Recent Investigations of Differential and Fractional Equations and Inclusions

Recent Investigations of Differential and Fractional Equations and Inclusions

Editor

Snezhana Hristova

MDPI • Basel • Beijing • Wuhan • Barcelona • Belgrade • Manchester • Tokyo • Cluj • Tianjin

Editor
Snezhana Hristova
University of Plovdiv
Bulgaria

Editorial Office
MDPI
St. Alban-Anlage 66
4052 Basel, Switzerland

This is a reprint of articles from the Special Issue published online in the open access journal *Mathematics* (ISSN 2227-7390) (available at: https://www.mdpi.com/journal/mathematics/special_issues/Recent_Investigations_Differential_Fractional_Equations_Inclusions).

For citation purposes, cite each article independently as indicated on the article page online and as indicated below:

LastName, A.A.; LastName, B.B.; LastName, C.C. Article Title. *Journal Name* **Year**, *Volume Number*, Page Range.

ISBN 978-3-0365-0074-4 (Hbk)
ISBN 978-3-0365-0075-1 (PDF)

Contents

About the Editor

Snezhana Hristova (Professor, PhD, DSc) is a Professor of Mathematics at University of Plovdiv "Paisii Hilendarski", Bulgaria. Her research interests are mainly in the field of differential equations and more precisely in functional differential equations, impulsive differential equations, fractional differential equations, and difference equations, as well as their applications. She is an author of 10 books (including one published by Springer) and more than 150 scientific papers (including ISI/Web of Science articles). She has an h-index of 17 (according to Scopus). Prof. Hristova is a Member of the Editorial Board of more than 10 journals (including journals indexed by ISI/Web of Science).

Preface to "Recent Investigations of Differential and Fractional Equations and Inclusions"

During the past decades, the subject of calculus of integrals and derivatives of any arbitrary real or complex order has gained considerable popularity and impact. This is mainly due to its demonstrated applications in numerous seemingly diverse and widespread fields of science and engineering. In connection with this, great importance is attached to the publication of results that focus on recent and novel developments in the theory of any types of differential and fractional differential equation and inclusions, especially covering analytical and numerical research for such kinds of equations.

This book is a compilation of articles from a Special Issue of Mathematics devoted to the topic of "Recent Investigations of Differential and Fractional Equations and Inclusions". It contains some theoretical works and approximate methods in fractional differential equations and inclusions as well as fuzzy integrodifferential equations. Many of the papers were supported by the Bulgarian National Science Fund under Project KP-06-N32/7.

Overall, the volume is an excellent witness of the relevance of the theory of fractional differential equations.

Snezhana Hristova
Editor

mathematics

MDPI

Article

A Note on the Topological Transversality Theorem for Weakly Upper Semicontinuous, Weakly Compact Maps on Locally Convex Topological Vector Spaces

Donal O'Regan

School of Mathematics, Statistics and Applied Mathematics, National University of Ireland, H91 TK33 Galway, Ireland; donal.oregan@nuigalway.ie

Received: 8 January 2020; Accepted: 18 February 2020; Published: 25 February 2020

Abstract: A simple theorem is presented that automatically generates the topological transversality theorem and Leray–Schauder alternatives for weakly upper semicontinuous, weakly compact maps. An application is given to illustrate our results.

Keywords: weakly upper semicontinuous; essential maps; homotopy

---

1. Introduction

Many problems arising in natural phenomena give rise to problems of the form $x \in F\,x$, for some map F. In applications for a complicated F, the intent is to attempt to relate it to a simpler (and solvable) problem $x \in G\,x$, where the map G is homotopic (in an appropriate way) to F, and then to hopefully deduce that $x \in F\,x$ is solvable. This approach was initiated by Leray and Schauder and extended to a very general formulation in, for example, [1,2]. The goal, to begin with, is to consider a class of maps that arise in applications and then to present the notion of homotopy for the class of maps that are fixed point free on the boundary of the considered set.

In this paper we consider weakly upper semicontinuous, weakly compact maps F and G, with $F \cong G$. We present the topological transversality theorem, which states that F is essential if, and only if, G is essential. The proof is based on a new result (Theorem 1) for weakly upper semicontinuous, weakly compact maps. Our topological transversality theorem will then immediately generate Leray–Schauder type alternatives (see Theorem 4 and Corollary 1). In addition, we note that these results are useful from an application viewpoint (see Theorem 5).

2. Topological Transversality Theorem

Let X be a Hausdorff locally convex topological vector space and U be a weakly open subset of C, where C is a closed convex subset of X. First we present the class of maps, M, that we will consider in this paper.

Definition 1. *We say $F \in M(\overline{U^w}, C)$ if $F : \overline{U^w} \to K(C)$ is a weakly upper semicontinuous, weakly compact map; here $\overline{U^w}$ denotes the weak closure of U in C and $K(C)$ denotes the family of nonempty, convex, weakly compact subsets of C.*

Definition 2. *We say $F \in M_{\partial U}(\overline{U^w}, C)$ if $F \in M(\overline{U^w}, C)$ and $x \notin F(x)$ for $x \in \partial U$; here ∂U denotes the weak boundary of U in C.*

Now we present the notion of homotopy for the class of maps, M, with the fixed point free on the boundary.

Definition 3. *Let $F, G \in M_{\partial U}(\overline{U^w}, C)$. We write $F \cong G$ in $M_{\partial U}(\overline{U^w}, C)$ if there exists a weakly upper semicontinuous, weakly compact map $\Psi : \overline{U^w} \times [0,1] \to K(C)$ with $x \notin \Psi_t(x)$ for $x \in \partial U$ and $t \in (0,1)$ (here $\Psi_t(x) = \Psi(x,t)$), $\Psi_0 = F$ and $\Psi_1 = G$.*

Definition 4. *Let $F \in M_{\partial U}(\overline{U^w}, C)$. We say that F is essential in $M_{\partial U}(\overline{U^w}, C)$ if, for every map $J \in M_{\partial U}(\overline{U^w}, C)$ with $J|_{\partial U} = F|_{\partial U}$, there exists a $x \in U$ with $x \in J(x)$.*

We present a simple theorem that will immediately yield the so called topological transversality theorem (motivated from [1]) for weakly upper semicontinuous, weakly compact maps (see Theorem 2). The topological transversality theorem essentially states that if a map F is essential and $F \cong G$ then the map G is essential (and so in particular has a fixed point).

Theorem 1. *Let X be a Hausdorff locally convex topological vector space, U be a weakly open subset of C, C be a closed convex subset of X, $F \in M_{\partial U}(\overline{U^w}, C)$ and $G \in M_{\partial U}(\overline{U^w}, C)$ is essential in $M_{\partial U}(\overline{U^w}, C)$. Also suppose*

$$\begin{cases} \text{for any map } J \in M_{\partial U}(\overline{U^w}, C) \text{ with } J|_{\partial U} = F|_{\partial U} \\ \text{we have } G \cong J \text{ in } M_{\partial U}(\overline{U^w}, C). \end{cases} \tag{1}$$

Then F is essential in $M_{\partial U}(\overline{U^w}, C)$.

Proof. Let $J \in M_{\partial U}(\overline{U^w}, C)$ with $J|_{\partial U} = F|_{\partial U}$. We must show there exists a $x \in U$ with $x \in J(x)$. Let $H^J : \overline{U^w} \times [0,1] \to K(C)$ be a weakly upper semicontinuous, weakly compact map with $x \notin H^J_t(x)$ for any $x \in \partial U$ and $t \in (0,1)$ (here $H^J_t(x) = H^J(x,t)$), $H^J_0 = G$ and $H^J_1 = J$ (this is guaranteed from (2.1)). Let

$$\Omega = \left\{ x \in \overline{U^w} : \ x \in H^J(x,t) \ \text{ for some } \ t \in [0,1] \right\}$$

and

$$D = \left\{ (x,t) \in \overline{U^w} \times [0,1] : \ x \in H^J(x,t) \right\}.$$

Now recall that $X = (X, w)$, the space X endowed with the weak topology, is completely regular. First, $D \neq \emptyset$ (note G is essential in $M_{\partial U}(\overline{U^w}, C)$) and D is weakly closed (note H^J is weakly upper semicontinuous) and so D is weakly compact (note H^J is a weakly compact map). Let $\pi : \overline{U^w} \times [0,1] \to \overline{U^w}$ be the projection. Now $\Omega = \pi(D)$ is weakly closed (see Kuratowski's theorem ([3] p. 126)) and so in fact weakly compact. Also note that $\Omega \cap \partial U = \emptyset$ (since $x \notin H^J_t(x)$ for any $x \in \partial U$ and $t \in [0,1]$). Thus there exists a weakly continuous map $\mu : \overline{U^w} \to [0,1]$ with $\mu(\partial U) = 0$ and $\mu(\Omega) = 1$. We define the map R by $R(x) = H^J(x, \mu(x)) = H^J \circ g(x)$, where $g : \overline{U^w} \to \overline{U^w} \times [0,1]$ is given by $g(x) = (x, \mu(x))$. Note that $R \in M_{\partial U}(\overline{U^w}, C)$ with $R|_{\partial U} = G|_{\partial U}$ (note, if $x \in \partial U$, then $R(x) = H^J(x,0) = G(x)$) so the essentiality of G guarantees a $x \in U$ with $x \in R(x)$ i.e., $x \in H^J_{\mu(x)}(x)$). Thus $x \in \Omega$ so $\mu(x) = 1$ and as a result $x \in H^J_1(x) = J(x)$. □

Before we state the topological transversality theorem we note two things:

(a). If Λ, $\Theta \in M_{\partial U}(\overline{U^w}, C)$ with $\Lambda|_{\partial U} = \Theta|_{\partial U}$ then $\Lambda \cong \Theta$ in $M_{\partial U}(\overline{U^w}, C)$. To see this let $\Psi(x,t) = (1-t)\Lambda(x) + t\Theta(x)$ and note that $\Psi : \overline{U^w} \times [0,1] \to K(C)$ is a weakly upper semicontinuous, weakly compact map [some authors prefer to assume (but it is not necessary) the following property:

$$\begin{cases} \text{if } W \text{ is a weakly compact subset of} \\ C \text{ then } \overline{co}\,(W) \text{ is weakly compact} \end{cases}$$

to guarantee that Ψ is weakly compact. Note, this property is a Krein–Šmulian type property [4,5], which we know is true if X is a quasicomplete locally convex linear topological space]. Note, $x \notin \Psi_t(x)$ for $x \in \partial U$ and $t \in [0,1]$ (note, $\Lambda|_{\partial U} = \Theta|_{\partial U}$).
(b). A standard argument guarantees that $\cong$ in $M_{\partial U}(\overline{U^w}, C)$ is an equivalence relation.

Theorem 2. *Let X be a Hausdorff locally convex topological vector space, U be a weakly open subset of C, and C be a closed convex subset of X. Suppose F and G are two maps in $M_{\partial U}(\overline{U^w}, C)$ with $F \cong G$ in $M_{\partial U}(\overline{U^w}, C)$. Then F is essential in $M_{\partial U}(\overline{U^w}, C)$ if, and only if, G is essential in $M_{\partial U}(\overline{U^w}, C)$.*

Proof. Assume G is essential in $M_{\partial U}(\overline{U^w}, C)$. To show that F is essential in $M_{\partial U}(\overline{U^w}, C)$ let $J \in M_{\partial U}(\overline{U^w}, C)$ with $J|_{\partial U} = F|_{\partial U}$. Now since $F \cong G$ in $M_{\partial U}(\overline{U^w}, C)$, then (a) and (b) above guarantees that $G \cong J$ in $M_{\partial U}(\overline{U^w}, C)$ i.e., (2.1) holds. Then Theorem 1 guarantees that F is essential in $M_{\partial U}(\overline{U^w}, C)$. A similar argument shows that if F is essential in $M_{\partial U}(\overline{U^w}, C)$, then G is essential in $M_{\partial U}(\overline{U^w}, C)$. □

Next, we present an example of an essential map in $M_{\partial U}(\overline{U^w}, C)$, which will be useful from an application viewpoint (see Corollary 1 and Theorem 5).

Theorem 3. *Let X be a Hausdorff locally convex topological vector space, U be a weakly open subset of C, $0 \in U$, and C be a closed convex subset of X. Then the zero map is essential in $M_{\partial U}(\overline{U^w}, C)$.*

Proof. Let $J \in M_{\partial U}(\overline{U^w}, C)$ with $J|_{\partial U} = \{0\}|_{\partial U}$. We must show there exists a $x \in U$ with $x \in J(x)$. Consider the map R given by

$$R(x) = \begin{cases} J(x), & x \in \overline{U^w} \\ \{0\}, & x \in C \setminus \overline{U^w}. \end{cases}$$

Note, $R : C \to K(C)$ is a weakly upper semicontinuous, weakly compact map, thus [6] guarantees that there exists a $x \in C$ with $x \in R(x)$. If $x \in C \setminus \overline{U^w}$ then since $R(x) = \{0\}$ and $0 \in U$ we have a contradiction. Thus $x \in U$ so $x \in R(x) = J(x)$. □

We combine Theorem 2 and Theorem 3 and we obtain:

Theorem 4. *Let X be a Hausdorff locally convex topological vector space, U be a weakly open subset of C, $0 \in U$, and C be a closed convex subset of X. Suppose $F \in M_{\partial U}(\overline{U^w}, C)$ with*

$$x \notin t\,F(x) \ \ for \ \ x \in \partial U \ \ and \ \ t \in (0,1). \tag{2}$$

Then F is essential in $M_{\partial U}(\overline{U^w}, C)$ (in particular there exists a $x \in U$ with $x \in F(x)$).

Proof. Note, Theorem 3 guarantees that the zero map is essential in $M_{\partial U}(\overline{U^w}, C)$. The result will follow from Theorem 2 if we note the usual homotopy between the zero map and F, namely, $\Psi(x,t) = t\,F(x)$ (note $x \notin \Psi_t(x)$ for $x \in \partial U$ and $t \in [0,1]$; see (2.2)). □

Corollary 1. *Let X be a Hausdorff locally convex topological vector space, U be a weakly open subset of C, $0 \in U$, C be a closed convex subset of X, and $\overline{U^w}$ be a Šmulian space (i.e., for any $\Omega \subseteq \overline{U^w}$ if $x \in \overline{\Omega^w}$ then there exists a sequence $\{x_n\}$ in Ω with $x_n \rightharpoonup x$). Suppose $F : \overline{U^w} \to K(C)$ is a weakly sequentially upper semicontinuous i.e., for any weakly closed set A of C we have that $F^{-1}(A) = \{x \in \overline{U^w} : F(x) \cap A \neq \emptyset\}$ is a weakly sequentially closed), weakly compact map with*

$$x \notin t\,F(x) \ \text{for} \ x \in \partial U \ \text{and} \ t \in (0,1]. \tag{3}$$

Then F is essential in $M_{\partial U}(\overline{U^w}, C)$ (in particular there exists a $x \in U$ with $x \in F(x)$).

Proof. The result follows from Theorem 4, as $F \in M_{\partial U}(\overline{U^w}, C)$. To see this we simply need to show that $F : \overline{U^w} \to K(C)$ is weakly upper semicontinuous. The argument is similar to that in [2,7]. Let A be a weakly closed subset of C and let $x \in \overline{F^{-1}(A)}^w$. As $\overline{U^w}$ is Šmulian then there exists a sequence $\{x_n\}$ in $F^{-1}(A)$ with $x_n \rightharpoonup x$. Now, $x \in F^{-1}(A)$ since $F^{-1}(A)$ is weakly sequentially closed. Thus, $\overline{F^{-1}(A)}^w = F^{-1}(A)$ so $F^{-1}(A)$ is weakly closed. □

We consider the second order differential inclusion

$$\begin{cases} y'' \in f(t,y,y') \ \text{a.e. on} \ [0,1] \\ y(0) = y(1) = 0 \end{cases} \tag{4}$$

where $f : [0,1] \times \mathbf{R}^2 \to CK(\mathbf{R})$ is a L^p–Carathéodory function (here $p > 1$ and $CK(\mathbf{R})$ denotes the family of nonempty, convex, compact subsets of $\mathbf{R}$); by this we mean
(a). $t \mapsto f(t,x,y)$ is measurable for every $(x,y) \in \mathbf{R}^2$,
(b). $(x,y) \mapsto f(t,x,y)$ is upper semicontinuous for a.e. $t \in [0,1]$,
and
(c). for each $r > 0$, $\exists\, h_r \in L^p[0,1]$ with $|f(t,x,y)| \le h_r(t)$ for a.e. $t \in [0,1]$ and every $(x,y) \in \mathbf{R}^2$ with $|x| \le r$ and $|y| \le r$.

We present an existence principle for (2.4) using Corollary 1. For notational purposes for appropriate functions u, let

$$\|u\|_0 = \sup_{[0,1]} |u(t)|, \ \|u\|_1 = \max\{\|u\|_0, \|u'\|_0\} \text{ and } \|u\|_{L^p} = \left(\int_0^1 |u(t)|^p\,dt\right)^{\frac{1}{p}}.$$

Recall that $W^{k,p}[0,1]$, $1 \le p < \infty$ denotes the space of functions $u : [0,1] \to \mathbf{R}^n$, with $u^{(k-1)} \in AC[0,1]$ and $u^{(k)} \in L^p[0,1]$. Note, $W^{k,p}[0,1]$ is reflexive if $1 < p < \infty$.

Theorem 5. *Let $f : [0,1] \times \mathbf{R}^2 \to CK(\mathbf{R})$ be a L^p–Carathéodory function ($1 < p < \infty$) and assume there exists a constant M_0 (independent of λ) with $\|y\|_1 \neq M_0$ for any solution $y \in W^{2,p}[0,1]$ to*

$$\begin{cases} y'' \in \lambda f(t,y,y') \ \text{a.e. on} \ [0,1] \\ y(0) = y(1) = 0 \end{cases}$$

for $0 < \lambda \le 1$. Then (2.4) has a solution in $W^{2,p}[0,1]$.

Proof. Since f is L^p–Carathéodory, there exists $h_{M_0} \in L^p[0,1]$ with

$$\begin{cases} |f(t,u,v)| \le h_{M_0}(t) \ \text{for a.e.} \ t \in [0,1] \ \text{and} \\ \text{every} \ (u,v) \in \mathbf{R}^2 \ \text{with} \ |u| \le M_0 \ \text{and} \ |v| \le M_0. \end{cases}$$

Let

$$G(t,s) = \begin{cases} (t-1)\,s, & 0 \le s \le t \le 1 \\ (s-1)\,t, & 0 \le t \le s \le 1 \end{cases}$$

and $N = \max\{N_0, N_1, M_0\}$ where (here $\frac{1}{p} + \frac{1}{q} = 1$),

$$N_0 = \|h_{M_0}\|_{L^p} \sup_{t\in[0,1]} \left(\int_0^1 |G(t,s)|^q \, ds \right)^{\frac{1}{q}}$$

and

$$N_1 = \|h_{M_0}\|_{L^p} \sup_{t\in[0,1]} \left(\int_0^1 |G_t(t,s)|^q \, ds \right)^{\frac{1}{q}}.$$

We also let

$$N_2 = \|h_{M_0}\|_{L^p}.$$

We will apply Corollary 1 with $X = W^{2,p}[0,1]$,

$$C = \left\{ u \in W^{2,p}[0,1] : \ \|u\|_1 \le N \ \text{ and } \ \|u''\|_{L^p} \le N_2 \right\}$$

and

$$U = \left\{ u \in W^{2,p}[0,1] : \ \|u\|_1 < M_0 \ \text{ and } \ \|u''\|_{L^p} \le N_2 \right\}.$$

Now, let

$$F = L \circ N_f : C \to 2^X$$

where $L : L^p[0,1] \to W^{2,p}[0,1]$ and $N_f : W^{2,p}[0,1] \to 2^{L^p[0,1]}$ are given by

$$Ly(t) = \int_0^1 G(t,s)\, y(s)\, ds$$

and

$$N_f u = \{ y \in L^p[0,1] : \ y(t) \in f(t, u(t), u'(t)) \ \text{ a.e. } \ t \in [0,1] \}.$$

Note, N_f is well defined, since if $x \in C$ then ([8] p. 26 or [9], p. 56) guarantees that $N_f x \ne \emptyset$.

Notice that C is a convex, closed, bounded subset of X. We first show that U is weakly open in C. To do this, we will show that $C \setminus U$ is weakly closed. Let $x \in \overline{C \setminus U}^w$. Then there exists $x_n \in C \setminus U$ (see [10] p. 81) with $x_n \rightharpoonup x$ (here $W^{2,p}[0,1]$ is endowed with the weak topology and $\rightharpoonup$ denotes weak convergence). We must show $x \in C \setminus U$. Now since the embedding $j : W^{2,p}[0,1] \to C^1[0,1]$ is completely continuous ([11], p. 144 or [12], p. 213), there is a subsequence S of integers with

$$x_n \to x \ \text{ in } \ C^1[0,1] \ \text{ and } \ x_n'' \rightharpoonup x'' \ \text{ in } \ L^p[0,1]$$

as $n \to \infty$ in S. Also

$$\|x\|_1 = \lim_{n\to\infty} \|x_n\|_1 \ \text{ and } \ \|x''\|_{L^p} \le \liminf \|x_n''\|_{L^p} \le N_2.$$

Note, $M_0 \le \|x\|_1 \le N$ since $M_0 \le \|x_n\|_1 \le N$ for all n. As a result, $x \in C \setminus U$, so $\overline{C \setminus U}^w = C \setminus U$. Thus, U is weakly open in C. Also,

$$\partial U = \{u \in C : \ \|u\|_1 = M_0\} \ \text{ and } \ \overline{U^w} = \{u \in C : \ \|u\|_1 \leq M_0\};$$

note, $\overline{U^w} = \overline{U}$ ([5] p. 66) since U is convex (alternatively take $x \in \overline{U^w}$ and follow a similar argument as above). Also note that $\overline{U^w}$ is weakly compact (note $W^{2,p}[0,1]$ is reflexive) so $\overline{U^w}$ is Šmulian. Notice also that $F : \overline{U^w} \to 2^C$ since if $y \in \overline{U^w}$ then from above we have

$$\|Fy\|_0 \leq \|h_{M_0}\|_{L^p} \sup_{t\in[0,1]} \left(\int_0^1 |G(t,s)|^q \, ds \right)^{\frac{1}{q}} = N_0,$$

$$\|(Fy)'\|_0 \leq \|h_{M_0}\|_{L^p} \sup_{t\in[0,1]} \left(\int_0^1 |G_t(t,s)|^q \, ds \right)^{\frac{1}{q}} = N_1,$$

and

$$\|(Fy)''\|_0 \leq \|h_{M_0}\|_{L^p} = N_2.$$

A standard argument (see for example ([13] p. 283)) guarantees that $F : \overline{U^w} \to K(C)$ is weakly sequentially upper semicontinuous.

Now we apply Corollary 1 to deduce our result: Note that (2.3) holds since, if there exists $x \in \partial U$ and $\lambda \in (0,1]$ with $x \in \lambda F x$, then $\|x\|_1 = M_0$ (since $x \in \partial U$) and $\|x\|_1 \neq M_0$ by assumption. Thus, F is essential in $M_{\partial U}(\overline{U^w}, C)$, so in particular, F has a fixed point in U. □

Conflicts of Interest: The author declares no conflict of interest.

References

1. Granas, A. Sur la méthode de continuité de Poincaré. *C.R. Acad. Sci. Paris* **1976**, *282*, 983–985.
2. O'Regan, D. Fixed point theory of Mönch type for weakly sequentially upper semicontinuous maps. *Bull. Aust. Math. Soc.* **2000**, *61*, 439–449. [CrossRef]
3. Engelking, R. *General Topology*; Heldermann Verlag: Berlin, Germany, 1989.
4. Edwards, R.E. *Functional Analysis, Theory and Applications*; Holt, Rinehart and Winston: New York, NY, USA, 1965.
5. Rudin, W. *Functional Analysis*; McGraw Hill: New York, NY, USA, 1991.
6. Himmelberg, C.J. Fixed points of compact multifunctions. *J. Math. Anal. Appl.* **1972**, *38*, 205–207. [CrossRef]
7. Arino, O.; Gautier, S.; Penot, J.P. A fixed point theorem for sequentially continuous mappings with applications to ordinary differential equations. *Funkc. Ekvac.* **1984**, *27*, 273–279.
8. Deimling, K. *Multivalued Differential Equations*; Walter de Gruyter: Berlin, Germany, 1992.
9. Frigon, M. *Existence Theorems for Solutions of Differential Inclusions*; Topological Methods in Differential Equations and Inclusions; Kluwer Academic Publishers: Dordrecht, The Netherlands, 1995; pp. 51–87.
10. Browder, F.E. Nonlinear operators and nonlinear equations of evolution in Banach spaces. *Proc. Sympos. Pure Math.* **1976**, *18*, 1–305. [CrossRef]
11. Adams, R.A. *Sobolev Spaces*; Academic Press: New York, NY, USA, 1975.
12. Brezis, H. *Functional Analysis, Sobolev Spaces and Partial Differential Equations*; Springer: New York, NY, USA, 2011.
13. Agarwal, R.P.; O'Regan, D. Fixed point theory for weakly sequentially upper semicontinuous maps with applications to differential inclusions. *Nonlinear Oscil.* **2002**, *5*, 277–286. [CrossRef]

mathematics

MDPI

Article

Computer Simulation and Iterative Algorithm for Approximate Solving of Initial Value Problem for Riemann-Liouville Fractional Delay Differential Equations

Snezhana Hristova [1,*], Kremena Stefanova [2] and Angel Golev [3]

1 Department of Applied Mathematics and Modeling, University of Plovdiv "Paisii Hilendarski", Plovdiv 4000, Bulgaria
2 Department of Computer Technologies, University of Plovdiv "Paisii Hilendarski", Plovdiv 4000, Bulgaria; kvstefanova@gmail.com
3 Department of Software Technologies, University of Plovdiv "Paisii Hilendarski", Plovdiv 4000, Bulgaria; angel.golev@gmail.com
* Correspondence: snehri@gmail.com

Received: 19 February 2020; Accepted: 29 March 2020; Published: 1 April 2020

Abstract: The main aim of this paper is to suggest an algorithm for constructing two monotone sequences of mild lower and upper solutions which are convergent to the mild solution of the initial value problem for Riemann-Liouville fractional delay differential equation. The iterative scheme is based on a monotone iterative technique. The suggested scheme is computerized and applied to solve approximately the initial value problem for scalar nonlinear Riemann-Liouville fractional differential equations with a constant delay on a finite interval. The suggested and well-grounded algorithm is applied to a particular problem and the practical usefulness is illustrated.

Keywords: Riemann-Liouville fractional differential equation; delay; lower and upper solutions; monotone-iterative technique

1. Introduction

Fractional differential operators are applied successfully to model various processes with anomalous dynamics in science and engineering [1,2]. At the same time, only a small number of fractional differential equations could be solved explicitly. It requires the application of different approximate methods for solving nonlinear factional equations.

This paper deals with an initial value problem for a nonlinear scalar Riemann-Liouville (RL) fractional differential equation with a delay on a closed interval is studied. Mild lower and mild upper solutions are defined. An algorithm for constructing two convergent monotone functional sequences $\{v^n\}$, $\{w^n\}$ are given. It is proved both sequences $\{(t-t_0)^{1-q}v^n\}$ and $\{(t-t_0)^{1-q}w^n\}$ are the mild minimal and the mild maximal solutions of the given problem. The uniform convergence of both sequences is proved. A special computer program is built and applied to solve particular problems and to illustrate the practical application of the suggested schemes.

Note the monotone iterative techniques combined with lower and upper solutions are applied in the literature to solve various problems in ordinary differential equations [3], differential equations with maxima [4], difference equations with maxima [5], Caputo fractional differential equations [6], Riemann-Liouville fractional differential equations [7–10].

In this paper, we consider an initial value problem for a scalar nonlinear Riemann-Liouville fractional differential equation with a constant delay on a finite interval. We apply the method of lower and upper solutions and monotone-iterative technique to suggest an algorithm for approximate solving of the studied problem. The suggested and well-grounded algorithm is used in an appropriate computer environment and it is applied to a particular problem to illustrate the practical usefulness.

2. Preliminary and Auxiliary Results

Let $m : [0, \infty) \to \mathbb{R}$ be a given function and $q \in (0, 1)$ be a fixed number. Then the Riemann-Liouville fractional derivative of order $q \in (0, 1)$ is defined by (see, for example, [2]

$$ {}_0^{RL}D_t^q m(t) = \frac{1}{\Gamma(1-q)} \frac{d}{dt} \Big(\int_0^t (t-s)^{-q} m(s) ds \Big), \quad t \geq 0. $$

We will give RL fractional derivatives of some elementary functions which will be used later:

Proposition 1. *Reference [2] the following equalities are true:*

$$ {}_0^{RL}D_t^q C = \frac{1}{\Gamma(1-q)} t^{-q}, $$
$$ {}_0^{RL}D_t^q t^{\beta} = \frac{\Gamma(1+\beta)}{\Gamma(1+\beta-q)} t^{\beta-q}. $$

Consider the initial value problem (IVP) for the nonlinear *Riemann-Liouville delay fractional differential equation* (FrDDE)

$$ \begin{aligned} &{}_0^{RL}D_t^q\, x(t) = F(t, x(t), x(t-\tau)) \quad \text{for } t \in (0, T] \\ &x(s) = \psi(s) \quad \text{for } s \in [-\tau, 0] \\ &t^{1-q}x(t)|_{t=0} = \lim_{t \to 0+} t^{1-q}x(t) = \psi(0), \end{aligned} \tag{1} $$

where $q \in (0, 1)$, $F : [0, T] \times \mathbb{R} \times \mathbb{R} \to \mathbb{R}$, $\psi : [-\tau, 0] \to \mathbb{R} :\ \psi(0) < \infty$ with $T \in ((N-1)\tau, N\tau]$, N is a natural number, and $\tau > 0$ is a given number.

The solution of the IVP (1) could have a discontinuity at $t = 0$.

Denote the interval $I = [-\tau, T]/\{0\}$.

Denote

$$ C_{1-q}([a,b]) = \{x(t) :\ [a,b] \to \mathbb{R} :\ (t-a)^{1-q}x(t) \in C([a,b], \mathbb{R})\}, $$

where $a, b,\ a < b$ are real numbers.

Define the norm in $C_{1-q}([a,b])$ by $||x||_{C_{1-q}[a,b]} = \max_{t \in [a,b]} |(t-a)^{1-q}x(t)|$.

Consider the linear scalar delay RL fractional equation of the type

$$ \begin{aligned} &{}_0^{RL}D_t^q x(t) = \lambda x(t) + \mu x(t-\tau) + f(t) \quad \text{for } t \in (0, T], \\ &x(t) = \psi(t) \text{ for } t \in [-\tau, 0], \qquad t^{1-q}x(t)|_{t=0} = \psi(0), \end{aligned} \tag{2} $$

where $\lambda,\ \mu$ are real constant, $f \in C([0, T], \mathbb{R})$. There exits an explicit formula for the solution of (2) given by see [11]:

$$ x(t) = \begin{cases} \psi(t) & \text{for } t \in [-\tau, 0], \\ \psi(0)\Gamma(q)E_{q,q}(\lambda t^q)t^{q-1} + \int_0^t (t-s)^{q-1}E_{q,q}(\lambda(t-s)^q)\big(f(s) + \mu x(s-\tau)\big)ds, & t \in (0, T] \end{cases} \tag{3} $$

where $E_{\alpha,\beta}(z) = \sum_{k=0}^{\infty} \frac{z^k}{\Gamma(\alpha k+\beta)}$ is the Mittag-Leffler function with two parameters.

Note that the solution in the simplest linear case is not easy to obtain. It requires the application of some approximate methods.

Similar to References [12], we have the following result:

Proposition 2. *Let $f \in C([0,T],\mathbb{R})$, $\psi \in C([-\tau,0],\mathbb{R})$, $\lambda \in \mathbb{R}$, $\mu \geq 0$ be constants and*

$$
\begin{aligned}
&{}_0^{RL}D_t^q v(t) \leq \lambda v(t) + \mu v(t-\tau) + f(t) \quad \text{for } t \in (0,T],\\
&v(t) = \psi(t) \quad \text{for } t \in [-\tau,0], \qquad t^{1-q}v(t)|_{t=0} = \psi(0).
\end{aligned}
$$

Then

$$
v(t) \leq \begin{cases} \psi(t), & t \in [-\tau,0],\\ \psi(0)\Gamma(q)E_{q,q}(\lambda t^q)t^{q-1} + \int_0^t (t-s)^{q-1}E_{q,q}(\lambda(t-s)^q)\Big(f(s) + \mu v(s-\tau)\Big)ds, & t \in (0,T]. \end{cases}
$$

Similar to References [7], we define the mild solutions:

Definition 1. *The function $x \in C(I,\mathbb{R})$ is a mild solution of the IVP for FrDDE (1), if it satisfies*

$$
x(t) = \begin{cases} \psi(t) \quad \text{for} \quad t \in [-\tau,0],\\ \psi(0)\Gamma(q)E_{q,q}(\lambda t^q)t^{q-1} + \int_0^t (t-s)^{q-1}E_{q,q}(\lambda(t-s)^q)f(s,x(s),x(s-\tau))ds, \ t \in (0,T]. \end{cases} \tag{4}
$$

Remark 1. *Note that the mild solution $x(t) \in C(I,\mathbb{R})$ of the IVP for FrDDE (1) might not be from $C_{1-q}([0,T])$ and it might not have the fractional derivative ${}_0^{RL}D_t^q\, x(t)$.*

Definition 2. *The function $x \in C(I,\mathbb{R})$ is a mild maximal solution (a mild minimal solution) of the IVP for FrDDE (1), if it is a mild solution of (1) and for any mild solution $u(t) \in C(I,\mathbb{R})$ of (1) the inequality $x(t) \leq u(t)$ ($x(t) \geq u(t)$) holds on I and $t^{1-q}x(t)|_{t=0} \leq (\geq) t^{1-q}u(t)|_{t=0}$.*

3. Mild Lower and Mild Upper Solutions of FrDDE

Definition 3. *The function $v(t) \in C(I,\mathbb{R})$ is a mild lower (a mild upper) solution of the IVP for FrDDE (1), if it satisfies the integral inequalities*

$$
v(t) \leq (\geq) \begin{cases} \psi(t) \quad \text{for} \quad t \in]-\tau,0],\\ \psi(0)\Gamma(q)E_{q,q}(\lambda t^q)t^{q-1} +\\ \qquad + \int_0^t (t-s)^{q-1}E_{q,q}(\lambda(t-s)^q)f(s,v(s),v(s-\tau))ds, \ t \in (0,T] \end{cases} \tag{5}
$$

and $t^{1-q}v(t)|_{t=0} = \psi(0)$.

Definition 4. *We say that the function $v(t) \in C_{1-q}(I,\mathbb{R})$ is a lower (an upper) solution of the IVP for FrDDE (1), if*

$$
\begin{aligned}
&{}_0^{RL}D_t^q v(t) \leq (\geq) F(t,v(t),v(t-\tau)) \quad \text{for} \quad t \in (0,T],\\
&v(t) = \psi(t) \quad \text{for } t \in [-\tau,0], \qquad t^{1-q}v(t)|_{t=0} = \psi(0).
\end{aligned}
$$

Remark 2. *A function could be a mild lower solution or a mild upper solution, respectively, of the IVP for FrDDE (1) but it could not be a lower solution or an upper solution, respectively, of the IVP for FrDDE (1).*

Remark 3. *Note that the mild lower solution (mild upper solution) is not unique. At the same time, because of the inequalities in (5) it is much easier to obtain at least one mild lower solution (mild upper solution) than a mild solution of the IVP for FrDDE (1).*

4. Monotone-Iterative Techniques for FrDDE

Now we will consider a nonlinear RL fractional differential equation with a constant delay. We will apply a monotone iterative technique to obtain approximate solution. The idea of the formulas for the successive approximations is based on linear RL-fractional differential equations of type (2) and its explicit formula for the solution obtained in [11].

For any two $u, v \in PC([-\tau, T], \mathbb{R})$ and the constants M, L define the operator (the values of the constants M, L will be defined later):

$$\Omega(u,v)(t) = \begin{cases} \psi(t), & t \in [-\tau, 0] \\ \psi(0)\Gamma(q)E_{q,q}(Mt^q)t^{q-1} + \int_0^t (t-s)^{q-1}E_{q,q}(M(t-s)^q)F(s,u(s),u(s-\tau))ds \\ \quad - \int_0^t (t-s)^{q-1}E_{q,q}(M(t-s)^q)\Big(Mu(s) + L(u(s-\tau) - v(s-\tau))\Big)ds, & t \in (0,T]. \end{cases}$$

Theorem 1. *Let the following conditions be fulfilled:*

1. *Let the functions $v, w \in C(I, \mathbb{R}) \cup C_{1-q}([0,T])$ be a lower solution and an upper solution, respectively, of the IVP for FrDDE (1) such that $v(t) \leq w(t)$ for $t \in [0,T]$ and $v(0) - \psi(0) \leq v(s) - \psi(s)$, $w(0) - \psi(0) \geq w(s) - \psi(s)$ for $s \in [-\tau, 0)$.*
2. *The function $F \in C([0,T] \times \mathbb{R} \times \mathbb{R}, \mathbb{R})$ and there exist constants $M \in \mathbb{R}$ and $L > 0$: $\frac{LT^q}{\Gamma(1_q)} < 1$ such that for any $t \in [0,T]$, $x, y, u, v \in \mathbb{R}$: $v(t) \leq x \leq y \leq w(t)$, $v(t-\tau) \leq u \leq v \leq w(t-\tau)$ the inequality $F(t,x,u) - F(t,y,v) \leq M(x-y) + L(u-v)$ holds.*

Then there exist two sequences of functions $\{v^{(n)}(t)\}_0^\infty$ and $\{w^{(n)}(t)\}_0^\infty$, $t \in [-\tau, T]$, such that:

a. *The sequences $\{v^{(n)}(t)\}$ and $\{w^{(n)}(t)\}$ are defined by $v^{(0)}(t) = v(t)$, $w^{(0)}(t) = w(t)$ and*

$$v^{(n)}(t) = \Omega\left(v^{(n-1)}, v^{(n)}\right)(t), \quad w^{(n)}(t) = \Omega\left(w^{(n-1)}, w^{(n)}\right)(t) \text{ for } n \geq 1,$$

that is,

$$v^{(n)}(t) = \begin{cases} \psi(t), & t \in [-\tau, 0], \\ \psi(0)\Gamma(q)E_{q,q}(Mt^q)t^{q-1} + \\ \quad + \int_0^t (t-s)^{q-1}E_{q,q}(M(t-s)^q)F(s, v^{(n-1)}(s), v^{(n-1)}(s-\tau))ds - \\ \quad - \int_0^t (t-s)^{q-1}E_{q,q}(M(t-s)^q) \times \\ \quad \times \Big(Mv^{(n-1)}(s) + L(v^{(n-1)}(s-\tau) - v^{(n)}(s-\tau))\Big)ds, & t \in (0,T], \end{cases}$$

$$w^{(n)}(t) = \begin{cases} \psi(t), & t \in [-\tau, 0], \\ \psi(0)\Gamma(q)E_{q,q}(Mt^q)t^{q-1} \\ \quad + \int_0^t (t-s)^{q-1}E_{q,q}(M(t-s)^q)F(s, w^{(n-1)}(s), w^{(n-1)}(s-\tau))ds \\ \quad - \int_0^t (t-s)^{q-1}E_{q,q}(M(t-s)^q) \times \\ \quad \times \Big(Mw^{(n-1)}(s) + L(w^{(n-1)}(s-\tau) - w^{(n)}(s-\tau))\Big)ds, & t \in (0,T], \end{cases}$$

where the constants M, L are defined in condition 2.

b. *The sequence $\{v^{(j)}(t)\}_{j=0}^\infty$ is increasing, that is, $v^{(j-1)}(t) \leq v^{(j)}(t)$ for $t \in (0,T]$, $j = 1, 2, \ldots$.*

c. *The sequence $\{w^{(j)}(t)\}_{j=0}^\infty$ is decreasing , that is, $w^{(j-1)}(t) \geq w^{(j)}(t)$ for $t \in (0,T]$, $j = 1, 2, \ldots$.*

d. *The inequality*

$$v^{(k)}(t) \leq w^{(k)}(t) \quad \text{for } t \in (0,T], \ k = 1, 2, \ldots \tag{6}$$

holds.

e. *The sequences* $\{t^{1-q}v^{(n)}(t)\}_0^\infty$ *and* $\{t^{1-q}w^{(n)}(t)\}_0^\infty$ *converge uniformly on* $[0,T]$ *and* $t^{1-q}V(t) = \lim_{k\to\infty} t^{1-q}v^{(n)}(t)$, $t^{1-q}W(t) = \lim_{k\to\infty} t^{1-q}w^{(n)}(t)$ *on* $[0,T]$.

f. *The limit functions* $V(t)$ *and* $W(t)$ *are mild solutions of the IVP for FrDDE (1) on* $[-\tau,T]$.

g. *The inequalities* $v^{(n)}(t) \le V(t) \le W(t) \le w^{(n)}(t)$ *hold on* $(0,T]$ *for any* $n = 0,1,2,\ldots$.

Proof of Theorem 1. Let $v(t)$ be a lower solution of the IVP for FrDDE (1), that is,

$$ {}_{\theta_0}^{RL}D_t^q\, v(t) \le Mv(t) + Lv(t-\tau) + G(t,v(t),v(t-\tau)), \tag{7} $$

where $G(t,u,v) = F(t,u,v) - Mu - Lv,\quad t\in[0,T],\quad u,v\in\mathbb{R}$.

According to Proposition 2, the inequality

$$ \begin{aligned} v(t) \le{}& \psi(0)\Gamma(q)E_{q,q}(Mt^q)t^{q-1} + \int_0^t (t-s)^{q-1}E_{q,q}(M(t-s)^q)G(s,v(s),v(s-\tau))ds \\ &+ L\int_0^t (t-s)^{q-1}E_{q,q}(M(t-s)^q)v(s-\tau)ds,\quad t\in(0,T] \end{aligned} \tag{8} $$

holds.

Let $v^{(0)}(t) = v(t)$ and $w^{(0)}(t) = w(t)$ for $t\in[-\tau,T]$.

We use induction w.r.t. the interval to prove properties of the sequences of successive approximations.

From the definition of the operator Ω and equality $E_{q,q}(0) = \frac{1}{\Gamma(q)}$ it follows that $t^{1-q}w^{(n)}(t)|_{t=0} = t^{1-q}v^{(n)}(t)|_{t=0} = \lim_{t\to 0+} t^{1-q}v^{(n)}(t) = \lim_{t\to 0+} \psi(0)\Gamma(q)E_{q,q}(Mt^q) = \psi(0)$ for all integers $n\ge 1$.

Let $t\in(0,\tau]$. From the definition of the operator Ω and inequalities (8) we obtain

$$ \begin{aligned} v^{(0)}(t) \le{}& \psi(0)\Gamma(q)E_{q,q}(Mt^q)t^{q-1} + \int_0^t (t-s)^{q-1}E_{q,q}(M(t-s)^q)G(s,v^{(0)}(s),v^{(0)}(s-\tau))ds \\ &+ L\int_0^t (t-s)^{q-1}E_{q,q}(M(t-s)^q)\psi(s-\tau)ds = v^{(1)}(t)\ \text{ for } t\in(0,\tau]. \end{aligned} \tag{9} $$

From the definition of the operator Ω, condition 2, the inequality (9) and the equality $v^{(1)}(s-\tau) - v^{(2)}(s-\tau) = 0$ for $s\in(0,\tau]$ we get

$$ \begin{aligned} v^{(1)}(t) ={}& \psi(0)\Gamma(q)E_{q,q}(Mt^q)t^{q-1} + L\int_0^t (t-s)^{q-1}E_{q,q}(M(t-s)^q)v^{(1)}(s-\tau)ds \\ &+ \int_0^t (t-s)^{q-1}E_{q,q}(M(t-s)^q)F(s,v^{(0)}(s),v^{(0)}(s-\tau))ds \\ &- \int_0^t (t-s)^{q-1}E_{q,q}(M(t-s)^q)\Big(Mv^{(0)}(s) + Lv^{(0)}(s-\tau)\Big)ds \\ \le{}& \psi(0)\Gamma(q)E_{q,q}(Mt^q)t^{q-1} + L\int_0^t (t-s)^{q-1}E_{q,q}(M(t-s)^q)v^{(2)}(s-\tau)ds \\ &+ \int_0^t (t-s)^{q-1}E_{q,q}(M(t-s)^q)F(s,v^{(1)}(s),v^{(1)}(s-\tau))ds \\ &- \int_0^t (t-s)^{q-1}E_{q,q}(M(t-s)^q)\Big(Mv^{(1)}(s) + Lv^{(1)}(s-\tau)\Big)ds \\ ={}& v^{(2)}(t)\ \text{ for } t\in(0,\tau]. \end{aligned} \tag{10} $$

Similarly, we can prove

$$ v^{(n)}(t) \le v^{(n+1)}(t),\ \text{ for } t\in(0,\tau],\ n=2,\ldots, $$

and

$$w^{(n)}(t) \geq w^{(n+1)}(t), \text{ for } t \in (0,\tau], \; n = 0,1,2,\ldots.$$

Let $t \in (\tau, 2\tau]$. From the definition of the operator Ω, the inequalities (8) and $v^{(0)}(t-\tau) \leq v^{(1)}(t-\tau)$ for $t \in (\tau, 2\tau]$ we obtain

$$\begin{aligned}
v^{(0)}(t) \leq\ & \psi(0)\Gamma(q)E_{q,q}(Mt^q)t^{q-1} + \int_0^t (t-s)^{q-1}E_{q,q}(M(t-s)^q)F(s, v^{(0)}(s), v^{(0)}(s-\tau))ds \\
& + L\int_0^\tau (t-s)^{q-1}E_{q,q}(M(t-s)^q)v^{(1)}(s-\tau)ds \\
& + L\int_\tau^t (t-s)^{q-1}E_{q,q}(M(t-s)^q)v^{(1)}(s-\tau)ds \\
& - \int_0^t (t-s)^{q-1}E_{q,q}(M(t-s)^q)\Big(Mv^{(0)}(s) + Lv^{(0)}(s-\tau)\Big)ds = v^{(1)}(t).
\end{aligned} \tag{11}$$

Also, from condition 2, the inequalities (8) and $v^{(1)}(t-\tau) \leq v^{(2)}(t-\tau)$ for $t \in (\tau, 2\tau]$ we get

$$\begin{aligned}
v^{(1)}(t) =\ & \psi(0)\Gamma(q)E_{q,q}(Mt^q)t^{q-1} + \int_0^t (t-s)^{q-1}E_{q,q}(M(t-s)^q)F(s, v^{(0)}(s), v^{(0)}(s-\tau))ds \\
& - \int_0^t (t-s)^{q-1}E_{q,q}(M(t-s)^q)\Big(Mv^{(0)}(s) + L(v^{(0)}(s-\tau) - v^{(1)}(s-\tau)\Big)ds \\
\leq\ & \psi(0)\Gamma(q)E_{q,q}(Mt^q)t^{q-1} + \int_0^t (t-s)^{q-1}E_{q,q}(M(t-s)^q)F(s, v^{(1)}(s), v^{(1)}(s-\tau))ds \\
& + \int_0^t (t-s)^{q-1}E_{q,q}(M(t-s)^q)\Big(M(v^{(0)}(s) - v^{(1)}(s)) + L(v^{(0)}(s-\tau) - v^{(1)}(s-\tau))\Big)ds \\
& - \int_0^t (t-s)^{q-1}E_{q,q}(M(t-s)^q)\Big(Mv^{(0)}(s) + L(v^{(0)}(s-\tau) - v^{(1)}(s-\tau))\Big)ds \\
=\ & \psi(0)\Gamma(q)E_{q,q}(Mt^q)t^{q-1} + \int_0^t (t-s)^{q-1}E_{q,q}(M(t-s)^q)F(s, v^{(1)}(s), v^{(1)}(s-\tau))ds \\
& - \int_0^t (t-s)^{q-1}E_{q,q}(M(t-s)^q)Mv^{(1)}(s)ds \\
\leq\ & \psi(0)\Gamma(q)E_{q,q}(Mt^q)t^{q-1} + \int_0^t (t-s)^{q-1}E_{q,q}(M(t-s)^q)F(s, v^{(1)}(s), v^{(1)}(s-\tau))ds \\
& - \int_0^t (t-s)^{q-1}E_{q,q}(M(t-s)^q)\Big(Mv^{(1)}(s) + L(v^{(1)}(s-\tau) - v^{(2)}(s-\tau))\Big)ds \\
=\ & v^{(2)}(t), \; t \in (\tau, 2\tau].
\end{aligned}$$

Similarly, we can prove

$$v^{(n)}(t) \leq v^{(n+1)}(t), \text{ for } t \in (\tau, 2\tau], \; n = 2,3\ldots,$$

and

$$w^{(n)}(t) \geq w^{(n+1)}(t), \text{ for } t \in (\tau, 2\tau], \; n = 0,1,2,\ldots.$$

Following the induction process w.r.t. the interval we prove the claims (b) and (c).

Now, we will prove the claim (d). Let $t \in (0, \tau]$. From the definition of the operator Ω, condition 2, the inequality (9) we get

$$\begin{aligned}
v^{(1)}(t) - w^{(1)}(t) = &\int_0^t (t-s)^{q-1} E_{q,q}(M(t-s)^q) F(s, v^{(0)}(s), v^{(0)}(s-\tau)) ds \\
&- \int_0^t (t-s)^{q-1} E_{q,q}(M(t-s)^q) F(s, w^{(0)}(s), w^{(0)}(s-\tau)) ds \\
&- \int_0^t (t-s)^{q-1} E_{q,q}(M(t-s)^q) \times \\
&\times \left(Mv^{(0)}(s) + L(v^{(0)}(s-\tau) - v^{(1)}(s-\tau)) \right) ds \\
&+ \int_0^t (t-s)^{q-1} E_{q,q}(M(t-s)^q) \times \\
&\times \left(Mw^{(0)}(s) + Lw^{(0)}(s-\tau) - Lw^{(1)}(s-\tau) \right) ds \\
&\leq L \int_0^t (t-s)^{q-1} E_{q,q}(M(t-s)^q)(v^{(1)}(s-\tau) - w^{(1)}(s-\tau)) ds \\
&= 0 \quad \text{for } t \in (0, \tau].
\end{aligned} \tag{12}$$

Similarly, we can prove

$$v^{(n)}(t) \leq w^{(n+1)}(t), \quad \text{for } t \in (0, \tau], \quad n = 2, 3, \ldots.$$

Let $t \in (\tau, 2\tau]$. From condition 2, the inequalities (8) and $v^{(1)}(t-\tau) \leq w^{(1)}(t-\tau)$ for $t \in (\tau, 2\tau]$ we obtain we get

$$\begin{aligned}
v^{(1)}(t) = &\psi(0)\Gamma(q) E_{q,q}(Mt^q) t^{q-1} + \int_0^t (t-s)^{q-1} E_{q,q}(M(t-s)^q) F(s, v^{(0)}(s), v^{(0)}(s-\tau)) ds \\
&- \int_0^t (t-s)^{q-1} E_{q,q}\Big(M(t-s)^q)(Mv^{(0)}(s) + L(v^{(0)}(s-\tau) - v^{(1)}(s-\tau)) \Big) ds \\
&\leq \psi(0)\Gamma(q) E_{q,q}(Mt^q) t^{q-1} + \int_0^t (t-s)^{q-1} E_{q,q}(M(t-s)^q) F(s, w^{(0)}(s), w^{(0)}(s-\tau)) ds \\
&+ \int_0^t (t-s)^{q-1} E_{q,q}(M(t-s)^q) \Big(M(v^{(0)}(s) - w^{(0)}(s)) + L(v^{(0)}(s-\tau) - w^{(0)}(s-\tau)) \Big) ds \\
&- \int_0^t (t-s)^{q-1} E_{q,q}(M(t-s)^q) \Big(Mv^{(0)}(s) + L(v^{(0)}(s-\tau) - v^{(1)}(s-\tau)) \Big) ds \\
&\leq \psi(0)\Gamma(q) E_{q,q}(Mt^q) t^{q-1} + \int_0^t (t-s)^{q-1} E_{q,q}(M(t-s)^q) F(s, w^{(0)}(s), w^{(0)}(s-\tau)) ds \\
&- \int_0^t (t-s)^{q-1} E_{q,q}(M(t-s)^q) Mw^{(0)}(s) \\
&\leq \psi(0)\Gamma(q) E_{q,q}(Mt^q) t^{q-1} + \int_0^t (t-s)^{q-1} E_{q,q}(M(t-s)^q) F(s, w^{(0)}(s), w^{(0)}(s-\tau)) ds \\
&- \int_0^t (t-s)^{q-1} E_{q,q}(M(t-s)^q) \Big(Mw^{(0)}(s) + L(w^{(0)}(s-\tau) - w^{(1)}(s-\tau)) \Big) ds \\
&= w^{(1)}(t), \quad t \in (\tau, 2\tau].
\end{aligned}$$

Similarly, we can prove

$$v^{(n)}(t) \leq w^{(n)}(t), \quad \text{for } t \in (\tau.2\tau], \quad n = 2, 3, \ldots.$$

Now consider the sequences $\{t^{1-q} v^{(n)}(t)\}_0^\infty$ and $\{t^{1-q} w^{(n)}(t)\}_0^\infty$. They are increasing and decreasing, respectively, and bounded. Thus, they are equicontinuous on $[0, T]$ (the proof is similar to that in [13] and we omit it). Therefore, they are uniformly convergent on $[0, T]$. Denote,

$\tilde{V}(t) = \lim_{n\to\infty} t^{1-q}v^{(n)}(t)$ and $\tilde{W}(t) = \lim_{n\to\infty} t^{1-q}w^{(n)}(t)$, $t \in [0,T]$. According to the above (b), (c) and (d) the inequalities

$$\begin{aligned} &t^{1-q}v^{(n)}(t) \le \tilde{V}(t),\ t\in[0,T], \quad \tilde{W}(t) \le t^{1-q}w^{(n)}(t),\ t\in[0,T],\ n=0,1,2,\dots, \\ &\tilde{V}(t) \le \tilde{W}(t),\ t\in[0,T]. \end{aligned} \tag{13}$$

hold.

From the uniform convergence of the sequences $\{t^{1-q}v^{(n)}(t)\}_0^\infty$ and $\{t^{1-q}w^{(n)}(t)\}_0^\infty$ we have the point-wise convergence of the sequences $\{v^{(n)}(t)\}_0^\infty$ and $\{w^{(n)}(t)\}_0^\infty$ on $(0,T]$ to $V(t) = \frac{\tilde{V}(t)}{t^{1-q}} \in C_{1-q}([0,T])$ and $W(t) = \frac{\tilde{W}(t)}{t^{1-q}} \in C_{1-q}([0,T])$, respectively.

Consider the continuous extension of the integral form of $t^{1-q}v^{(n+1)}(t)$ on $[0,T]$:

$$\begin{aligned} t^{1-q}v^{(n)}(t) &= \psi(0)\Gamma(q)E_{q,q}(Mt^q) \\ &\quad + t^{1-q}\int_0^t (t-s)^{q-1}E_{q,q}(M(t-s)^q)F(s,v^{(n-1)}(s),v^{(n-1)}(s-\tau))ds \\ &\quad - t^{1-q}\int_0^t (t-s)^{q-1}E_{q,q}(M(t-s)^q)\times \\ &\qquad \times\Big(Mv^{(n-1)}(s) + L(v^{(n-1)}(s-\tau) - v^{(n)}(s-\tau))\Big)\,ds. \end{aligned} \tag{14}$$

Take the limit in (14) and we obtain the Volterra fractional integral equation

$$\begin{aligned} \tilde{V}(t) &= \psi(0)\Gamma(q)E_{q,q}(Mt^q) \\ &\quad + t^{1-q}\int_0^t (t-s)^{q-1}E_{q,q}(M(t-s)^q)\Big(F(s,V(s),V(s-\tau)) - MV(s)\Big)ds, \quad \text{for } t\in(0,T], \end{aligned} \tag{15}$$

or

$$\begin{aligned} V(t) &= \psi(0)\Gamma(q)E_{q,q}(Mt^q)t^{1-q} \\ &\quad + \int_0^t (t-s)^{q-1}E_{q,q}(M(t-s)^q)\Big(F(s,V(s),V(s-\tau)) - MV(s)\Big)ds, \quad \text{for } t\in(0,T]. \end{aligned} \tag{16}$$

From equalities $t^{1-q}V(t)|_{t=0} = \tilde{V}(t)|_{t=0} = \lim_{n\to\infty}(t-0)^{1-q}v^{(n)}(t)|_{t=0} = \lim_{n\to\infty}\psi(0) = \psi(0)$ according to Proposition 2 applied to Equation (16) the limit function $V(t)$ is a solution of the linear FrDDE

$$ {}_{t_0}^{RL}D_t^q\, v(t) = Mv(t) - \Big(F(t,v(t),v(t-\tau)) - Mv(t)\Big) = F(t,v(t),v(t-\tau)), \quad t\in(0,T]. $$

Therefore, the function $v(t)$ is a solution of the IVP for FrDDE (1).

The proof about $w(t)$ is similar.

Proof of claim g). From claim (d) and the inequality (6) it follows that $t^{1-q}v^{(k)}(t) \le t^{1-q}w^{(k)}(t)$ for any fixed $t\in(0,T]$ and $k=1,2,\dots$. Then applying claim (e) we get $t^{1-q}v^{(k)}(t) \le t^{1-q}V(t) \le t^{1-q}W(t) \le t^{1-q}w^{(k)}(t)$ for any fixed $t\in(0,T]$. Therefore, $v^{(k)}(t) \le V(t) \le W(t) \le w^{(k)}(t)$ for any on $t\in(0,T]$. □

5. Application of the Suggested Algorithm

Now we will apply the algorithm suggested in Theorem 1 for approximate obtaining of the solution of nonlinear RL fractional differential equation with a delay. We will use computer realization of this algorithm to obtain the values of the approximate solutions and to graph them.

Example 1. *Let $\tau = 0.5$, $T = 1$ and consider the IVP for scalar nonlinear Riemann-Liouville FrDDE*

$$\begin{aligned} &{}_{0}^{RL}D_t^{0.5}x(t) = (x^2(t) + 0.05)\left(-0.5 + \frac{x(t-0.5)}{t+1}\right) \quad \text{for } t \in (0,1], \\ &x(t) = 0.5t \quad \text{for } t \in [-0.5, 0], \\ &t^{0.5}x(t)|_{t=0} = 0 \end{aligned} \tag{17}$$

with $\psi(t) = 0.5t$, $t \in [-0.5, 0]$, and $F(t, x, y) = (x^2 + 0.05)(-0.5 + \frac{y}{t+1})$.

The function

$$w(t) = \begin{cases} 0.5t, & t \in [-0.5, 0] \\ t^2, & t \in (0, 1] \end{cases}$$

is an upper solution on $[-0.5, 1]$ of the IVP for FrDE (17) since $t^{0.5}t^2|_{t=0} = 0$ and according to Proposition 1 with $q = 0.5$ and $\beta = 2$ the following inequalities

$${}_{0}^{RL}D_t^{0.5}t^2 = \frac{\Gamma(3)}{\Gamma(2.5)}t^{1.5} \geq \begin{cases} (t^4 + 0.05)\left(-0.5 + \frac{0.5(t-0.5)}{t+1}\right), & t \in (0, 0.5] \\ (t^4 + 0.05)\left(-0.5 + \frac{(t-0.5)^2}{t+1}\right), & t \in (0.5, 1] \end{cases}$$

are satisfied (see Figure 1).

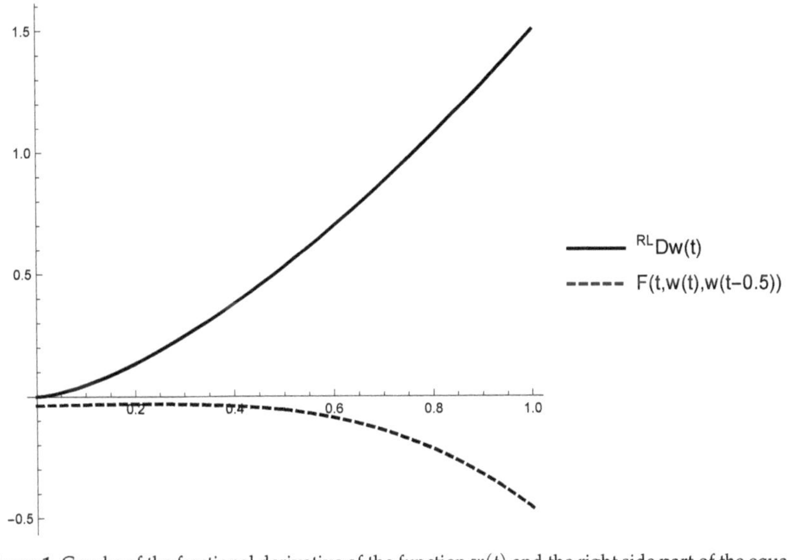

Figure 1. Graphs of the fractional derivative of the function $w(t)$ and the right side part of the equation on $[0, 1]$.

The function

$$v(t) = \begin{cases} 0.5t, & t \in [-0.5, 0] \\ -0.2t^{0.5}, & t \in (0, 1] \end{cases}$$

is a lower solution on $[-0.5, 1]$ *of the IVP for FrDDE (17) because* $t^{0.5}(-5t^{0.5})|_{t=0} = 0$ *holds and according to Proposition 1 with* $q = \beta = 0.5$ *the following inequalities*

$$ {}_0^{RL}D_t^{0.5}(-0.2t^{0.5}) = -0.2\Gamma(1.5) \leq \begin{cases} \left((0.2t^{0.5})^2 + 0.05\right)\left(-0.5 + 0.5\frac{t-0.5}{t+1}\right), & t \in (0, 0.5] \\ \left((-0.2t^{0.5})^2 + 0.05\right)\left(-0.5 - 0.2\frac{(t-0.5)^{0.5}}{t+1}\right), & t \in (0.5, 1] \end{cases} $$

are satisfied (see Figure 2).

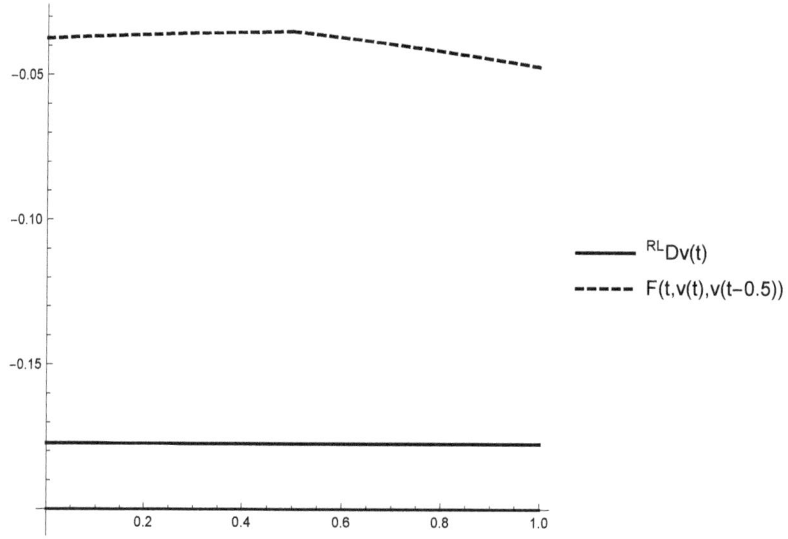

Figure 2. Graphs of the fractional derivative of the functions $v(t)$ and the right side part of the equation on $[0, 1]$.

Note that the lower and upper solutions $v(t)$ *and* $w(t)$ *are not unique. For example the function*

$$ w(t) = \begin{cases} 0.5t, & t \in [-0.5, 0] \\ t^3, & t \in (0, 1] \end{cases} $$

is also an upper solution. But we take just one lower (upper) solution to start the procedure.

Also, the inequality $v(t) \leq w(t)$ *on* $[-0.5, 1]$ *holds.*

For any $t \in [0, 1]$, $x, y, u, v \in \mathbb{R}$ *we have* $-0.2 \leq -0.2t^{0.5} = v(t) \leq x \leq y \leq w(t) = t^2 \leq 1$, $-0.25 \leq v(t-0.5) \leq u \leq v \leq w(t-0.5) \leq \sqrt{0.5}$ *and therefore,*

$$ \begin{aligned} F(t,x,u) - F(t,y,v) &= (x^2 + 0.05)\left(-0.5 + \frac{u}{t+1}\right) - (y^2 + 0.05)\left(-0.5 + \frac{v}{t+1}\right) \\ &= -0.5(x^2 + 0.05 - y^2 - 0.05) + (x^2 + 0.05)\frac{u}{t+1} \\ &\quad - (y^2 + 0.05)\frac{u}{t+1} + (y^2 + 0.05)\frac{u}{t+1} - (y^2 + 0.05)\frac{v}{t+1} \\ &= \left(\frac{u}{t+1} - 0.5\right)(x+y)(x-y) + \frac{y^2 + 0.05}{t+1}(u - v). \end{aligned} $$

Applying the inequalities $-0.5 \leq (\frac{u}{t+1} - 0.5) \leq \sqrt{0.5} - 0.5$, $-0.4 \leq x + y \leq 2$, *and* $(\frac{u}{t+1} - 0.5)(x + y) \geq -1$, $\frac{y^2+0.05}{t+1} \geq 0.05$, *we get the inequality* $F(t,x,u) - F(t,y,v) \leq M(x-y) + L(u-v)$ *with* $M = -1$, $L = 0.05 > 0$. *Therefore, all conditions of Theorem 1 are fulfilled.*

We apply the iterative scheme, suggested in Theorem 1, to obtain the successive approximations to the mild solution and to illustrate the claims of Theorem 1.

Define the zero approximation by $v^{(0)}(t) = v(t)$ *and* $w^{(0)}(t) = w(t)$ *for* $t \in [-0.5, 1]$.

Starting from the function $v^{(0)}(t)$ *we obtain the first lower approximation*

$$v^{(1)}(t) = \begin{cases} 0.5t, \quad t \in [-0.5, 0] \\ \int_0^t (t-s)^{-0.5} E_{0.5,0.5}(-(t-s)^{0.5})((v^{(0)}(s))^2 + 0.05)(-0.5 + \dfrac{v^{(0)}(s-0.5)}{s+1}))ds \\ \quad - \int_0^t (t-s)^{-0.5} E_{0.5,0.5}(-(t-s)^{0.5}) \times \\ \quad \times (-(v^{(0)}(s)) + 0.05)(v^{(0)}(s-0.5) - v^{(1)}(s-0.5))ds, \quad t \in (0, 1], \end{cases} \tag{18}$$

the second lower approximation

$$v^{(2)}(t) = \begin{cases} 0.5t, \quad t \in [-0.5, 0] \\ \int_0^t (t-s)^{-0.5} E_{0.5,0.5}(-(t-s)^{0.5})((v^{(1)}(s))^2 + 0.05)(-0.5 + \dfrac{v^{(1)}(s-0.5)}{s+1}))ds \\ \quad - \int_0^t (t-s)^{-0.5} E_{0.5,0.5}(-(t-s)^{0.5}) \times \\ \quad \times (-(v^{(1)}(s)) + 0.05)(v^{(1)}(s-0.5) - v^{(2)}(s-0.5))ds, \quad t \in (0, 1] \end{cases} \tag{19}$$

and so on.

About the upper approximations we start from $w^{(0)}(t)$ *and obtain the first upper approximation*

$$w^{(1)}(t) = \begin{cases} 0.5t, \quad t \in [-0.5, 0] \\ \int_0^t (t-s)^{-0.5} E_{0.5,0.5}(-(t-s)^{0.5})((w^{(0)}(s))^2 + 0.05)(-0.5 + \dfrac{w^{(0)}(s-0.5)}{s+1}))ds \\ \quad - \int_0^t (t-s)^{-0.5} E_{0.5,0.5}(-(t-s)^{0.5}) \times \\ \quad \times (-(w^{(0)}(s)) + 0.05)(w^{(0)}(s-0.5) - w^{(1)}(s-0.5))ds, \quad t \in (0, 1], \end{cases}$$

the second upper approximation

$$w^{(2)}(t) = \begin{cases} 0.5t, \quad t \in [-0.5, 0] \\ \int_0^t (t-s)^{-0.5} E_{0.5,0.5}(-(t-s)^{0.5})((w^{(1)}(s))^2 + 0.05)(-0.5 + \dfrac{w^{(1)}(s-0.5)}{s+1}))ds \\ \quad - \int_0^t (t-s)^{-0.5} E_{0.5,0.5}(-(t-s)^{0.5}) \times \\ \quad \times (-(w^{(1)}(s)) + 0.05)(w^{(1)}(s-0.5) - w^{(2)}(s-0.5))ds, \quad t \in (0, 1], \end{cases}$$

and so on.

The numerical values of the lower/upper approximations, given analytically above, are obtained by a computer program written in C#. We will briefly describe the computerized algorithm for obtaining these successive approximations:

The numerical values of the sequences of successive approximations $v^{(k)}(t)$ *and* $w^{(k)}(t)$, $k = 0, 1, 2, 3, \ldots$, $t \in [-0.5, 1]$, *are written in two dimensional arrays. The length of any of these arrays depends on the step in the interval* $[-0.5, 1]$.

We calculate in advance the values of the Mittag-Leffler function $E_{0.5,0.5}(-t^{0.5})$, $t \in [0,1)$, *in the points* t, *which will be used for numerical solving of the integrals* $\int_0^t ...ds, t \in (0,1)$ *(see Equations (18) and (19)). In the same points we also obtain the values of* $(t)^{-0.5}$. *These results are written in arrays with lengths, depending on the step on interval* $(0,1)$. *Note that the values of the Mittag-Leffler function are calculated by the help of the main definition (as an infinite sum) with an initially given error.*

We use the trapezoid method with an initially given error to solve numerically the integrals of the type $\int_0^t (t-s)^{-0.5} E_{0.5,0.5}(-(t-s)^{0.5}) \ldots ds$ *for each approximation k and any fixed* $t \in (0,1]$. *The values of both multipliers* $(t-s)^{-0.5}$ *and* $E_{0.5,0.5}(-(t-s)^{0.5})$ *are taken from initially formed arrays. Note that it could be used another numerical method for solving the required definite integrals.*

For example, to calculate the values of $v^{(k)}(t)$ *we use the following function*

```
private double Calc_v_t(double[,] v, int k, long it)
{
double f, pf, sum = 0, s = 0, q = 0.5;
long shift = (long)(0.5/eps)+1;
long sh2 = shift/2;
long i = shift, ie = it;

pf = PowTmS[ie] * Eqq[ie] *
((v[k-1,i]*v[k-1,i]+0.05) * (-q+v[k-1,i-sh2]/(s+1)) -
(-v[k-1,i]+0.05*(v[k-1,i-sh2]-v[k,i-sh2])));

while (s < tval)
{
i++; ie--; s += eps;
f = PowTmS[ie] * Eqq[ie] *
((v[k-1,i]*v[k-1,i]+0.05) * (-q+v[k-1,i-sh2]/(s+1)) -
(-v[k-1,i]+0.05*(v[k-1,i-sh2]-v[k,i-sh2])));
sum += (pf + f) * eps;
pf = f;
}
return sum / 2;
}
```

A part of the obtained numerical values of the successive approximations are given in Table 1 and they are used to generate the graphs on Figures 3–6).

Table 1 and Figures 3–6 illustrate the claims of Theorem 1 for the obtained successive approximations:

- *claim (b) - the sequence of lower approximate solutions* $v^{(n)}(t)$, $n = 0, 1, 2, 3$ *is increasing (see Figure 4 and the last four columns of Table 1);*
- *claim (c)- the sequence of upper approximate solutions* $w^{(n)}(t)$, $n = 0, 1, 2, 3$ *is decreasing (see Figure 5 and the first four columns of Table 1);*
- *claim (d) - the inequality* $v^{(3)}(t) \le w^{(3)}(t)$, $t \in [0,1]$ *holds (see Figure 6 and the 5-th and 6-th columns of Table 1).*

According to the claim (g) of Theorem 1 the mild solutions $V(t)$ *and* $W(t)$ *of the FrDDE (17) are between the last obtained lower solution* $v^{(3)}(t)$ *and upper solution* $w^{(3)}(t)$. *So, practically the suggested algorithm for the approximate solving of IVP for FrDDE gives us a lower and upper bounds of the unknown exact solution.*

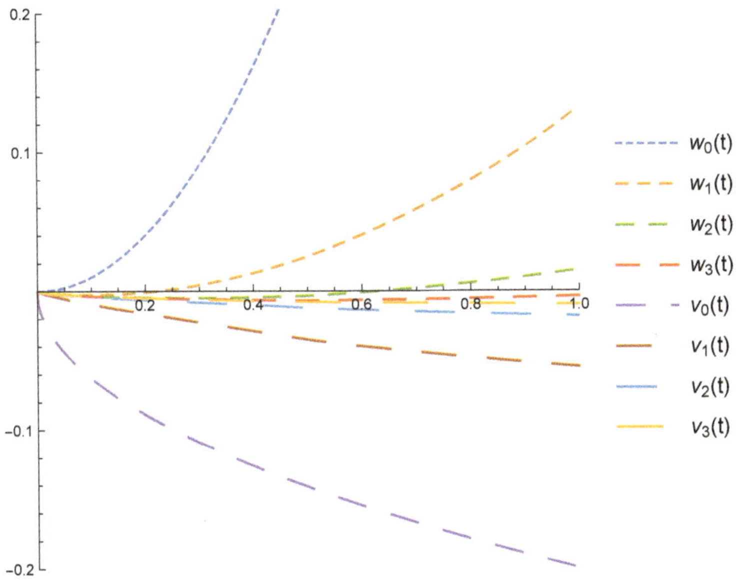

Figure 3. Graphs of the upper/lpwer successive approximations $v^{(n)}(t)$ and $w^{(n)}(t)$, $n = 0, 1, 2, 3$, on the interval $[0, 1]$.

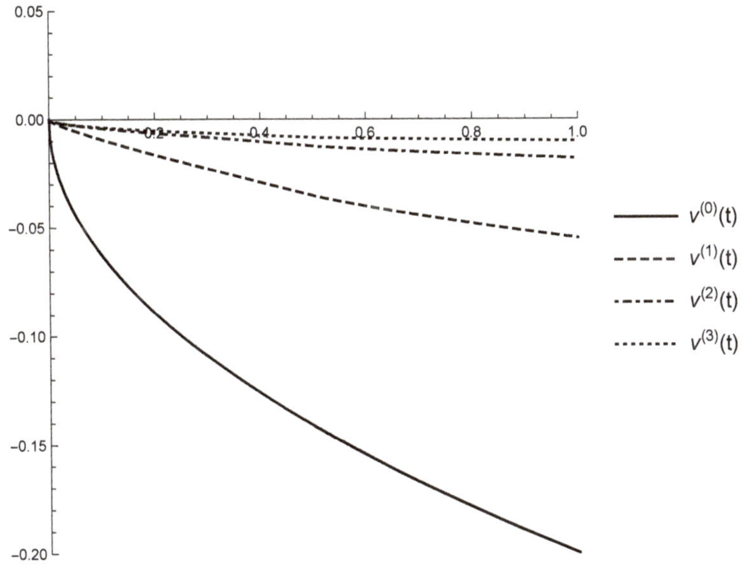

Figure 4. Graphs of the successive lower approximations $v^{(n)}(t)$, $n = 0, 1, 2, 3$, on the interval $[0, 1]$.

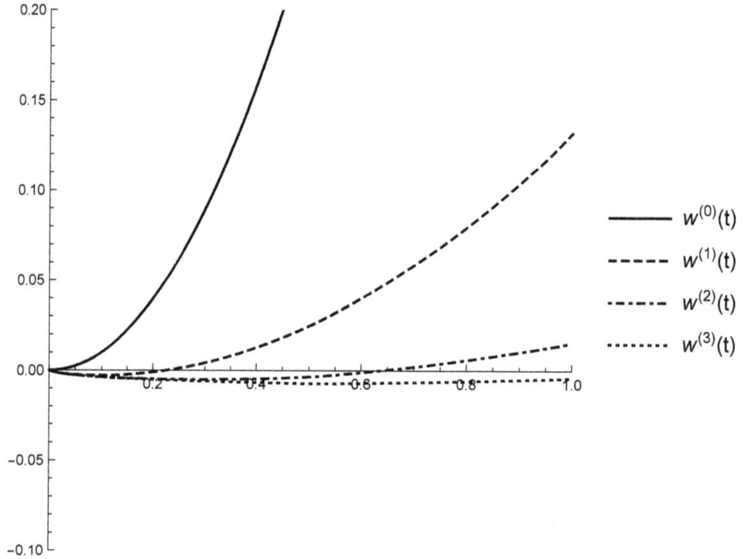

Figure 5. Graphs of the successive upper approximations $w^{(n)}(t)$, $n = 0, 1, 2, 3$, on the interval $[0, 1]$.

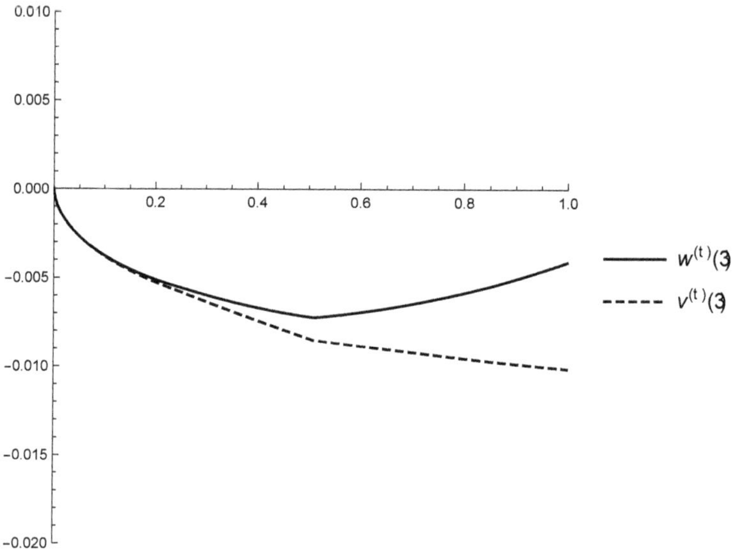

Figure 6. Graphs of the successive approximations $v^{(3)}(t)$ and $w^{(3)}(t)$ on the interval $[0, 1]$.

Table 1. Values of successive approximations $w^{(n)}(t)$ and $v^{(n)}(t)$, $n = 0, 1, 2, 3$ for $t \in [0,1]$.

t	$w^{(0)}(t)$	$w^{(1)}(t)$	$w^{(2)}(t)$	$w^{(3)}(t)$	$v^{(3)}(t)$	$v^{(2)}(t)$	$v^{(1)}(t)$	$v^{(0)}(t)$
0	0	0	0		0	0	0	0
...	...	...	...	...	...	...	...	...
0.05	0.0025000	−0.0024014	−0.002669875	−0.0026830	−0.0026910	−0.0028370	−0.0054822	−0.0447213
...	...	...	...	...	...	...	...	...
0.1	0.0100000	−0.0028030	−0.003707587	−0.0037665	−0.0038006	−0.0042116	−0.0094143	−0.0632455
...	...	...	...	...	...	...	...	...
0.15	0.0225000	−0.0023273	−0.004371989	−0.0045293	−0.0046100	−0.0053537	−0.0130148	−0.0774596
...	...	...	...	...	...	...	...	...
0.2	0.0400000	−0.0010073	−0.004782033	−0.0051102	−0.0052606	−0.0063855	−0.0164157	−0.0894427
...	...	...	...	...	...	...	...	...
0.25	0.0625000	0.0011752	−0.004974678	−0.0055640	−0.0058104	−0.0073536	−0.0196715	−0.1000000
...	...	...	...	...	...	...	...	...
0.3	0.0900000	0.0041728	−0.005025551	−0.0059818	−0.0063535	−0.0083442	−0.0228876	−0.1095445
...	...	...	...	...	...	...	...	...
0.35	0.1225000	0.0080102	−0.004920663	−0.0063617	−0.0068915	−0.0093546	−0.0260671	−0.1183215
...	...	...	...	...	...	...	...	...
0.4	0.1600000	0.0127018	−0.004640016	−0.0066932	−0.0074170	−0.0103733	−0.0291992	−0.1264911
...	...	...	...	...	...	...	...	...
0.45	0.2025000	0.0182403	−0.004175051	−0.0069743	−0.0079311	−0.0113985	−0.0322849	−0.1341640
...	...	...	...	...	...	...	...	...
0.5	0.2500000	0.0246100	−0.003520597	−0.0072042	−0.0084358	−0.0124294	−0.0353267	−0.1414213
...	...	...	...	...	...	...	...	...
0.55	0.3025000	0.0320137	−0.002426030	−0.0071326	−0.0086820	−0.0131993	−0.0378996	−0.1483239
...	...	...	...	...	...	...	...	...
0.6	0.3600000	0.0402038	−0.001086952	−0.0069496	−0.0088569	−0.0138649	−0.0401436	−0.1549193
...	...	...	...	...	...	...	...	...
0.65	0.4225000	0.0490747	0.000417801	−0.0067284	−0.0090315	−0.0145008	−0.0422464	−0.1612451
...	...	...	...	...	...	...	...	...
0.7	0.4900000	0.0585998	0.002078399	−0.0064716	−0.0092072	−0.0151118	−0.0442469	−0.1673320
...	...	...	...	...	...	...	...	...
0.75	0.5625000	0.0687781	0.003887027	−0.0061807	−0.0093841	−0.0157009	−0.0461668	−0.1732050
...	...	...	...	...	...	...	...	...
0.8	0.6400000	0.0796462	0.005842443	−0.0058527	−0.0095575	−0.0162662	−0.0480153	−0.1788854
...	...	...	...	...	...	...	...	...
0.85	0.7225000	0.0912844	0.007949841	−0.0054815	−0.0097203	−0.0168024	−0.0497938	−0.1843908
...	...	...	...	...	...	...	...	...
0.9	0.8100000	0.1038155	0.010211677	−0.0050684	−0.0098728	−0.0173114	−0.0515115	−0.1897366
...	...	...	...	...	...	...	...	...
0.95	0.9025000	0.1174206	0.012636895	−0.0046141	−0.0100154	−0.0177952	−0.0531754	−0.1949358
...	...	...	...	...	...	...	...	...
0.999	0.9980010	0.1320345	0.015188142	−0.0041287	−0.0101456	−0.0182462	−0.0547594	−0.1998999

6. Conclusions

The main aim of the paper is to suggest a scheme for the approximate solving of the initial value problem for scalar nonlinear Riemann-Liouville fractional differential equations with a constant delay on a finite interval. The iterative scheme is based on the method of lower and upper solutions. In connection with this, mild lower and mild upper solutions are defined. An algorithm for constructing two monotone sequences of mild lower and mild upper solutions, respectively, is given. It is proved both sequences are convergent to the exact solution of the studied problem. The iterative scheme is used in a computer environment to illustrate its application for solving a particular nonlinear problem. The suggested and computerized algorithm can be applied to solve approximately and to study the behavior of scalar models with RL fractional derives and delays. The practical application requires the next step in the investigations, more exactly to obtain an algorithm for approximate solving of systems with RL derivatives and delays.

Author Contributions: Conceptualization, S.H.; Methodology, S.H.; Software, A.G. and K.S.; Validation, A.G. and K.S.; Formal Analysis, S.H.; Writing—Original Draft Preparation, S.H., A.G. and K.S.; Writing—Review and Editing, S.H., A.G. and K.S.; Visualization, A.G. and K.S.; Supervision, S.H.; Funding Acquisition, S.H. All authors have read and agreed to the published version of the manuscript.

Funding: S.H. is partially supported by the Bulgarian National Science Fund under Project KP-06-N32/7.

Conflicts of Interest: The authors declare no conflict of interest.

References

1. Das, S. *Functional Fractional Calculus*; Springer: Berlin/Heidelberg, Germany, 2011.
2. Podlubny, I. *Fractional Differential Equations*; Academic Press: San Diego, CA, USA, 1999.
3. Ladde, G.; Lakshmikantham, V.; Vatsala, A. *Monotone Iterative Techniques for Nonlinear Differential Equations*; Pitman: Boston, MA, USA, 1985 .
4. Hristova, S.; Golev, A.; Stefanova, K. Approximate method for boundary value problems of anti-periodic type for differential equations with "maxima". *Bound. Value Probl.* **2013**, *2013*, 12. [CrossRef]
5. Agarwal, R.; Hristova, S.; Golev, A.; Stefanova, K. Monotone-iterative method for mixed boundary value problems for generalized difference equations with "maxima". *J. Appl. Math. Comput.* **2013**, *43*, 213–233. [CrossRef]
6. Agarwal, R.; Golev, A.; Hristova, S.; O'Regan, D.; Stefanova, K. Iterative techniques with computer realization for the initial value problem for Caputo fractional differential equations. *J. Appl. Math. Comput.* **2017**. [CrossRef]
7. Agarwal, R.; Golev, A.; Hristova, S.; O'Regan, D. Iterative techniques with computer realization for initial value problems for Riemann–Liouville fractional differential equations. *J. Appl. Anal.* **2020**. [CrossRef]
8. Bai, Z.; Zhang, S.; Sun, S.; Yin, C. Monotone iterative method for fractional differential equations. *Electron. J. Differ. Equ.* **2016**, *2016*, 1–8.
9. Denton, Z. Monotone method for Riemann-Liouville multi-order fractional differential systems. *Opusc. Math.* **2016**, *36*, 189–206. [CrossRef]
10. Wang, G.; Baleanu, D.; Zhang, L. Monotone iterative method for a class of nonlinear fractional differential equations. *Fract. Calc. Appl. Anal.* **2012**, *15*, 244–252. [CrossRef]
11. Agarwal, R.; Hristova, S.; O'Regan, D. Explicit solutions of initial value problems for linear scalar Riemann-Liouville fractional differential equations with a constant delay. *Mathematics* **2020**, *8*, 32. [CrossRef]
12. Denton, Z.; Ng, P.; Vatsala, A. Quasilinearization method via lower and upper solutions for Riemann-Liouville fractional differential equations. *Nonlinear Dyn. Syst. Theory* **2011**, *11*, 239–251.
13. Ramırez, J.; Vatsala, A. Monotone iterative technique for fractional differential equations with periodic boundary conditions. *Opusc. Math.* **2009**, *29*, 289–304. [CrossRef]

mathematics

MDPI

Article

Infinitely Many Homoclinic Solutions for Fourth Order p-Laplacian Differential Equations

Stepan Tersian

Institute of Mathematics and Informatics, Bulgarian Academy of Sciences (BAS), 1113 Sofia, Bulgaria; sterzian@uni-ruse.bg

Received: 17 March 2020; Accepted: 1 April 2020; Published: 2 April 2020

Abstract: The existence of infinitely many homoclinic solutions for the fourth-order differential equation $(\varphi_p(u''(t)))'' + w(\varphi_p(u'(t)))' + V(t)\varphi_p(u(t)) = a(t)f(t,u(t)), t \in \mathbb{R}$ is studied in the paper. Here $\varphi_p(t) = |t|^{p-2}t, p \geq 2$, w is a constant, V and a are positive functions, f satisfies some extended growth conditions. Homoclinic solutions u are such that $u(t) \to 0, |t| \to \infty, u \neq 0$, known in physical models as ground states or pulses. The variational approach is applied based on multiple critical point theorem due to Liu and Wang.

Keywords: homoclinic solutions; fourth-order p-Laplacian differential equations; minimization theorem; Clark's theorem

---

1. Introduction

In this paper, we study the existence of infinitely many nonzero solutions homoclinic solutions for the fourth-order p-Laplacian differential equation

$$(\varphi_p(u''(t)))'' + w(\varphi_p(u'(t)))' + V(t)\varphi_p(u(t)) = a(t)f(t,u(t)), \tag{1}$$

where $t \in \mathbb{R}$, w is a constant, $\varphi_p(t) = |t|^{p-2}t$, for $p \geq 2$, V is a positive bounded function, a is a positive continuous function and $f \in C^1(\mathbb{R}, \mathbb{R})$ satisfies some growth conditions with respect to p. As usual, we say that a solution u of (1) is a nontrivial homoclinic solution to zero solution of (1) if

$$u \neq 0, u(t) \to 0, \qquad |t| \to \infty. \tag{2}$$

They are known in phase transitions models as ground states or pulses (see [1]). The existence of homoclinic and heteroclinic solutions of fourth-order equations is studied by various authors (see [2–12] and references therein). Sun and Wu [4] obtained existence of two homoclinic solutions for a class of fourth-order differential equations:

$$u^{(4)} + wu'' + a(t)u = f(t,u) + \lambda h(t)|u|^{p-2}u, \ t \in \mathbb{R},$$

where w is a constant, $\lambda > 0, 1 \leq p < 2$, $a \in C(\mathbb{R}, \mathbb{R}^+)$ and $h \in L^{\frac{2}{2-p}}(\mathbb{R})$ by using mountain pass theorem.

Yang [8] studies the existence of infinitely many homoclinic solutions for a the fourth-order differential equation:

$$u^{(4)} + wu'' + a(t)u = f(t,u), \ t \in \mathbb{R},$$

where w is a constant, $a \in C(\mathbb{R})$ and $f \in C(\mathbb{R} \times \mathbb{R}, \mathbb{R})$. A critical point theorem, formulated in the terms of Krasnoselskii's genus (see [13], Remark 7.3), is applied, which ensures the existence of infinitely many homoclinic solutions.

We suppose the following conditions on the functions a, f and V.

(A) $a \in C(\mathbb{R}, \mathbb{R}^+)$ and $a(t) \to 0$ as $|t| \to +\infty$.

(F_1) There are numbers p and q s.t. $1 < q < 2 \leq p$ and for $f \in C^1(\mathbb{R}, \mathbb{R})$

$$uf(t,u) \leq qF(t,u),\ \forall u \in \mathbb{R}, u \neq 0,$$

where $F(t,u) = \int_0^u f(t,x)dx$.

(F_2) $|f(t,u)| \leq b(t)|u|^{q-1}$, $\forall (t,u) \in \mathbb{R} \times \mathbb{R}$, where b is a positive function, s.t. $b \in L^r(\mathbb{R}) \bigcap L^{\frac{p}{2-q}}(\mathbb{R})$, where $r = \frac{p}{p-q}$.

(F_3) There exists an interval $J \subset \mathbb{R}$ and a constant $c > 0$ s. t. $F(t,u) \geq c|u|^q$, $\forall (t,u) \in J \times \mathbb{R}$.

(F_4) $F(t,-u) = F(t,u)$ for all $(t,u) \in \mathbb{R} \times \mathbb{R}$.

(V) There exist positive constants v_1 and v_2 such that $0 < v_1 \leq V(t) \leq v_2$, $\forall t \in \mathbb{R}$.

Let

$$w^* = \inf_{u \neq 0} \frac{\int_{\mathbb{R}} \left(|u''(t)|^p + |u(t)|^p\right) dt}{\int_{\mathbb{R}} |u'(t)|^p\, dt}.$$

Denote by X the Sobolev's space

$$X := W^{2,p}(\mathbb{R}) = \{u \in L^p(\mathbb{R}) : u' \in L^p(\mathbb{R}), u'' \in L^p(\mathbb{R})\},$$

equipped by the usual norm

$$||u||_X := \left(\int_{\mathbb{R}} \left(|u''(t)|^p + |u'(t)|^p + |u(t)|^p\right) dt\right)^{1/p}.$$

The functional $I : X \to \mathbb{R}$ is defined as follows

$$I(u) = \int_{\mathbb{R}} (\Phi_p(u''(t)) - w\Phi_p(u'(t)) + V(t)\Phi_p(u(t)))dt - \int_{\mathbb{R}} a(t)F(t,u(t)dt, \tag{3}$$

where $\Phi(t) = \frac{|t|^p}{p}$ for $p \geq 2$.

Under conditions $(A), (F_1) - (F_3)$ and V the functional I is differentiable and for all $u, v \in X$ we have

$$\begin{aligned} \langle I'(u), v\rangle &= \int_{\mathbb{R}} \left(\varphi_p\left(u''(t)\right) v''(t) - w\varphi_p\left(u'(t)\right) v'(k)\right) dt + V(t)\varphi_p\left(u(t)\right) v(t)dt \\ &\quad - \int_{\mathbb{R}} a(t)f\left(t, u(t)\right) v(t)dt. \end{aligned}$$

where $\langle .,. \rangle$ means the duality pairing between X and it's dual space X^*. The homoclinic solutions of the Equation (1) are the critical points of the functional I, i.e., u_0 is a homoclinic solution of the problem if $\langle I'(u_0), v\rangle = 0$ for every $v \in X$ (see [6,11,12]).

Let $v_0 = \min\{1, v_1\}$, where v_1 is the positive constant from condition (V). Our main result is:

Theorem 1. *Let $p \geq 2$, $w < v_0 w^*$ and the functions a, f and V satisfy the assumptions (A), $(F_1) - (F_3)$ and (V). Then the Equation (1) has at least one nonzero homoclinic solution $u_0 \in X$. Additionally if (F_4) holds, the Equation (1) has infinitely many nonzero solutions u_j such that $||u_j||_\infty \to 0$ as $j \to \infty$.*

Remark 1. *An example of a function $f(t,u)$, which satisfies the assumptions $(F_1) - (F_4)$ is as follows. Let $p = 3, q = \frac{3}{2}$ and $f(t,u) = \alpha(t)|u|^{1/2}u$, where*

$$\alpha(t) = \begin{cases} \frac{3-t^2}{2}, & |t| \leq 1, \\ \frac{1}{|t|}, & |t| \geq 1. \end{cases}$$

We have that $r = \frac{p}{p-q} = 2$, $\frac{p}{2-q} = 6$ *and* $b(t) = \alpha(t) \in L^2(\mathbb{R}) \cap L^6(\mathbb{R})$, *because* $\int\limits_1^\infty \frac{1}{t^2} dt = 1$ *and* $\int\limits_1^\infty \frac{1}{t^6} dt = \frac{1}{5}$. *Moreover* $\alpha(t) \geq 1$ *if* $t \in (-1,1) = J$. *Next, we have*

$$\begin{aligned} |f(t,u)| &= \alpha(t)|u|^{3/2}, \\ F(t,u) &= \frac{2}{5}\alpha(t)|u|^{5/2}, \end{aligned}$$

and $F(t,u) \geq \frac{2}{5}|u|^{5/2}$, $t \in J = (-1,1)$.

As an open problem we state the existence of weak solutions of the problem when $1 < q < p < 2$.

This paper is organized as follows. In Section 2 we present the variational formulation of the problem and critical point theorems used in the proof of the main result. In Section 3, we give the proof of Theorem 1.

2. Preliminaries

In this section we give the variational formulation of the problem and present two critical point theorems.

Let X_1 be the Sobolev's space

$$X_1 := \{u \in X : \int_{\mathbb{R}} \left(|u''(t)|^p - w\,|u'(t)|^p + V(t)\,|u(t)|^p\right) dt < \infty\},$$

equipped by the norm

$$||u|| := \left(\int_{\mathbb{R}} \left(|u''(t)|^p - w\,|u'(t)|^p + V(t)\,|u(t)|^p\right) dt\right)^{1/p}.$$

Denote

$$w^* = \inf_{u \neq 0} \frac{\int_{\mathbb{R}} \left(|u''(t)|^p + |u(t)|^p\right) dt}{\int_{\mathbb{R}} |u'(t)|^p\, dt}.$$

and $v_0 = \min\{1, v_1\}$. The next lemma shows that under condition (V) for $w < v_0 w^*$ the norms $||.||$ and $||.||_X$ are equivalent and $X = X_1$.

Lemma 1. *Let* $w < v_0 w^*$. *Then, there exists a constant* $C > 0$ *such that*

$$\int_{\mathbb{R}} \left(|u''(t)|^p - w\,|u'(t)|^p + V(t)\,|u(t)|^p\right) dt \geq C\,||u||_X^p\,, \ \forall u \in X. \tag{4}$$

Proof of Lemma 1. In view of Lemma 4.10 in [14], there exists a positive constant $K = K(p)$ depending only on p such that

$$\int_{\mathbb{R}} |u'(t)|^p\, dt \leq K \int_{\mathbb{R}} \left(|u''(t)|^p + |u(t)|^p\right) dt.$$

Then

$$\frac{1}{K} \leq w^* = \inf_{u \neq 0} \frac{\int_{\mathbb{R}} \left(|u''(t)|^p + |u(t)|^p\right) dt}{\int_{\mathbb{R}} |u'(t)|^p\, dt}.$$

Let

$$C_0 = \frac{v_0 w^* - w}{(K+1) v_0 w^*}$$

and $C = v_0 C_0$. We have

$$\begin{aligned}
&\int_{\mathbb{R}}\left(|u''(t)|^p - w\,|u'(t)|^p + V(t)\,|u(t)|^p\right)dt \\
\geq\ & v_0\int_{\mathbb{R}}\left(|u''(t)|^p - \frac{w}{v_0}\,|u'(t)|^p + |u(t)|^p\right)dt \\
=\ & v_0((1-\frac{w}{v_0w^*})\int_{\mathbb{R}}\left(|u''(t)|^p + |u(t)|^p\right)dt \\
& +\frac{w}{v_0w^*}\int_{\mathbb{R}}\left(|u''(t)|^p - w^*\,|u'(t)|^p + |u(t)|^p\right)dt) \\
\geq\ & v_0(1-\frac{w}{v_0w^*})\int_{\mathbb{R}}\left(|u''(t)|^p + |u(t)|^p\right)dt \\
=\ & v_0C_0(K+1)\int_{\mathbb{R}}\left(|u''(t)|^p + |u(t)|^p\right)dt \\
\geq\ & v_0C_0\int_{\mathbb{R}}\left(|u''(t)|^p + |u'(t)|^p + |u(t)|^p\right)dt = C||u||_X^p,
\end{aligned}$$

which completes the proof. □

By Brezis [15], Theorem 8.8 and Corollary 8.9 for $u \in X$ and $s > p$

$$\begin{aligned}
||u||_\infty\ &: = ||u||_{L^\infty(\mathbb{R})} \leq C_1||u||_X, \\
\int_{\mathbb{R}}|u(t)|^s dt\ &\leq\ ||u||_\infty^{s-p}||u||_X^p,
\end{aligned}$$

and $\lim\limits_{|t|\to\infty} u(t) = 0.$

We consider the functional $I : X \to \mathbb{R}$

$$I(u) = \int_{\mathbb{R}}(\Phi_p(u''(t)) - w\Phi_p(u'(t)) + V(t)\Phi_p(u(t)))dt - \int_{\mathbb{R}} a(t)F(t,u(t)dt, \tag{5}$$

where $\Phi(t) = \frac{|t|^p}{p}$ for $p \geq 2$.

One can show that under conditions $(A), (F_1) - (F_3)$ and V the functional I is differentiable and for all $u, v \in X$ we have

$$\begin{aligned}
\langle I'(u), v\rangle = \int_{\mathbb{R}}\left(\varphi_p\left(u''\left(t\right)\right)v''(t) - w\varphi_p\left(u'\left(t\right)\right)v'(k)\right)dt + V(t)\varphi_p\left(u(t)\right)v(t)dt \\
- \int_{\mathbb{R}} a(t)f\left(t, u(t)\right)v(t)dt. \qquad (6)
\end{aligned}$$

Let $L_a^p(\mathbb{R}), p \geq 1$ be the weighted Lebesque space of functions $u : \mathbb{R} \to \mathbb{R}$ with norm $||u||_{p,a} := \left(\int\limits_{\mathbb{R}} a(t)|u(t)|^p dt\right)^{1/p}$. We have

Lemma 2. *Assume that the assumptions (A) and (V) hold. Then, the inclusion $X \subset L_a^p(\mathbb{R})$ is continuous and compact.*

Proof of Lemma 2. The embedding $X \subset L_a^p(\mathbb{R})$ is continuous by the boundedness of the function a by (A). We show that the inclusion is compact. Let $\{u_j\} \subset X$ be a sequence such that $||u_j|| \leq M$ and $u_j \rightharpoonup u$ weakly in X. We'll show that $u_j \to u$ strongly in $L_a^p(\mathbb{R})$. Without loss of generality we can assume that $u = 0$, considering the sequence $\{u_j - u\}$. By (A) for any $\varepsilon > 0$, there exists $R > 0$, such that for $|t| \geq R$

$$0 \leq a(t) \leq \frac{\varepsilon}{2(1+M^p)}.$$

Then

$$\int_{|t|\geq R} a(t)|u_j(t)|^p dt \leq \frac{\varepsilon M^p}{2(1+M^p)}.$$

By Sobolev's imbedding theorem $u_j \to 0$ strongly in $C([-R,R])$ and there exists j_0 such that for $j > j_0$:

$$\int_{|t|\leq R} a(t)|u_j(t)|^p dt < \frac{\varepsilon}{2(1+M^p)}.$$

Then, for $j > j_0$ we have $\int_{\mathbb{R}} a(t)|u_j(t)|^p dt < \varepsilon$, which shows that $u_j \to 0$ strongly in $L^p_a(\mathbb{R})$. □

Lemma 3. *Let assumptions* $(A), (F_1) - (F_3)$ *and* (V) *hold. If* $u_j \rightharpoonup u$ *weakly in* X*, there exists a subsequence of the sequence* $\{u_j\}$ *, still denoted by* $\{u_j\}$ *such that* $f(t,u_j) \to f(t,u)$ *in* $L^p_a(\mathbb{R})$.

Proof of Lemma 3. Let $u_j \rightharpoonup u$ weakly in X. By Banach-Steinhaus theorem there exists $M_1 > 0$, such that $||u_j|| \leq M_1$ and $||u|| \leq M_1$. By the elementary inequality for $a > 0, b > 0, p > 1$

$$(a+b)^p \leq 2^{p-1}(a^p + b^p),$$

and (F_2) we have

$$\begin{aligned}|f(t,u_j) - f(t,u)|^p &\leq 2^{p-1}(|f(t,u_j)|^p + |f(t,u)|^p) \\ &\leq 2^{p-1}|b(t)|^p(|u_j|^{p(q-1)} + |u|^{p(q-1)}).\end{aligned}$$

Let $0 < a(t) \leq A$. Then, by Hölder inequality and $b \in L^{\frac{p}{2-q}}(\mathbb{R})$ it follows that

$$\begin{aligned}&\int_{\mathbb{R}} a(t)|f(t,u_j(t)) - f(t,u(t))|^p dt \\ \leq\ & 2^{p-1}A\int_{\mathbb{R}} |b(t)|^p(|u_j|^{p(q-1)} + |u|^{p(q-1)})dt \\ \leq\ & 2^{p-1}A(\int_{\mathbb{R}} |b(t)|^{\frac{p}{2-q}})^{2-q}((\int_{\mathbb{R}} |u_j(t)|^p dt)^{q-1} + (\int_{\mathbb{R}} |u(t)|^p dt)^{q-1}) \\ \leq\ & 2^p A||b||^p_{L^{\frac{p}{2-q}}(\mathbb{R})} M_1^{p(q-1)}.\end{aligned}$$

By Lemma 2, $u_j \rightharpoonup u$ weakly in X implies that there exists a subsequence $\{u_j\}$, such that $u_j \to u$ strongly in $L^p_a(\mathbb{R})$. By analogous way as above we have that there exists $B > 0$, such that

$$\int_{\mathbb{R}} |f(t,u_j(t)) - f(t,u(t))|^p dt \leq B.$$

Let $\varepsilon > 0, R > 0$ are s.t. $0 < a(t) < \frac{\varepsilon}{2B}$ for $|t| \geq R$ by (A). Then

$$\int_{|t|\geq R} a(t)|f(t,u_j(t)) - f(t,u(t))|^p dt < \frac{\varepsilon}{2}. \tag{7}$$

Let $0 < a_R < a(t) \leq A$ for $|t| \leq R$. By $u_j \to u$ strongly in $L^p_a(\mathbb{R})$ it follows that

$$\int_{|t|\leq R} a(t)|u_j(t) - u(t)|^p dt \geq a_R \int_{|t|\leq R} |u_j(t) - u(t)|^p dt \to 0$$

and $u_j(t) - u(t) \to 0$ a.e. in $|t| \leq R$. Then, by Lebesque's dominated convergence theorem

$$I_R := \int_{|t| \leq R} a(t)|f(t, u_j(t)) - f(t, u(t))|^p dt \to 0.$$

Let j_0 is sufficiently large, such that for $j > j_0, 0 \leq I_R < \frac{\varepsilon}{2}$. Then by (7) for $j > j_0$ we have

$$\int_{\mathbb{R}} a(t)|f(t, u_j(t)) - f(t, u(t))|^p dt < \varepsilon,$$

which completes the proof. □

Next we have:

Lemma 4. *Under assumptions $(A), (F_1) - (F_3), (V)$ the functional $I \in C^1(X, \mathbb{R})$ and the identity (6) holds for all $u, v \in X$. holds.*

It can be proved in a standard way using Lemma 3 (see Yang [8], Tersian, Chaparova [6]).

Lemma 5. *Under assumptions $(A), (F_1) - (F_3)$ and (V) the functional I satisfies the (PS) condition.*

Proof of Lemma 5. Let $\{u_j\}$ be a sequence such that $\{I(u_j)\}$ is bounded in X and $I'(u_j) \to 0$ in X^*. Then, there exists a constant $C_1 > 0$, s.t.

$$||I(u_j)|| \leq C_1, \quad ||I'(u_j)||_{X^*} \leq C_1.$$

By (F_2) we have

$$\begin{aligned} C_1 + \frac{C_1}{q}||u_j|| &\geq \frac{1}{q} < I'(u_j), u_j > -I(u_j) \\ &= \left(\frac{1}{q} - \frac{1}{p}\right)||u_j||^p + \int_{\mathbb{R}} a(t)(F(t, u_j(t)) - \frac{1}{q} f(t, u_j(t))u_j(t))dt \\ &\geq \left(\frac{1}{q} - \frac{1}{p}\right)||u_j||^p. \end{aligned}$$

Then, $\{u_j\}$ is a bounded sequence in X and up to a subsequence, still denoted by $\{u_j\}$, $u_j \rightharpoonup u$ weakly in X. There exists $M_2 > 0$, such that $||u_j|| \leq M_2, ||u|| \leq M_2$. By Lemma 2, $u_m \to u$ in $L^2_a(\mathbb{R})$ and by Lemma 3, $f(t, u_m(t)) \to f(t, u(t))$ in $L^2_a(\mathbb{R})$. By Hölder inequality we have:

$$\begin{aligned} I_j &:= \int_{\mathbb{R}} a(t)(f(t, u_j(t)) - f(t, u(t)))(u_j(t) - u(t))dt \\ &= \int_{\mathbb{R}} a^{\frac{p-1}{p}}(t)(f(t, u_j(t)) - f(t, u(t)))a^{\frac{1}{p}}(t)(u_j(t) - u(t))dt \\ &\leq A^{\frac{p-1}{p}} \int_{\mathbb{R}} a(t)|u_j(t) - u(t)|^p dt \left(\int_{\mathbb{R}} |f(t, u_j(t)) - f(t, u(t))|^{\frac{p}{p-1}} dt\right)^{\frac{p-1}{p}}. \end{aligned}$$

As in the proof of Lemma 3, by assumption (F_2), $b \in L^{\frac{p}{p-q}}(\mathbb{R})$ and Hölder inequality we have for $p_1 = \frac{p}{p-1} > 1$:

$$\begin{aligned}
&\int_{\mathbb{R}} |f(t,u_j(t)) - f(t,u(t))|^{p_1} dt \\
\leq\ & 2^{p_1-1} \int_{\mathbb{R}} |b(t)|^{p_1} | \left(|u_j(t)|^{(q-1)p_1} + |u(t)|^{(q-1)p_1} \right) dt \\
\leq\ & 2^{p_1-1} \left(\int_{\mathbb{R}} |b|^{\frac{p}{p-q}} dt \right)^{\frac{p-q}{p-1}} \left(\left(\int_{\mathbb{R}} |u_j|^p dt \right)^{\frac{q-1}{p-1}} + \left(\int_{\mathbb{R}} |u|^p dt \right)^{\frac{q-1}{p-1}} \right) \\
\leq\ & 2^{p_1} ||b||^{p_1}_{L^{\frac{p}{p-q}}} M_2^{(q-1)p_1}.
\end{aligned}$$

Then, by $u_j \to u$ in $L^2_a(\mathbb{R})$ it follows that $I_j \to 0$ as $j \to \infty$. Next, we have

$$||u_j - u||^p \leq < I'(u_j) - I'(u), u_j - u > + I_j,$$

which shows that $u_j \to u$ in X. □

Next, we recall a minimization theorem which will be used in the proof of Theorem 1. (see [16], Theorem 2.7 of [13]).

Theorem 2. *(Minimization theorem) Let E be a real Banach space and $J \in C^1(E, \mathbb{R})$ satisfying (PS) condition. If J is bounded below, then $c = \inf_E I$ is a critical value of J.*

We will use also the following generalization of Clark's theorem (see Rabinowitz [13], p. 53) due to Z. Liu and Z. Wang [17]:

Theorem 3. *(Generalized Clark's theorem, [17]) Let E be a Banach spa ce, $J \in C^1(E, \mathbb{R})$. Assume that J satisfies the (PS) condition, it is even, bounded from below and $J(0) = 0$. If for any $k \in \mathbb{N}$, there exists a $k-$dimensional subspace E^k of E and $\rho_k > 0$ such that $\sup_{E^k \cap S_{\rho_k}} J < 0$, where $S_\rho = \{u \in E,\ \|u\|_E = \rho\}$, then at least one of the following conclusions holds:*

1. *There exists a sequence of critical points $\{u_k\}$ satisfying $J(u_k) < 0$ for all k and $\lim_{k\to\infty} \|u_k\|_E = 0$.*
2. *There exists $r > 0$ such that for any $0 < \alpha < r$ there exists a critical point u such that $\|u\|_E = \alpha$ and $J(u) = 0$.*

Note that Theorem 3 implies the existence of infinitely many pairs of critical points $(u_k, -u_k)$, $u_k \neq 0$ of J, s.t. $J(u_k) \leq 0$, $\lim_{k\to+\infty} J(u_k) = 0$ and $\lim_{k\to+\infty} \|u_k\|_E = 0$.

Lemma 6. *Assume that assumptions (A), (F_2) and (V) hold. Then the functional I is bounded from below.*

Proof of Lemma 6. By (F_2) and the proof of Lemma 3 we have

$$|F(t,u)| \leq \frac{1}{q} b(t) |u|^q.$$

and

$$\begin{aligned}
I(u) &= \frac{1}{p}||u||^p - \int_{\mathbb{R}} a(t) F(t,u(t)) dt \\
&\geq \frac{1}{p}||u||^p - \frac{A}{q} \int_{\mathbb{R}} b(t) |u(t)|^q dt \\
&\geq \frac{1}{p}||u||^p - \frac{A}{q} \left(\int_{\mathbb{R}} |b(t)|^{\frac{p}{p-q}} dt \right)^{\frac{p-q}{p}} \left(\int_{\mathbb{R}} |u(t)|^p dt \right)^{\frac{q}{p}} \\
&\geq \frac{1}{p}||u||^p - \frac{A}{q} ||b||_{L^{\frac{p}{p-q}}} ||u||^q.
\end{aligned}$$

By $p > q$ it follows that I is bounded from below functional. □

3. Proof of the Main Result

In this section we prove Theorem 1. The proof is based on the minimization Theorem 2 and multiplicity result Theorem 3. Their conditions are satisfied according to Lemmas 1–6.

Proof of Theorem 1. The functional I satisfies the assumptions of minimization Theorem 2. Let u_0 be the minimizer of I. Since $I(0) = 0$ to show that $u_0 \neq 0$, let us take $v \in W_0^{2,p}(J)$, where J is the interval from condition (F_3). Suppose that $||v||_\infty \leq 1$. Then for $\lambda > 0$ by (F_3)

$$\begin{aligned} I(\lambda v) &= \frac{\lambda^p}{p}||v||^p - \int_J a(t)F(t,\lambda v(t))dt \\ &\leq \frac{\lambda^p}{p}||v||^p - c\lambda^q \int_J a(t)|v(t)|^q dt. \end{aligned}$$

By $1 < q < p$ and the last inequality it follows for λ_0 sufficiently small and $\lambda_0 > \lambda > 0$ $I(\lambda v) < 0$. Then $I(u_0) = \min\{I(u) : u \in X\} < I(\lambda v) < 0$ and u_0 is a nonzero weak solution. Let the condition (F_4) holds additionally. We show that the functional I satisfies the assumptions of Theorem 3. We construct a sequence of finite dimensional subspaces $X_n \subset X$ and spheres $S_{r_n}^{n-1} \subset X_n$ with sufficiently small radius $r_n > 0$ such that $sup\{I(u) : u \in S_{r_n}^{n-1}\} < 0$. Let $J = (a,b) \subset \mathbb{R}$ and for $k \in \{1,2,...,n\}$ $J_k = (x_{k-1}, x_k)$, where $x_k = a + \frac{k}{n}(b-a)$. Next, we choose functions $v_k \in C_0^2(J_k)$ such that $||v_k||_\infty < \infty$ and $||v_k||_X = 1$.

Let X_n be the $n-$ dimensional subspace $X_n := span\{v_1, ..., v_k\} \subset X$ and

$$S_\rho^{n-1} := \{u \in X_n : ||u||_X = \rho\}.$$

For $u = \sum_{k=1}^n c_k v_k \in X_n$ we have

$$\begin{aligned} ||u||^p &= \int_{\mathbb{R}} \left(|u''(t)|^p - w\,|u'(t)|^p + V(t)\,|u(t)|^p\right) dt \\ &= \sum_{j=k}^n |c_k|^p \int_{J_k} (|v_k''(t)|^p - w|v_k'(t)|^p + V(t)|v_k(t)|^p))dt \\ &= \sum_{k=1}^n |c_k|^p. \end{aligned}$$

By analogous way for $\gamma_k = \int_{J_k} (|v_k(t)|^q dt > 0$ we have

$$||u||_n^q = \sum_{k=1}^n \gamma_k |c_k|^q \tag{8}$$

The space X_n is n-dimensional and the norms $||.||$ and $||.||_n$ are equivalent. There are positive constants d_{1n} and d_{2n} s.t.

$$d_{1n}||u|| \leq ||u||_n \leq d_{2n}||u||, \quad \forall u \in X_n. \tag{9}$$

Then, for $u \in X_n \bigcap S_1^{n-1}$

$$\begin{aligned} I(\lambda u) &= \frac{\lambda^p}{p}||u||^p - \sum_{k=1}^n \int_{J_k} a(t)F(t,\lambda c_k v_k(t))dt \\ &\leq \frac{\lambda^p}{p}||u||^p - c\lambda^q \sum_{k=1}^n |c_k|^q \int_{J_k} a(t)|v_k(t)|^q dt \\ &\leq \frac{\lambda^p}{p}||u||^p - c\lambda^q d_{1n}||u||^q \end{aligned}$$

By $1 < q < p$ and the last inequality it follows that $I(v) < 0$ for $v \in S^{\rho}_{n-1} := \{u \in X_n : ||u|| = \rho\}$. Finally, all assumptions of Theorem 3 are satisfied and by Remark 1 there exist infinitely many weak solutions $\{u_j\}$ of the problem (1), such that $I(\{u_j\}) \leq 0$ and $||u_j|| \to 0$. By imbedding $X \subset L^{\infty}(\mathbb{R})$ it follows that $||u_j||_{\infty} \to 0$ as $j \to \infty$ which completes the proof. □

4. Conlusions

In this paper, we obtained the existence of infinitely many homoclinic solutions of Equation (1) under conditions $(A), (F_1) - (F_4), (V)$ in the case $1 < q < 2 \leq p$. The equation is an extension of the stationary Fisher-Kolmogorov equation which appears in the phase transition models. The variational approach is applied based on the multiple critical point theorem due to Liu and Wang. It will be interesting to extend the result to the case $1 < q < p < 2$.

Author Contributions: Conceptualization, S.T.; methodology, S.T.; software, S.T.; validation, S.T.; formal Analysis, S.T.; writing—original draft preparation, S.T.; writing—review and editing, S.T.; visualization, S.T.; supervision, S.T.; funding acquisition, S.T. The author has read and agreed to the published version of the manuscript.

Funding: S.T. is partially supported by the Bulgarian National Science Fund under Project KP-06-N32/7 and bilateral agreement between BAS and Serbian Academy of Sciences and Arts (SASA), 2020-2022.

Acknowledgments: The author is thankful to the reviewer's remarks.

Conflicts of Interest: The author declares no conflict of interest.

References

1. Peletier, L.A.; Troy, W.C. *Spatial Patterns: Higher Order Models in Physics and Mechnics*; Birkhauser: Boston, MA, USA, 2001.
2. Dimitrov, N.D.; Tersian, S.A. Homoclinic solutions for a class of nonlinear fourth order p-Laplacian differential equations. *Appl. Math. Lett.* **2019**, *96*, 208–215. [CrossRef]
3. Li, T.; Sun, J.; Wu, T.F. Existence of homoclinic solutions for a fourth order differential equation with a parameter. *Appl. Math. Comput.* **2015**, *251*, 499–506. [CrossRef]
4. Sun, J.; Wu, T.F. Two homoclinic solutions for a nonperiodic fourth order differential equation with a perturbation. *J. Math. Anal. Appl.* **2014**, *413*, 622–632. [CrossRef]
5. Sun, J.; Wu, T.; Li, F. Concentration of homoclinic solutions for some fourth-order equations with sublinear indefinite nonlinearities. *Appl. Math. Lett.* **2014**, *38*, 1–6. [CrossRef]
6. Tersian, S.; Chaparova, J. Periodic and homoclinic solutions of extended Fisher-Kolmogorov equations. *J. Math. Anal. Appl.* **2001**, *260*, 490–506. [CrossRef]
7. Timoumi, M. Multiple homoclinic solutions for a class of superquadratic fourth-order differential equations. *Gen. Lett. Math.* **2017**, 3, 154–163. [CrossRef]
8. Yang, L. Infnitely many homoclinic solutions for nonperiodic fourth order differential equations with general potentials. *Abstr. Appl. Anal.* **2014**, *2014*, 435125. [CrossRef]
9. Yeun, Y.L. Heteroclinic solutions of extended Fisher–Kolmogorov equations. *J. Math. Anal. Appl.* **2013**, *407*, 119–129. [CrossRef]
10. Zhang, Z.; Liu, Z. Homoclinic solutions for fourthorder differential equations with superlinear nonlinearitie. *J. Appl. Anal. Comput.* **2018**, *8*, 66–80.
11. Zhang, Z.H.; Yuan, R. Homoclinic solutions for a nonperiodic fourth order differential equations without coercive conditions. *Qual. Theory Dyn. Syst.* **2015**. [CrossRef]
12. Zhang, Z.H.; Yuan, R. Homoclinic Solutions for p-Laplacian Hamiltonian Systems with Combined Nonlinearities. *Mediterr. J. Math.* **2016**, *13*, 1589–1611. [CrossRef]
13. Rabinowitz, P. *Minimax Methods in Critical Point Theory with Applications to Differential Equations*; CBMS Regional Conference Series in Mathematics 65; AMS: Providence, RI, USA, 1986.
14. Adams, R. *Sobolev Spaces*; Academic Press: New York, NY, USA, 1975.
15. Brezis, H. *Functional Analysis, Sobolev Spaces and Partial Differential Equations*; Springer: Berlin, Germany, 2011; ISBN 978-0-387-70913-0.

16. Mawhin, J.; Willem, M. *Critical Point Theory and Hamiltonian Systems;* Springer: New York, NY, USA, 1989.
17. Liu, Z.; Wang, Z. On Clark's theorem and its applications to partially sublinear problems. *Ann. Inst. Henri Poincaré Anal. Non Linéaire* **2015**, *32*, 1015–1037. [CrossRef]

mathematics

MDPI

Article

On the Exponents of Exponential Dichotomies

Flaviano Battelli [1] **and Michal Fečkan** [2,3,*]

[1] Department of Industrial Engineering and Mathematics, Marche Polytecnic University, 60121 Ancona, Italy; battelli@dipmat.univpm.it
[2] Department of Mathematical Analysis and Numerical Mathematics, Faculty of Mathematics, Physics and Informatics, Comenius University in Bratislava, Mlynská dolina, 842 48 Bratislava, Slovakia
[3] Mathematical Institute, Slovak Academy of Sciences, Štefánikova 49, 814 73 Bratislava, Slovakia
* Correspondence: Michal.Feckan@fmph.uniba.sk

Received: 22 March 2020; Accepted: 18 April 2020; Published: 23 April 2020

Abstract: An exponential dichotomy is studied for linear differential equations. A constructive method is presented to derive a roughness result for perturbations giving exponents of the dichotomy as well as an estimate of the norm of the difference between the corresponding two dichotomy projections. This roughness result is crucial in developing a Melnikov bifurcation method for either discontinuous or implicit perturbed nonlinear differential equations.

Keywords: exponential dichotomy; roughness; asymptotically constant matrices

MSC: 34D09

1. Introduction

Exponential dichotomy of a linear system of differential equations is a type of conditional stability that goes back to an idea in Perron [1]. It was revealed to be a very important tool for the study of nonlinear systems because of its roughness. Indeed, it has been used to show the existence of chaotic behaviour in non autonomous perturbations of autonomous nonlinear equations having a homoclinic solution, since transverse intersection of stable and unstable manifolds along a homoclinic solution corresponds to the fact that the linearization of the nonlinear system along it has an exponential dichotomy on $\mathbb{R}$ [2]. Exponential dichotomies are also related with the so called *reducibillty*. A linear system of differential equations $\dot{x} = A(t)x$ is said to be reducible if there exists an invertible C^1 matrix $S(t)$ such that the change of variables $x = S(t)y$ transforms the system into a block diagonal system

$$\dot{y} = \begin{pmatrix} B_1(t) & 0 \\ 0 & B_2(t) \end{pmatrix} y.$$

In [3], it is proven that a system is reducible if and only if the original system has an exponential or ordinary dichotomy. The difference between the two cases is that in ordinary dichotomy the exponents are equal to zero. Another interesting property is the following (see [3]). The linear system $\dot{x} = A(t)x$ has an exponential dichotomy on $\mathbb{R}_+$ if and only if for every locally integrable function $f(t)$, $t \in \mathbb{R}_+$, such that

$$\sup_{t\geq 0} \int_t^{t+1} f(s)ds < \infty,$$

the inhomogeneous linear system $\dot{x} = A(t)x + f(t)$ has a bounded solution. Exponential dichotomies have also relations with such notions as integral separation or spectral theory, see for example [4,5]. Recently, it has been proved in [6] that if a bounded linear Hamiltonian system is exponentially separated into two subspaces of the same dimension, then it must have an exponential dichotomy.

Let us start with the definition of exponential dichotomy. A linear system

$$\dot{x} = A(t)x \tag{1}$$

where $A(t)$ is a piecewise continuous $n \times n$ matrix, is said to have an exponential dichotomy on an interval $I \subset \mathbb{R}$ (usually $\mathbb{R}, \mathbb{R}_+, \mathbb{R}_-$) with projection P, constant $k \geq 1$ and exponents $\alpha, \beta > 0$ if the fundamental matrix $X(t)$ of the Equation (1) (with $X(0) = \mathbb{I}$) satisfies the following conditions:

$$\begin{array}{ll} |X(t)PX(s)^{-1}| \leq ke^{-\alpha(t-s)} & \text{for } s \leq t,\, s,t \in I \\ |X(s)(\mathbb{I} - P)X(t)^{-1}| \leq ke^{-\beta(t-s)} & \text{for } s \leq t,\, s,t \in I. \end{array} \tag{2}$$

Here $\mathbb{R} = (-\infty, \infty)$, $\mathbb{R}_+ = [0, \infty)$ and $\mathbb{R}_- = (-\infty, 0]$. It follows immediately from the definition that, if $\alpha' \leq \alpha$ and $\beta' \leq \beta$ then α' and β' are also exponents of the dichotomy with the same projection P and constant k and also that the linear system (1) has an exponential dichotomy on an interval $J \subseteq I$ if it has one in the interval I. Next, from Gronwall inequality it follows that, on a compact interval, the linear system (1) has an exponential dichotomy with any projection P and exponents α and β (but the constant may change).

We give few examples of systems having an exponential dichotomy. An autonomous system $\dot{x} = Ax$ has an exponential dichotomy on $\mathbb{R}$ if and only if all the eigenvalues of A have nonzero real parts. A periodic system $\dot{x} = A(t)x$ has an exponential dichotomy on $\mathbb{R}$ if and only if all the Floquet exponents have nonzero real parts. A scalar equation $\dot{x} = a(t)x$ has an exponential dichotomy on $I = \mathbb{R}_+$ or $I = \mathbb{R}_-$, if and only if

$$\liminf_{t-s \to \infty} \frac{1}{t-s} \int_s^t a(\tau)d\tau > 0 \quad \text{or} \quad \limsup_{t-s \to \infty} \frac{1}{t-s} \int_s^t a(\tau)d\tau < 0.$$

where the limits are taken as $t - s \to \pm\infty$ in case $I = \mathbb{R}_\pm$ respectively.

Suppose the linear system $\dot{x} = A(t)x$ has an exponential dichotomy on $\mathbb{R}_+$ with exponents α and β. The result that motivates this paper is the following, see [3] (Proposition 1, p. 34).

Theorem 1 (Roughness). *Let $\dot{x} = A(t)x$ have an exponential dichotomy on $\mathbb{R}_+$ with exponents α and β. Given $0 < \tilde{\alpha} < \alpha$ and $0 < \tilde{\beta} < \beta$ there exists $\varepsilon > 0$ such that if $B(t)$ is a piecewise continuous matrix such that $\sup_{t \in \mathbb{R}_+} |B(t)| < \varepsilon$ then the linear system $\dot{x} = [A(t) + B(t)]x$ has an exponential dichotomy on $\mathbb{R}_+$ with exponents $\tilde{\alpha}, \tilde{\beta}$ (but the constant may be larger).*

As a matter of fact in [3] (Proposition 1, p. 34), an estimate on the size of ε is also given, showing that, if $\beta = \alpha$ and $\varepsilon < \frac{\alpha}{4k^2}$ then $\dot{x} = [A(t) + B(t)]x$ has an exponential dichotomy on $\mathbb{R}_+$ with exponent $\alpha - 2k\varepsilon$. So if $\beta = \alpha$ and $\tilde{\alpha} = \tilde{\beta} < \alpha$ we have $\varepsilon = \frac{\alpha - \tilde{\alpha}}{2k}$. We emphasize the fact that in [7] the assumptions on $B(t)$ have been weakened to obtain a roughness result valid also for unbounded perturbations.

However, the exponents of the dichotomy determine the rate of convergence to zero of bounded solution either at ∞ (when the dichotomy is in $\mathbb{R}_+$) or at $-\infty$ (when the dichotomy is in $\mathbb{R}_-$). Sometimes it becomes important to determine this rate of convergence, and hence the exponents of the dichotomy, for example when studying chaotic behaviour of discontinuous systems [8] or developing Melnikov theory for implicit nonlinear differential equations [9]. As a matter of fact in [8] the following result has been proved.

Theorem 2. *Let $\dot{x} = A(t)x$ have an exponential dichotomy on $\mathbb{R}_+$ with exponents α, β. Then there exists $\varepsilon > 0$ such that if $B(t)$ is a piecewise continuous function such that, for some $T > 0$, $\sup_{t \geq T} |B(t)| < \varepsilon$ and*

$$\int_{\tilde{T}}^{\infty} |B(t)|dt < \frac{1}{k}$$

then the linear system $\dot{x} = [A(t) + B(t)]x$ has an exponential dichotomy on $[T, \infty)$ (and hence also on $\mathbb{R}_+$) with the same exponents α, β.

Of course Theorems 1 and 2 hold equally well when the dichotomy of $\dot{x} = A(t)x$ is on $\mathbb{R}_-$.

The proof given in [8] follows an idea in [3] where an exponential estimate is derived for bounded solutions of certain integral inequalities. In this paper we want to give another, more direct, proof of the same result. As a matter of fact we work directly in the space of continuous functions decaying to zero as $t \to \infty$ at a certain given rate. This approach leads us to derive the first of the two exponential estimates given in (2). The second is derived passing to the adjoint system and using the fact that one has a certain freedom in choosing the projection of the dichotomy (see Proposition 2).

Our method has also the advantage that relates the projection of the dichotomy of the perturbed system with the one of the unperturbed. As a matter of fact, we will give an estimate of the norm of the difference between the two projections in term of $\sup_{t\in I}|B(t)|$, where $I = \mathbb{R}_+, \mathbb{R}_-$ is the interval where the exponential dichotomy is considered. This estimate allows us to prove the same result also when the dichotomy of the unperturbed system is on $\mathbb{R}$, a fact that was not noted in [8].

We now briefly resume the content of this paper. In Section 2 we recall basic properties of exponential dichotomy, stable and unstable spaces, roughness, freedom in the choice of the projection etc. Section 3 is devoted to the proof of our main result. Finally, Section 4 contains applications to asymptotically constant matrices and to the linearization of nonlinear systems.

We conclude this section by giving some notations used in the paper. For a linear map L from a Banach space into another, we denote by $\mathcal{R}L$ and $\mathcal{N}L$ its range, resp. its kernel. Next $C_b^0(I)$ denotes the Banach space of bounded continuous functions $x(t)$ on the interval I with the norm $\|x\| = \sup_{t\in I}|x(t)|$. When $I = \mathbb{R}_+$ or $\mathbb{R}_-$ we omit I and write C_b^0 instead of $C_b^0(\mathbb{R}_+)$ or $C_b^0(\mathbb{R}_-)$.

2. Properties of Exponential Dichotomies

First we start with a remark. Let $\nu \in \mathbb{R}$ be a real number. Then $Y(t) = X(t)e^{\nu t}$ is a fundamental matrix of the linear system

$$\dot{x} = [A(t) + \nu\mathbb{I}]x. \tag{3}$$

Assuming that (1) has an exponential dichotomy on I with exponents α, β, we have, for $s, t \in I$, with $s \le t$:

$$|Y(t)PY(s)^{-1}| \le ke^{-(\alpha-\nu)(t-s)}$$
$$|Y(s)(\mathbb{I} - P)Y(t)^{-1}| \le ke^{-(\beta+\nu)(t-s)}$$

that is (3) has an exponential dichotomy on I with the same projection P, constant k and exponents $(\alpha - \nu)$ and $(\beta + \nu)$. Viceversa, if (3) has an exponential dichotomy on I with projections P, constant k and exponents $\tilde{\alpha}, \tilde{\beta}$, then (1) has an exponential dichotomy on I with the same projections P and constant k, and exponents $\alpha = \tilde{\alpha} + \nu$, $\beta = \tilde{\beta} - \nu$. Taking $\nu = \frac{\alpha-\beta}{2}$ the exponents of the dichotomy of (3) are then

$$\tilde{\alpha} = \alpha - \frac{\alpha - \beta}{2} = \frac{\alpha + \beta}{2} = \beta + \frac{\alpha - \beta}{2} = \tilde{\beta}.$$

So, starting from a linear system with an exponential dichotomy, shifting the coefficient matrix by $\nu\mathbb{I}$, $\nu = \frac{\alpha-\beta}{2}$, we can assume that the exponents are the same.

Proposition 1. *Suppose that (1) has an exponential dichotomy of the intervals I_1 and I_2 with the same projection and exponents. Suppose, also, that $I_1 \cap I_2 \neq \emptyset$. Then (1) has an exponential dichotomy of the interval $I_1 \cup I_2$ with the same projection and exponents but possibly different constant.*

Proof. If $I_1 \subset I_2$ or $I_2 \subset I_1$ there is nothing to prove. So we assume that $I = I_1 \cap I_2$ is different from both I_1 and I_2. We can also assume that I_1 is on *the left* and I_2 is on *the right* that is: if $t_1 \in I_1 \setminus I$ and $t_2 \in I_2 \setminus I$ then $t_1 < t_2$.

It is clear that (2) holds if $s, t \in I_1$ or $s, t \in I_2$. So, let $s \in I_1 \setminus I$ and $t \in I_2 \setminus I$. Take $\bar{t} \in I$. Then $s < \bar{t} < t$ and we have:

$$\begin{aligned}
|X(t)PX(s)^{-1}| &\leq |X(t)PX(\bar{t})^{-1}|\,|X(\bar{t})PX(s)^{-1}| \\
&\leq ke^{-\alpha(t-\bar{t})}ke^{-\alpha(\bar{t}-s)} = k^2e^{-\alpha(t-s)} \\
|X(s)(I-P)X(t)^{-1}| &\leq |X(s)(\mathbb{I}-P)X(\bar{t})^{-1}|\,|X(\bar{t})(\mathbb{I}-P)X(t)^{-1}| \\
&\leq ke^{\beta(s-\bar{t})}ke^{\beta(\bar{t}-t)} = k^2e^{\beta(s-t)}
\end{aligned}$$

the proof is complete. □

Since in compact intervals $I = [a, b]$ a linear system (1) has an exponential dichotomy with any projection and any exponents, it follows from Proposition 1 that if a linear system has an exponential dichotomy on an interval $[T, \infty)$ (resp. $(-\infty, -T]$) then it has an exponential dichotomy with the same exponents and projection on $\mathbb{R}_+ = [0, \infty)$ (resp. $\mathbb{R}_- = (-\infty, 0]$). Hence, in the following we will only consider $I = \mathbb{R}_+$ or $I = \mathbb{R}_-$.

When the dichotomy is on $\mathbb{R}_+$ (or on $\mathbb{R}_-$) we have some freedom in the choice of the projection. Indeed we have the following

Proposition 2. [3] (p. 16–17). *Suppose* (1) *has an exponential dichotomy on $\mathbb{R}_+$ with projection P. Then*

$$\mathcal{R}P = \{\xi \in \mathbb{R}^n : \sup_{t\geq 0} |X(t)\xi|e^{\alpha t} < \infty\} = \{\xi \in \mathbb{R}^n : \sup_{t\geq 0} |X(t)\xi| < \infty\}$$

but the kernel of P, $\mathcal{N}P$, can be any complement of $\mathcal{R}P$. Moreover if $Q : \mathbb{R}^n \to \mathbb{R}^n$ is another projection such that $\mathcal{R}Q = \mathcal{R}P$ then there exist a constant k_Q such that (2) *holds with Q and k_Q instead of P and k (with the same exponents). If the dichotomy is on $\mathbb{R}_-$ then it is $\mathcal{N}P$ which is uniquely defined being*

$$\mathcal{N}P = \{\xi \in \mathbb{R}^n : \sup_{t\leq 0} |X(t)\xi|e^{-\beta t} < \infty\} = \{\xi \in \mathbb{R}^n : \sup_{t\leq 0} |X(t)\xi| < \infty\}.$$

Moreover $\mathcal{R}P$ can be any complement of $\mathcal{N}P$ and if $Q : \mathbb{R}^n \to \mathbb{R}^n$ is another projection such that $\mathcal{N}Q = \mathcal{N}P$ then there exist a constant k_Q such that (2) *holds with Q and k_Q instead of P and k (with the same exponents).*

A consequence of the roughness Theorem 1 is the following.

Corollary 1. *Suppose the linear system* (1) *has an exponential dichotomy on $\mathbb{R}_+$ [resp. $\mathbb{R}_-$] with projection P and exponents α and β. Let $B(t)$ be a matrix such that*

$$\lim_{t\to\infty} |B(t)| = 0$$

where the limit is taken at $+\infty$ if $I = \mathbb{R}_+$ and at $-\infty$ when $I = \mathbb{R}_-$. Then, given $\tilde{\alpha} < \alpha$ and $\tilde{\beta} < \beta$, the linear system $\dot{x} = [A(t) + B(t)]x$ has an exponential dichotomy on $\mathbb{R}_+$ [resp. $\mathbb{R}_-$] with exponent $\tilde{\alpha}$ and $\tilde{\beta}$ and projection $\tilde{P}$ such that $\mathcal{N}\tilde{P} = \mathcal{N}P$ [resp. $\mathcal{R}\tilde{P} = \mathcal{R}P$].

Proof. Let $\tilde{\alpha}$ and $\tilde{\beta}$ be as in the statement of the theorem and let $\varepsilon > 0$ be as in Theorem 1. It follows from the assumption the existence of T such that for $t \geq T$ we have $|B(t)| \leq \varepsilon$ and the linear system $\dot{x} = A(t)x$ has an exponential dichotomy on $[T, +\infty)$ with projection P and exponents α and β. Then from Theorem 1 it follows that $\dot{x} = [A(t) + B(t)]x$ has an exponential dichotomy on $[T, \infty)$ with exponent $\tilde{\alpha}$ and $\tilde{\beta}$ and projection as in the statement of the Corollary. However, we have already observed that on $[0, T]$, $\dot{x} = [A(t) + B(t)]x$ has an exponential dichotomy with the same projection and exponents. Then the conclusion follows from Proposition 1. □

Example. Consider the scalar equation $\dot{x} = \left(-1 + \frac{1}{t+1}\right) x$. The unperturbed equation $\dot{x} = -x$ has an exponential dichotomy on $\mathbb{R}$ (and hence on both $\mathbb{R}_+$ and $\mathbb{R}_-$) with $k = 1$, $\alpha = 1$ and projection $P = \mathbb{I}$. The solution of the perturbed equation with $x(0) = 1$ is $x(t) = (t+1)e^{-t}$ and

$$\left|\frac{x(t)}{x(s)}\right| = \frac{t+1}{s+1}e^{-(t-s)}.$$

Let $\alpha < 1$. The function $(t+1)e^{-(1-\alpha)t}$ is increasing on $\left[0, \frac{\alpha}{1-\alpha}\right]$ and decreasing on $\left[\frac{\alpha}{1-\alpha}, \infty\right)$ hence

$$(t-s+1)e^{-(1-\alpha)(t-s)} \leq \frac{e^{-\alpha}}{1-\alpha}$$

for any $s \leq t$. Next, observe that for $0 \leq s \leq t$ we have

$$\frac{1}{s+1} \leq 1 \Leftrightarrow \frac{t-s}{s+1} \leq t-s \Leftrightarrow \frac{t+1}{s+1} \leq t-s+1$$

hence

$$\frac{t+1}{s+1}e^{-(t-s)} \leq \frac{e^{-\alpha}}{1-\alpha}e^{-\alpha(t-s)}.$$

So the equation $\dot{x} = \left(-1 + \frac{1}{t+1}\right) x$ has an exponential dichotomy on $\mathbb{R}_+$ with exponent $\alpha < 1$ but not with exponent $= 1$ since otherwise there should exists $k \geq 1$ such that

$$\frac{t+1}{s+1} \leq k$$

for any $0 \leq s \leq t$ which is absurd. However, the fundamental solution of scalar equation $\dot{x} = \left(-1 + \frac{1}{t^2+1}\right) x$ is

$$x(t) = e^{-t+\arctan t}$$

and

$$\left|\frac{x(t)}{x(s)}\right| = \frac{e^{\arctan t}}{e^{\arctan s}}e^{-(t-s)} \leq e^{\frac{\pi}{2}}e^{-(t-s)}.$$

for any $0 \leq s \leq t$. So, the scalar equation $\dot{x} = \left(-1 + \frac{1}{t^2+1}\right) x$ has an exponential dichotomy on $\mathbb{R}_+$ with exponent $\alpha = -1$.

The difference between the two examples is that the integral of $\frac{1}{t+1}$ in $[0, \infty)$ is divergent whereas the integral of $\frac{1}{t^2+1}$ in $[0, \infty)$ is convergent. Thus we guess that that the statement of Theorem 1 can be improved when

$$\int_0^\infty |B(t)|dt < \infty.$$

3. The Main Result

In this section we prove the following result.

Theorem 3. *Suppose the linear system $\dot{x} = A(t)x$ has an exponential dichotomy on $\mathbb{R}_+$ with exponents α, β. Then there exists $\varepsilon > 0$ such that if $B(t)$ is a piecewise continuous function such that $\sup_{t\in\mathbb{R}_+} |B(t)| < \varepsilon$ and*

$$\int_0^\infty |B(t)|dt < \infty$$

then the linear system $\dot{x} = [A(t) + B(t)]x$ has an exponential dichotomy on $\mathbb{R}_+$ with the same exponents α, β and projection Q such that

$$|Q - P| = O(\varepsilon).$$

A similar result holds when the dichotomies are considered on $\mathbb{R}_-$ and on $\mathbb{R}$.

Proof. First, replacing $A(t)$ with $A(t) = A(t) + \nu\mathbb{I}$, $\nu = \frac{\alpha-\beta}{2}$, we may assume that the exponents are equal. Denote them by δ. Next, consider the perturbed system

$$\dot{x} = [A(t) + B(t)]x. \tag{4}$$

Let $\tilde{\delta} < \delta$ and take $\varepsilon > 0$ as in Theorem 1. Then (4) has an exponential dichotomy on $\mathbb{R}_+$ with projection, say, $\tilde{P}$ and exponent $\tilde{\delta}$. We now follow the approach in [10] to construct a suitable projection for the dichotomy of the perturbed equation.

Let $X_B(t)$ be the fundamental matrix of system (4) and $X(t)$ be the fundamental matrix of $\dot{x} = A(t)x$. A well known standard argument shows that a bounded solution of (4) satisfies the fixed point equation

$$\hat{x}(t) = X(t)P\xi + \int_0^t X(t)PX(s)^{-1}B(s)x(s)ds - \int_t^\infty X(t)(\mathbb{I} - P)X(s)^{-1}B(s)x(s)ds.$$

for some $\xi \in \mathbb{R}^n$. It is easy to see that if $x(t)$, $x_1(t)$ and $x_2(t)$ are bounded functions then

$$|\hat{x}(t)| \le k|\xi| + \frac{2k}{\delta}\|B\|\,\|x\|_b$$

and

$$|\hat{x}_1(t) - \hat{x}_2(t)| \le \frac{2k}{\delta}\|B\|\,\|x_1 - x_2\|_b$$

So taking $\varepsilon > 0$ such that $2k\varepsilon < \delta$, we see that the map $x(t) \mapsto \hat{x}(t)$ is a uniform contraction (with respect to ξ) on the space $C_b^0(\mathbb{R}_+)$ of bounded continuous functions of $\mathbb{R}_+$. So, for any $\xi \in \mathbb{R}^n$ the map $x(t) \mapsto \hat{x}(t)$ has a unique fixed point $x(t,\xi)$ such that

$$\|x(\cdot,\xi)\|_b \le k(1 - 2k\varepsilon\delta^{-1})^{-1}|\xi|. \tag{5}$$

Note that $x(t, P\xi)$ is the unique fixed point of

$$\hat{x}(t) = X(t)P^2\xi + \int_0^t X(t)PX(s)^{-1}B(s)x(s)ds - \int_t^\infty X(t)(\mathbb{I} - P)X(s)^{-1}B(s)x(s)ds$$

and then $x(t, P\xi) = x(t,\xi)$, because of $P^2 = P$ and the uniqueness of the fixed point.

It is straightforward to see that such a fixed point is a solution of (4) and that it is linear with respect to ξ. So

$$x(t,\xi) = X_B(t)Q\xi.$$

where

$$Q\xi = x(0,\xi) = P\xi - \int_0^\infty (\mathbb{I} - P)X(s)^{-1}B(s)x(s,\xi)ds.$$

We pause for a moment to observe that

$$|(Q-P)\xi| \le \int_0^\infty ke^{-\delta s}|B(s)||x(s,\xi)|ds \le k^2(\delta - 2k\varepsilon)^{-1}\varepsilon|\xi|$$

that is

$$|Q - P| \le k^2(\delta - 2k\varepsilon)^{-1}\varepsilon. \tag{6}$$

From the previous considerations it follows that $X_B(t)\xi$ is a bounded solution of (4) if and only if $\xi = Q\xi$. Moreover, we have

$$PQ = P \text{ and}$$
$$QP\xi = x(0, P\xi) = x(0,\xi) = Q\xi$$

So

$$Q^2 = [QP]Q = Q[PQ] = QP = Q$$

that is Q is a projection. Next, if $\xi \in \mathcal{N}Q$ then $P\xi = PQ\xi = 0$ and if $\xi \in \mathcal{N}P$ then $Q\xi = QP\xi = 0$. So

$$\mathcal{N}P = \mathcal{N}Q.$$

Finally, $\xi \in \mathcal{R}Q$ if and only if $X_B(t)\xi = X_B(t)Q\xi$ is a bounded solution of (4). From Proposition 2 it follows, then, that Q is a projection for the dichotomy of (4). So

$$\begin{aligned} |X_B(t)QX_B(s)^{-1}| &\le Ke^{-\tilde{\delta}(t-s)}, & 0 \le s \le t \\ |X_B(s)(\mathbb{I}-Q)X_B(t)^{-1}| &\le Ke^{-\tilde{\delta}(t-s)}, & 0 \le s \le t \end{aligned}$$

for some $K \ge 1$, or, if we go back to the original system with $A(t)$ instead of $A(t) + \nu\mathbb{I}$:

$$\begin{aligned} |X_B(t)QX_B(s)^{-1}| &\le Ke^{-\tilde{\alpha}(t-s)}, & 0 \le s \le t \\ |X_B(s)(\mathbb{I}-Q)X_B(t)^{-1}| &\le Ke^{-\tilde{\beta}(t-s)}, & 0 \le s \le t. \end{aligned}$$

Now assume that $\int_0^\infty |B(t)|dt < \infty$ and let $T > 0$ be such that

$$\Delta := \int_T^\infty |B(t)|dt < \frac{1}{2k}$$

together with $\sup_{t\ge 0} |B(t)| \le \varepsilon$ and $\alpha = \beta = \delta$. Let $t \ge s \ge T$. From the previous part we know that $x(t,s,\xi) = X_B(t)QX_B(s)^{-1}\xi$ is a solution of $\dot{x} = [A(t) + B(t)]x$ which is bounded for $t \ge s \ge T$. Actually we have

$$|x(t,s,\xi)| \le k|\xi|e^{-\tilde{\delta}(t-s)}.$$

We want to show that $\tilde{\delta}$ can be replaced by δ. To this end we consider the map $x(t) \mapsto \hat{x}(t)$:

$$\begin{aligned} \hat{x}(t) = X(t)PX(s)^{-1}\xi &+ \int_s^t X(t)PX(\sigma)^{-1}B(\sigma)x(\sigma)d\sigma \\ &- \int_t^\infty X(t)(\mathbb{I}-P)X(\sigma)^{-1}B(\sigma)x(\sigma)ds \end{aligned} \tag{7}$$

in the space $C^0_\delta([s,\infty))$, $s \ge T$, of functions $x(t)$ such that

$$\sup_{t\ge s} |x(t)|e^{\delta(t-s)} < \infty$$

with norm $\|x(\cdot)\| = \sup_{t\ge s} |x(t)|e^{\delta(t-s)}$. We have

$$\begin{aligned} |\hat{x}(t)| &\le ke^{-\delta(t-s)}|\xi| + \int_s^t ke^{-\delta(t-s)}|B(\sigma)|\|x(\cdot)\|ds + \int_t^\infty ke^{\delta(t+s-2\sigma)}|B(\sigma)|\|x(\cdot)\|ds \\ &\le ke^{-\delta(t-s)}\left(|\xi| + \int_s^\infty |B(\sigma)|d\sigma\|x(\cdot)\|\right) \le ke^{-\delta(t-s)}\left(|\xi| + \Delta\|x(\cdot)\|\right). \end{aligned}$$

or else

$$\|\hat{x}(\cdot)\| \le k\left(|\xi| + \Delta\|x(\cdot)\|\right)$$

and similarly

$$\|\hat{x}_1(\cdot) - \hat{x}_2(\cdot)\| \le k\Delta\|x_1(\cdot) - x_2(\cdot)\|.$$

So we have proved the following

Proposition 3. *Suppose the linear system* (1) *has an exponential dichotomy on* $\mathbb{R}_+$ *with projection* P *constant* k *and exponents* $\alpha = \beta = \delta$. *Let* $B(t)$ *be a matrix and suppose there exists* $T \geq 0$ *such that such that*

$$\|B\| = \sup_{t \geq T} |B(t)| \leq \varepsilon,$$

and

$$\int_T^\infty |B(t)|dt = \Delta < \infty$$

where $\varepsilon > 0$ *is sufficiently small and* Δ *satisfies* $2k\Delta \leq 1$. *Then for any* $s \geq T$ *the map* (7) *is a contraction on the set* $C_\delta^0([s,\infty))$ *and contraction constant* $= \frac{1}{2}$. *Thus its unique fixed point* $x(t,s,\xi)$ *belongs to* $C_\delta^0([s,\infty))$ *and*

$$\|x(t,s,\xi)\| \leq 2k|\xi|.$$

Hence we proved that

$$|X_B(t)QX_B(s)^{-1}| \leq 2ke^{-\delta(t-s)}$$

for any $t \geq s \geq T$, and we extend this inequality for any $t \geq s \geq 0$ provided we change $2k$ with a possibly larger constant K_1. Next, from Proposition 2, we also know that

$$|X_B(s)(\mathbb{I} - Q)X_B(t)^{-1}| \leq K_2 e^{-\tilde{\delta}(t-s)}$$

for $0 \leq s \leq t$ and possibly another constant K_2, since we know that Q can be taken as a projection of the dichotomy of the perturbed system. Thus:

$$\begin{aligned} &|X_B(t)QX_B(s)^{-1}| \leq Ke^{-\delta(t-s)} \\ &|X_B(s)(\mathbb{I} - Q)X_B(t)^{-1}| \leq Ke^{-\tilde{\delta}(t-s)} \end{aligned} \tag{8}$$

(where $\tilde{\delta} < \delta$) for any $t \geq s \geq 0$ and $K = \max\{K_1, K_2\}$.

To complete the proof we still have to prove that, for $T \leq s \leq t$, it results

$$|X_B(s)(\mathbb{I} - Q)X_B(t)^{-1}| \leq Ke^{-\delta(t-s)} \tag{9}$$

for possibly another constant K. The fundamental matrix $Y(t) = X(t)^{-1*}$ of the adjoint system

$$\dot{x} = -A(t)^* x$$

has an exponential dichotomy on $\mathbb{R}_+$ with projection $(\mathbb{I} - P^*)$. Indeed:

$$\begin{aligned} &|Y(t)(\mathbb{I} - P^*)Y(s)^{-1}| \leq ke^{-\delta(t-s)} \\ &|Y(t)P^*Y(s)^{-1}| \leq ke^{-\delta(t-s)} \end{aligned}$$

for any $0 \leq s \leq t$. From the previous part applied to the system $\dot{x} = -[A(t) + B(t)]^* x$ we see that a projection $\tilde{Q}^*$ exists such that $\mathcal{R}\tilde{Q}^* = \mathcal{R}P^*$ and

$$\begin{aligned} &|Y_B(t)(\mathbb{I} - \tilde{Q}^*)Y_B(s)^{-1}| \leq Ke^{-\delta(t-s)} \\ &|Y_B(s)\tilde{Q}^* Y_B(t)^{-1}| \leq Ke^{-\tilde{\delta}(t-s)} \end{aligned}$$

for $0 \leq s \leq t$, where $Y_B(t) = X_B(t)^{-1*}$ is the fundamental matrix of the perturbed system $\dot{x} = -[A(t) + B(t)]^* x$. Going back to $X_B(t)$ we see that

$$\begin{aligned} &|X_B(t)\tilde{Q}X_B(s)^{-1}| \leq Ke^{-\tilde{\delta}(t-s)} \\ &|X_B(s)(\mathbb{I} - \tilde{Q})X_B(t)^{-1}| \leq Ke^{-\delta(t-s)} \end{aligned}$$

for $0 \le s \le t$ and a certain constant K (possibly different from the previous one, however we do not introduce other notations for these constants since at the end we can take the larger of all). From the first inequality it follows that, if $\xi \in \mathcal{R}\tilde{Q}$ then $|X_B(t)\xi| \le Ke^{-\tilde{\delta}t}$ and hence $\mathcal{R}\tilde{Q} \subset \mathcal{R}Q$. So

$$\mathcal{R}\tilde{Q} = \mathcal{R}Q$$

since $\text{rank}\tilde{Q} = \text{rank}\tilde{Q}^* = \text{rank}P^* = \text{rank}P = \text{rank}Q$. Next:

$$\begin{aligned}&\xi \in \mathcal{N}\tilde{Q} \Leftrightarrow \tilde{Q}\xi = 0 \Leftrightarrow \langle \tilde{Q}\xi, \eta\rangle = 0, \forall \eta \\ &\Leftrightarrow \langle \xi, \tilde{Q}^*\eta\rangle = 0, \forall\eta \Leftrightarrow \xi \in [\mathcal{R}\tilde{Q}^*]^\perp = [\mathcal{R}P^*]^\perp.\end{aligned}$$

But in the same way we see that $[\mathcal{R}P^*]^\perp = \mathcal{N}P$ and hence

$$\mathcal{N}\tilde{Q} = \mathcal{N}P = \mathcal{N}Q.$$

As a consequence $Q = \tilde{Q}$ and we have

$$|X_B(t)QX_B(s)^{-1}| \le Ke^{-\delta(t-s)}$$

and

$$|X_B(s)(\mathbb{I} - Q)X_B(t)^{-1}| = |X_B(s)(\mathbb{I} - \tilde{Q})X_B(t)^{-1}| \le Ke^{-\delta(t-s)}$$

for $0 \le s \le t$.

Going back to the original system (that is before the shifting from $A(t)$ to $A(t) + \nu\mathbb{I}$) we see that

$$\begin{aligned}&|X_B(t)QX_B(s)^{-1}| \le 2Ke^{-\alpha(t-s)} \\ &|X_B(s)(\mathbb{I} - Q)X_B(t)^{-1}| \le 2Ke^{-\beta(t-s)}.\end{aligned}$$

for $0 \le s \le t$. This completes the proof when the dichotomy is on $\mathbb{R}_+$.

When the dichotomy is on $\mathbb{R}_-$, we reduce to the case of $\mathbb{R}_+$ by changing t with $-t$, $X(t)$ with $X(-t)$ and $A(t)$ with $-A(-t)$. When $\dot{x} = A(t)x$ has an exponential dichotomy on $\mathbb{R}$, we apply the previous result to see that $\dot{x} = [A(t) + B(t)]x$ has an exponential dichotomy on $\mathbb{R}_+$ with projection Q_+ and on $\mathbb{R}_-$ with projection Q_-. Then $\mathcal{R}Q_+ \cap \mathcal{N}Q_- = \{0\}$ because both projections are close to P and $\mathcal{R}P \cap \mathcal{N}P = \{0\}$ since $\dot{x} = A(t)x$ has an exponential dichotomy on $\mathbb{R}$. The conclusion follows from [3] [p. 19], (see also [2] (Proposition 2.1)). □

4. Asymptotically Constant Matrices

Let $A(t)$ be a piecewise continuous $n \times n$ matrix, $t \in \mathbb{R}_+$ and assume that a constant matrix A exists such that

(A1) $\lim\limits_{t\to\infty} A(t) = A$ and $\displaystyle\int_0^\infty |A(t) - A|dt < \infty$;

(A2) A has two semi-simple eigenvalues $-\alpha < 0$ and $\beta > 0$;

(A3) there exists $\mu > 0$ such that all others eigenvalues λ of A satisfy either $\Re\lambda \le -(\alpha + \mu)$ or $\Re\lambda \ge \beta + \mu$.

Let $X_0(t)$ be the fundamental matrix of $\dot{x} = Ax$ such that $X_0(0) = \mathbb{I}$. Since $-\alpha$ and β are semi-simple eigenvalues, their generalized eigenspaces, that we denote with V_s and V_u, consist of eigenvectors of $-\alpha$ and β, that is for any $v \in V_s$ (resp. $v \in V_u$) we have $X_0(t)v = ve^{-\alpha t}$ (resp. $X_0(t)v = ve^{\beta t}$). Write

$$\mathbb{R}^n = V_{ss} \oplus V_s \oplus V_u \oplus V_{uu}$$

where V_{ss} is the generalized eigenspace of the eigenvalues of A with real parts less than $-\alpha - \mu$ and V_{uu} is the generalized eigenspace of the eigenvalues of A with real parts greater than $\alpha + \mu$. Let d_{ss}, d_s, d_u d_{uu} be the dimensions of V_{ss}, V_s V_u, V_{uu} respectively.

Let $P_{ss} : \mathbb{R}^n \to \mathbb{R}^n$ be the projection onto V_{ss} with kernel $V_s \oplus V_u \oplus V_{uu}$, $P_s : \mathbb{R}^n \to \mathbb{R}^n$ be the projection onto V_s with kernel $V_{ss} \oplus V_u \oplus V_{uu}$, $P_u : \mathbb{R}^n \to \mathbb{R}^n$ be the projection onto V_u with kernel $V_{ss} \oplus V_s \oplus V_{uu}$, and $P_{uu} : \mathbb{R}^n \to \mathbb{R}^n$ be the projection onto V_{uu} with kernel $V_{ss} \oplus V_s \oplus V_u$.

Let $\{v_1^{ss}, \dots, v_{d_{ss}}^{ss}\}$ be a orthonormal basis of V_{ss}, $\{v_1^s, \dots, v_{d_s}^s\}$ be a orthonormal basis of V_s, $\{v_1^u, \dots, v_{d_u}^u\}$ be a orthonormal basis of V_u and $\{v_1^{uu}, \dots, v_{d_{uu}}^{uu}\}$ be a orthonormal basis of V_{uu}. For any $\xi \in \mathbb{R}^n$ we have

$$P_{ss}\xi = \sum_{j=1}^{d_{ss}} c_j^{ss} v_j^{ss} \qquad P_s\xi = \textstyle\sum_{j=1}^{d_s} c_j^s v_j^s$$
$$P_{uu}\xi = \sum_{j=1}^{d_{uu}} c_j^{uu} v_j^{uu} \quad P_u\xi = \textstyle\sum_{j=1}^{d_u} c_j^u v_j^u$$

Hence

$$\left\{ \sum_{j=1}^{d_{ss}} |c_j^{ss}|^2 \right\}^{\frac{1}{2}} = |P_{ss}\xi| \le |P_{ss}|\,|\xi|.$$

Similarly

$$\left\{ \sum_{j=1}^{d_s} |c_j^s|^2 \right\}^{\frac{1}{2}} \le |P_s|\,|\xi|$$
$$\left\{ \sum_{j=1}^{d_u} |c_j^u|^2 \right\}^{\frac{1}{2}} \le |P_u|\,|\xi|$$
$$\left\{ \sum_{j=1}^{d_{uu}} |c_j^{uu}|^2 \right\}^{\frac{1}{2}} \le |P_{uu}|\,|\xi|.$$

Next, V_{ss}, V_s, V_u and V_{uu} are all invariant under $X_0(s)$, that is

$$\xi \in V \Rightarrow X_0(s)\xi \in V$$

for $V = V_{ss}, V_s, V_u, V_{uu}$. So we have, for example

$$X_0(t)P_{ss}\xi = P_{ss}X_0(t)P_{ss}\xi$$

and

$$P_{ss}X_0(t)(\mathbb{I} - P_{ss}) = 0$$

because $X_0(t)(\mathbb{I} - P_{ss}) = X_0(t)P_s + X_0(t)P_u + X_0(t)P_{uu} \in V_s + V_u + V_{uu}$. So

$$X_0(t)P_{ss} = P_{ss}X_0(t).$$

Similarly:

$$X_0(t)P_s = P_sX_0(t)$$
$$X_0(t)P_u = P_uX_0(t)$$
$$X_0(t)P_{uu} = P_{uu}X_0(t).$$

Now we observe that

$$X_0(t)X_0(s)^{-1}P_s\xi = X_0(t-s)P_s\xi = X_0(t-s)\sum_{j=1}^{d_s} c_j^s v_j^s = \sum_{j=1}^{d_s} c_j^s X_0(t-s) v_j^s$$
$$= e^{-\alpha(t-s)} \sum_{j=1}^{d_s} c_j^s v_j^s = e^{-\alpha(t-s)} P_s\xi$$

then

$$|X_0(t)X_0(s)^{-1}P_s| \le |P_s|e^{-\alpha(t-s)}$$

for any $0 \le s, t$. Similarly,

$$|X_0(t)X_0(s)^{-1}P_u| \le |P_u|e^{\beta(t-s)}$$

for any $0 \le s, t$. A slightly different estimate occurs when considering $Y_0(t)Y_0(s)^{-1}P_{ss}$ and $Y_0(t)Y_0(s)^{-1}P_{uu}$. Indeed in this case the eigenvalues may not be simple so that, for example

$$X_0(t)X_0(s)^{-1}v_i^{ss} = X_0(t-s)v_i^{ss} = \sum_{j=1}^{d_{ss}} q_{ij}(t-s)e^{\lambda_j(t-s)}v_j^{ss}$$

where $q_{ij}(t)$ is a polynomial that may have positive degree (but less than the multiplicity of λ_i as an eigenvalue of A.) Since $\Re\lambda_i \le -\alpha - \mu$, for any $i = 1, \dots, d_{ss}$ in this case we have then

$$|X_0(t)X_0(s)^{-1}v_i^{ss}| = |X_0(t-s)v_i^{ss}| \le c_i e^{-(\alpha+\frac{\mu}{2})(t-s)}$$

for some $c_i > 0$. As a consequence

$$\begin{aligned}|X_0(t)X_0(s)^{-1}P_{ss}\xi| &\le \sum_{i=1}^{d_{ss}} |c_i^{ss}|\,|X_0(t-s)v_i^{ss}| \le c_i \sum_{i=1}^{d_{ss}} |c_i^{ss}|e^{-(\alpha+\frac{\mu}{2})(t-s)} \\ &\le k_1|P_{ss}|e^{-(\alpha+\frac{\mu}{2})(t-s)}|\xi|\end{aligned}$$

for any $t \ge s \ge 0$. Similarly:

$$|X_0(t)X_0(s)^{-1}P_{uu}\xi| \le k_2|P_{uu}|e^{(\beta+\frac{\mu}{2})(t-s)}|\xi|$$

for some k_2 and any $s \ge t \ge 0$. Summarising we see that $k \ge 1$ exists such that:

$$\begin{array}{ll} |X_0(t)X_0(s)^{-1}P_{ss}| \le k|P_{ss}|e^{-(\alpha+\frac{\mu}{2})(t-s)} & \text{for any } 0 \le s \le t \\ |X_0(t)X_0(s)^{-1}P_s| \le |P_s|e^{-\alpha(t-s)} & \text{for any } 0 \le s, t \\ |X_0(t)X_0(s)^{-1}P_u| \le |P_u|e^{\beta(t-s)} & \text{for any } 0 \le s, t \\ |X_0(t)X_0(s)^{-1}P_{uu}| \le k|P_{uu}|e^{(\beta+frac\mu 2)(t-s)} & \text{for any } 0 \le t \le s. \end{array}$$

and hence, using the commutativity of $X_0(s)$ with the projections

$$\begin{array}{ll} |X_0(t)P_{ss}X_0(s)^{-1}| \le k|P_{ss}|e^{-(\alpha+\frac{\mu}{2})(t-s)} & \text{for any } 0 \le s \le t \\ |X_0(t)P_sX_0(s)^{-1}| \le |P_s|e^{-\alpha(t-s)} & \text{for any } 0 \le s, t \\ |X_0(t)P_uX_0(s)^{-1}| \le |P_u|e^{\beta(t-s)} & \text{for any } 0 \le s, t \\ |X_0(t)P_{uu}X_0(s)^{-1}| \le k|P_{uu}|e^{(\beta+\frac{\mu}{2})(t-s)} & \text{for any } 0 \le t \le s. \end{array}$$

Setting

$$P = P_{ss} + P_s$$

and then $I - P = P_{uu} + P_u$ we get

$$\begin{array}{ll} |X_0(t)PX_0(s)^{-1}| \le (|P_s| + k|P_{ss}|)e^{-\alpha(t-s)} & \text{for any } t \ge s \ge 0 \\ |X_0(s)(\mathbb{I} - P)X_0(t)^{-1}| \le (|P_u| + k|P_{uu}|)e^{\beta(t-s)} & \text{for any } s \ge t \ge 0 \end{array}$$

From Theorem 3 we obtain the following result.

Proposition 4. *Suppose conditions (A1)–(A3) hold. Then the linear system $\dot{x} = A(t)x$ has an exponential dichotomy on both $\mathbb{R}_+$ and $\mathbb{R}_-$ with exponents α and β.*

We conclude this Section with an application of Proposition 4 to nonlinear systems. Let $f(x)$ be a C^1-function such that $f'(x)$ is Lipschitz with L_f as Lipschitz constant. Suppose the system $\dot{x} = f(x)$ has two hyperbolic fixed points $x = x_-$ and x_+ (that may coincide, i.e., $x_- = x_+$) together with a heteroclinic orbit $u(t)$, i.e., a bounded solution such that

$$\lim_{t\to\pm\infty} u(t) = x_\pm.$$

The fixed points being hyperbolic means that the matrices $f'(x_-)$ and $f'(x_+)$ have no eigenvalues with zero real part. Then both systems

$$\dot{x} = f'(x_\pm)x$$

have an exponential dichotomy on $\mathbb{R}$ with projections, say, P_+ and P_-. It is known that rank$P_\pm$ equals the number of eigenvalues of $f'(x_\pm)$ having negative real parts counted with multiplicities. Let $\alpha_\pm$ and $\beta_\pm$ be the exponents of the dichotomy of $\dot{x} = f'(x_\pm)x$ respectively. First we observe that $\dot{u}(t) = f(u(t))$ is bounded and it is also a solution of $\dot{x} = f'(u(t))x$. From the roughness theorem we know that this system has an exponential dichotomy on $\mathbb{R}_+$ with exponents $\tilde{\alpha}_+$ and $\tilde{\beta}_+$ slightly less that α_+ and β_+ respectively. Hence we get $|u(t) - x_+| \le Ke^{-\tilde{\alpha}_+ t}$ for $t \ge 0$. Similarly we get $|u(t) - x_-| \le Ke^{\tilde{\beta}_- t}$, for $t \le 0$. So we see that $f'(u(t)) - f'(x_+) \in L^1(\mathbb{R}_+)$ and $f'(u(t)) - f'(x_-) \in L^1(\mathbb{R}_-)$. A simple application of Theorem 3 gives then the following

Theorem 4. *Let $f(x)$ be a C^1-function with Lipschitz continuous derivative. Suppose there exists $x = x_-$ and x_+ such that $f(x_-) = f(x_+) = 0$ and $f'(x_-)$, $f'(x_+)$ has no eigenvalues with zero real parts. Then both linear systems $\dot{x} = f'(x_-)x$ and $\dot{x} = f'(x_+)x$ have an exponential dichotomy on $\mathbb{R}$. Let $\alpha_\pm$, $\beta_\pm$ and $P_\pm$ be the corresponding exponents and projections. Suppose further that the (nonlinear) equation $\dot{x} = f(x)$ has a solution $u(t)$ such that*

$$\lim_{t\to\pm\infty} u(t) = x_\pm$$

Then the linear equation $\dot{x} = f'(u(t))x$ has an exponential dichotomy on both $\mathbb{R}_-$ and $\mathbb{R}_+$ with exponents α_-, β_- and α_+, β_+ respectively, and projections $Q_\pm$ such that

$$\operatorname{rank} Q_\pm = \operatorname{rank} P_\pm.$$

5. Conclusions

We have given a new proof of a roughness result for linear systems with an exponential dichotomy different than the one in [8]. This new proof has the advantage that it is is more direct, can be easily extended to system having an exponential dichotomy on the whole line and gives a precise estimate on the norm of the difference of the projections of the dichotomies of the perturbed and the unperturbed system. Moreover it extends also to more general situations. Indeed the assumptions that $\sup_{t\in I}|B(t)| < \varepsilon$ is used just to prove that the map

$$x(t) \mapsto X(t)P\xi + \int_0^\infty \Gamma(t,s)B(s)x(s)ds,$$

where

$$\Gamma(t,s) = \begin{cases} X(t)PX(s)^{-1} & \text{if } 0 \le s \le t \\ -X(t)(\mathbb{I} - P)X(s)^{-1} & \text{if } 0 \le t < s, \end{cases}$$

is a contraction on C_b^0. According to [7] this holds also under the weaker assumption that

$$\inf_{T>0}\sup_{t\ge 0}\int_t^{t+T}|B(s)|ds\left(\frac{k}{1-e^{-\alpha T}} + \frac{k}{1-e^{-\beta T}}\right) < 1 \tag{10}$$

and the fixed point $x(t,s)$ satisfies again $\|x\| \leq C|\xi|$, for a suitable constant C. The remaining part of the proof showing that this fixed point indeed belongs to C^0_δ just depends on the fact that $|B(t)| \in L^1(\mathbb{R}_+)$. Hence Theorem 3 holds also under the weaker condition (10) instead of $\sup_{t\geq 0}|B(t)| < \varepsilon$.

Author Contributions: Investigation, M.F.; Methodology, F.B. The contributions of all authors are equal. All authors have read and agreed to the published version of the manuscript .

Funding: Partially supported by the Slovak Research and Development Agency under the contract No. APVV-18-0308 and by the Slovak Grant Agency VEGA No. 1/0358/20 and No. 2/0127/20.

Conflicts of Interest: The authors declare no conflict of interest.

References

1. Perron, O. Die Stabilitätsfrage bei Differentialgleichungen. *Math. Z.* **1930**, *32*, 703–728. [CrossRef]
2. Palmer, K.J. Exponential dichotomies and transversal homoclinic points. *J. Diff. Equ.* **1984**, *55*, 225–256. [CrossRef]
3. Coppel, W.A. *Dichotomies in Stability Theory*; Lecture Notes in Mathematics *629*; Springer: Berlin/Heidelberg, Germany, 1978.
4. Palmer, K.J. Exponential dichotomy, integral separation and diagonalizability of linear systems of ordinary differential equations. *J. Diff. Equ.* **1982**, *43*, 184–203. [CrossRef]
5. Palmer, K.J. Exponential dichotomy, exponential separation and spectral theory for linear systems of ordinary differential equations. *J. Diff. Equ.* **1982**, *46*, 324–345. [CrossRef]
6. Battelli, F.; Palmer, K.J. Strongly exponentially separated linear systems. *J. Dyn. Diff. Equ.* **2019**, *31*, 573–600. [CrossRef]
7. Ju, N.; Wiggins, S. On roughness of exponential dichotomy. *J. Math. Anal. Appl.* **2001**, *262*, 39–49. [CrossRef]
8. Calamai, A.; Franca, M. Mel'nikov methods and homoclinic orbits in discontinuous systems. *J. Dyn. Diff. Equ.* **2013**, *25*, 733–764. [CrossRef]
9. Battelli, F.; Fečkan, M. Melnikov theory for nonlinear implicit ODEs. *J. Diff. Equ.* **2014**, *256*, 1157–1190. [CrossRef]
10. Hale, J. Introductions to dynamic bifurcation. In *Bifurcation Theory and Applications*; Lecture Notes in Mathematics *1057*; Springer: Berlin, Germany, 1984; pp. 106–151.

mathematics MDPI

Article

Double Fuzzy Sumudu Transform to Solve Partial Volterra Fuzzy Integro-Differential Equations

Atanaska Georgieva

Department of Mathematical Analysis, University of Plovdiv Paisii Hilendarski, 24 Tzar Asen, 4003 Plovdiv, Bulgaria; afi2000@abv.bg

Received: 24 March 2020; Accepted: 24 April 2020; Published: 2 May 2020

Abstract: In this paper, the double fuzzy Sumudu transform (DFST) method was used to find the solution to partial Volterra fuzzy integro-differential equations (PVFIDE) with convolution kernel under Hukuhara differentiability. Fundamental results of the double fuzzy Sumudu transform for double fuzzy convolution and fuzzy partial derivatives of the n-th order are provided. By using these results the solution of PVFIDE is constructed. It is shown that DFST method is a simple and reliable approach for solving such equations analytically. Finally, the method is demonstrated with examples to show the capability of the proposed method.

Keywords: double fuzzy Sumudu transform; partial Volterra fuzzy integro-differential equations; n-th order fuzzy partial H-derivative

1. Introduction

Modeling of different physical systems gives us different differential, integral and integro-differential equations. We are not always sure that the models obtained are perfect. The fuzzy set theory is one of the most popular theories for describing this situation. The fuzzy logic is introduced with the proposal of fuzzy set theory by Zadeh and is applied when the observational parameters are imprecise or unclear. The neutrosophic logic is considered as the extension of the fuzzy logic and the measure of indeterminacy is added to the measures of truthiness and falseness. The theory of neutrosophic statistics can be applied when to observation indeterminate, imprecise, vague, and incomplete parameters. For more details [1–4].

The concept of fuzzy sets, fuzzy numbers and arithmetic operations firstly introduced by Zadeh [5]. In [6] Seikkala defined fuzzy derivatives. The concept of fuzzy integration was given by Dubois and Prade [7]. One of the first applications of fuzzy integration was given by Wu and Ma who investigated the fuzzy Fredholm integral equation of the second kind. The idea of fuzzy partial differential equations was introduced by Buckley in [8]. Allahveranloo proposed the difference method for solving this equations in [9].

In recent years, many mathematicians have studied the solution of fuzzy differential equations [10–12], fuzzy integral equations [13–17], and fuzzy integro-differential equations [18–21], which play a key role in engineering [22,23]. These equations in a fuzzy setting are a natural way to model the ambiguity of dynamic systems in different scientific fields such as physics, geography, medicine, and biology [24,25].

The fundamental tool in operational calculus are integral transforms. They are used in solving many practical problems in applied mathematics, physics and engineering. The integral transforms be very useful in solving partial differential equations. They convert the original function to a function that is simpler to solve. The Fourier transform is the precursor of the integral transforms. This transform is used to express functions in a finite interval. Similar integral transforms are Laplace, Mellin and Hankel transforms. In the 1990s Watugala [26,27] has introduced a new integral transform called the Sumudu transform. Later, Weerakon [28] used the Sumudu transform for solving partial differential equations. Some fundamental theorems and properties for Sumudu transform can be seen in [8,29]. Furthermore, in [30] the Sumudu transform is applied for Bessel functions and equations.

One of the recent methods in handling problems modelled under fuzzy environment is fuzzy Sumudu transform proposed by Ahmad and Abdul Rahman [31]. This transform is used for solving of fuzzy differential equations , fuzzy integral equations and fuzzy integro-differential equations as the problem is reduced to problem which is much simpler to be solved.

In [29] Ahmad and Abdul Rahman proposed the idea of the fuzzy method of transformation of Sumudu to solve fuzzy partial differential equations. The technique of the fuzzy Sumudu transform method for solving a fuzzy convolution Volterra integral equations and the fuzzy integro-differential equation was developed in [32,33]. The studies are the followed by the application of fuzzy Sumudu transform on fuzzy fractional differential equations and fuzzy Volterra integral equations in [34]. In [35] is introduced double fuzzy Sumudu transform (DFST) method and is applied to solve fuzzy convolution Volterra integral equation of two variable.

In [36] the solution of classical partial integro-differential equations was discussed using classical double Elzaki transform method. In the present paper we investigate the solution of PVFIDE with convolution kernel under Hukuhara differentiability using DFST method. The main difficulties overcome in solving the this problem are related to the application of the DFST for fuzzy partial H-derivative of the n-th order. So, we obtain a new results on the Sumudu transform for fuzzy partial H-derivative of the n-th order . After, the studied equation we convert to a nonlinear system of partial Volterra integro-differential equations in a crisp case. To be find the lower and upper functions of the solution we us DFST and we convert this system to system of algebraic equations.

The paper is organized as follows: In Section 2, some definitions and results of fuzzy numbers, fuzzy functions and fuzzy partial derivative of the n-th order is given. In Section 3, the definition of DFST is recalled, double fuzzy convolution theorem is stated. New results on DFST for fuzzy partial derivative of the n-th order are proposed. In Section 4, the DFST is applied to fuzzy partial convolution Volterra fuzzy integro-differential equation to construct the general technique. In Section 5, an example is provided to demonstrate the proposed method and finally in Section 6, conclusions are drawn.

2. Premilinaries and Notations

In this section, we give some basics definitions and theorems for fuzzy number, fuzzy-valued function and derivative of fuzzy-valued function.

Definition 1 ([37]). *A fuzzy number is defined as the mapping $u : \mathbb{R} \to [0,1]$ satisfying the following four properties:*

(i) *u is upper semi-continuous on $\mathbb{R}$;*
(ii) *$u(x) = 0$ outside of some interval $[c,d]$;*
(iii) *there are the real numbers a and b with $c \leq a \leq b \leq d$, such that u is increasing on $[c,a]$, decreasing on $[b,d]$ and $u(x) = 1$ for each $x \in [a,b]$;*
(iv) *$u(rx + (1-r)y) \geq \min\{u(x), u(y)\}$ for any $x, y \in \mathbb{R}$, $r \in [0,1]$.*

needed throughout the paper such

Denote E^1 the set of all fuzzy numbers and $D = \mathbb{R}_+ \times \mathbb{R}_+$. Any real number $a \in \mathbb{R}$ can be interpreted as a fuzzy number $\tilde{a} = \chi(a)$ and therefore $\mathbb{R} \subset E^1$.

Definition 2 ([38]). *Let $u \in E^1$ and $r \in (0,1]$. The r-level set of u is the crisp set*

$$[u]^r = \{x \in \mathbb{R} : u(x) \geq r\},$$

where $[u]^r$ denotes r-level set of fuzzy number u.

It can be concluded that any r-level set is bounded and closed interval and denoted by $[\underline{u}(r), \overline{u}(r)]$ for all $r \in [0,1]$, where the functions $\underline{u}$, $\overline{u}$: $[0,1] \to \mathbb{R}$ are the lower and upper bound of $[u]^r$, respectively.

Definition 3 ([38]). *A fuzzy number in parametric form is given as an order pair of the form $u = (\underline{u}(r), \overline{u}(r))$, where $0 \leq r \leq 1$ satisfying the following conditions:*

(i) $\underline{u}(r)$ *is a bounded left continuous monotonic increasing function in* $[0,1]$;
(ii) $\overline{u}(r)$ *is a bounded left continuous monotonic decreasing function in* $[0,1]$;
(iii) $\underline{u}(r) \leq \overline{u}(r)$.

For arbitrary fuzzy number $u = (\underline{u}(r), \overline{u}(r)), v = (\underline{v}(r), \overline{v}(r))$ and an arbitrary crisp number $k \in \mathbb{R}$ the addition and the scalar multiplication are defined by $[u \oplus v]^r = [u]^r + [v]^r = [\underline{u}(r) + \underline{v}(r), \overline{u}(r) + \overline{v}(r)]$ and

$$[k \odot u]^r = k.[u]^r = \begin{cases} [k\underline{u}(r), k\overline{u}(r)], & k \geq 0 \\ [k\overline{u}(r), k\underline{u}(r)], & k < 0. \end{cases}$$

The neutral element with respect to $\oplus$ in E^1 is denoted by $\tilde{0} = \chi_{\{0\}}$.
For basic algebraic properties of fuzzy numbers, please see ([37]).

Definition 4 ([39]). *Let $x, y \in E^1$ and exists $z \in E^1$, such that $x = y \oplus z$. Then z is called the H-difference of x and y and is given by $x \ominus y$.*

We use the Hausdorff metric as a distance between fuzzy numbers.

Definition 5 ([37]). *For arbitrary fuzzy numbers $u = (\underline{u}(r), \overline{u}(r))$ and $v = (\underline{v}(r), \overline{v}(r))$ the quantity*

$$d(u, v) = \sup_{r \in [0,1]} \max\{|\underline{u}(r) - \underline{v}(r)|, |\overline{u}(r) - \overline{v}(r)|\}$$

is the distance between u, v.

The metric d is a complete metric space in E^1.
For any fuzzy-number-valued function $w : D \to E^1$ we define the functions $\underline{w}(.,.,r)$, $\overline{w}(.,.,r)$: $D \to \mathbb{R}$, for all $r \in [0,1]$. These functions are called the left and right $r-$ level functions of w.

Definition 6 ([15]). *A fuzzy-number-valued function $w : D \to E^1$ is said to be continuous at $(s_0, t_0) \in D$ if for each $\varepsilon > 0$ there is $\delta > 0$ such that $d(f(s,t), f(s_0, t_0)) < \varepsilon$ whenever $|s - s_0| + |t - t_0| < \delta$. If w be continuous for each $(s,t) \in D$ then we say that w is continuous on D.*

Let $R > 0$. Denote $D_R = D \cap \overline{U}(0, R)$, where

$$\overline{U}(0, R) = \{(x, y) : x^2 + y^2 \leq R^2\}$$

is the closed circle with radius R.
Let $w : D \to E^1$ be fuzzy-valued function with parametric form $(\underline{w}(x,y,r), \overline{w}(x,y,r))$ for all $r \in [0,1]$.

Theorem 1. *Let for all $r \in [0,1]$*

1. *the functions $\underline{w}(x,y,r)$ and $\overline{w}(x,y,r)$ are Riemann-integrable on D_R,*

2. *there are constants* $\underline{M}(r) > 0$ *and* $\overline{M}(r) > 0$, *such that*

$$\iint_{D_R} |\underline{w}(x,y,r)|dxdy \leq \underline{M}(r), \quad \iint_{D_R} |\overline{w}(x,y,r)|dxdy \leq \overline{M}(r)$$

for every $R > 0$.

Then the function $w(x,y)$ *is improper fuzzy Riemann-integrable on D and*

$$(FR)\int_0^\infty (FR)\int_0^\infty w(x,y)dxdy = \left(\int_0^\infty\int_0^\infty \underline{w}(x,y,r)dxdy, \int_0^\infty\int_0^\infty \overline{w}(x,y,r)dxdy\right).$$

Proof. Define the function $\underline{I} : (0,\infty) \to \mathbb{R}_+$ by

$$\underline{I}(R) = \iint_{D_R} |\underline{w}(x,y,r)|dxdy \;\; forall \;\; r \in [0,1].$$

From condition 2, it follows that $\underline{I}$ is bounded and monotonically increasing. Hence, there exists

$$\lim_{R\to\infty} \underline{I}(R) = \int_0^\infty\int_0^\infty \underline{w}(x,y,r)dxdy.$$

□

For fuzzy valued functions $w = w(x,y)$ we define the n-th order partial H-derivatives with respect to x and y as given in [11].

Definition 7. *Let* $w : (a,b) \times (c,d) \to E^1$ *be a fuzzy function. We call that* w *is H-differentiable of the n-th order at* $x_0 \in (a,b)$, *with respect to x, if there exists an element* $\frac{\partial^n w(x_0,y)}{\partial x^n} \in E^1$ *such that*

1. *for all* $h > 0$ *sufficiently small the H-differences*

$$\frac{\partial^{n-1}w(x_0+h,y)}{\partial x^{n-1}} \ominus \frac{\partial^{n-1}w(x_0,y)}{\partial x^{n-1}}, \quad \frac{\partial^{n-1}w(x_0,y)}{\partial x^{n-1}} \ominus \frac{\partial^{n-1}w(x_0-h,y)}{\partial x^{n-1}},$$

exist and the following limits hold (in the metric d)

$$\lim_{h\to 0}\frac{1}{h}\left(\frac{\partial^{n-1}w(x_0+h,y)}{\partial x^{n-1}} \ominus \frac{\partial^{n-1}w(x_0,y)}{\partial x^{n-1}}\right)$$

$$= \lim_{h\to 0}\frac{1}{h}\left(\frac{\partial^{n-1}w(x_0,y)}{\partial x^{n-1}} \ominus \frac{\partial^{n-1}w(x_0-h,y)}{\partial x^{n-1}}\right) = \frac{\partial^n w(x_0,y)}{\partial x^n}$$

or

2. *for all* $h > 0$ *sufficiently small the H-differences*

$$\frac{\partial^{n-1}w(x_0,y)}{\partial x^{n-1}} \ominus \frac{\partial^{n-1}w(x_0+h,y)}{\partial x^{n-1}}, \quad \frac{\partial^{n-1}w(x_0-h,y)}{\partial x^{n-1}} \ominus \frac{\partial^{n-1}w(x_0,y)}{\partial x^{n-1}},$$

exist and the following limits hold (in the metric d)

$$\lim_{h\to 0}\frac{-1}{h}\left(\frac{\partial^{n-1}w(x_0,y)}{\partial x^{n-1}} \ominus \frac{\partial^{n-1}w(x_0+h,y)}{\partial x^{n-1}}\right)$$

$$= \lim_{h\to 0}\frac{-1}{h}\left(\frac{\partial^{n-1}w(x_0-h,y)}{\partial x^{n-1}} \ominus \frac{\partial^{n-1}w(x_0,y)}{\partial x^{n-1}}\right) = \frac{\partial^n w(x_0,y)}{\partial x^n}.$$

Similarly,

Definition 8. *Let $w : (a,b) \times (c,d) \to E^1$ be a fuzzy function. We call that w is H-differentiable of the n-th order at $y_0 \in (c,d)$, with respect to y, if there exists an element $\frac{\partial^n w(x,y_0)}{\partial y^n} \in E^1$ such that*

1. *For $h > 0$ sufficiently small the H-differences*

$$\frac{\partial^{n-1}w(x,y_0+h)}{\partial y^{n-1}} \ominus \frac{\partial^{n-1}w(x,y_0)}{\partial y^{n-1}}, \quad \frac{\partial^{n-1}w(x,y_0)}{\partial y^{n-1}} \ominus \frac{\partial^{n-1}w(x,y_0-h)}{\partial y^{n-1}},$$

exist and the following limits hold (in the metric d)

$$\lim_{h\to 0}\frac{1}{h}\left(\frac{\partial^{n-1}w(x,y_0+h)}{\partial y^{n-1}} \ominus \frac{\partial^{n-1}w(x,y_0)}{\partial y^{n-1}}\right)$$

$$= \lim_{h\to 0}\frac{1}{h}\left(\frac{\partial^{n-1}w(x,y_0)}{\partial y^{n-1}} \ominus \frac{\partial^{n-1}w(x,y_0-h)}{\partial y^{n-1}}\right) = \frac{\partial^{n}w(x,y_0)}{\partial y^{n}}$$

or

2. *For $h > 0$ sufficiently small the H-differences*

$$\frac{\partial^{n-1}w(x,y_0)}{\partial y^{n-1}} \ominus \frac{\partial^{n-1}w(x,y_0+h)}{\partial y^{n-1}}, \quad \frac{\partial^{n-1}w(x,y_0-h)}{\partial y^{n-1}} \ominus \frac{\partial^{n-1}w(x,y_0)}{\partial y^{n-1}},$$

exist and the following limits hold (in the metric d)

$$\lim_{h\to 0}\frac{-1}{h}\left(\frac{\partial^{n-1}w(x,y_0)}{\partial y^{n-1}} \ominus \frac{\partial^{n-1}w(x,y_0+h)}{\partial y^{n-1}}\right)$$

$$= \lim_{h\to 0}\frac{-1}{h}\left(\frac{\partial^{n-1}w(x,y_0-h)}{\partial y^{n-1}} \ominus \frac{\partial^{n-1}w(x,y_0)}{\partial y^{n-1}}\right) = \frac{\partial^{n}w(x,y_0)}{\partial y^{n}}.$$

The first type of differentiability as in Definition 7 and Definition 8 are referred as (i)-differentiable, while the second type as (ii)-differentiable.

Theorem 2 ([11])**.** *Let $w : \mathbb{R} \times \mathbb{R} \to E^1$ be a continuous fuzzy-valued function and $w(x,y) = (\underline{w}(x,y,r), \overline{w}(x,y,r))$ for all $r \in [0,1]$. Then*

1. *if $w(x,y)$ is (i)-differentiable of the n-th order with respect to x, then $\underline{w}(x,y,r)$ and $\overline{w}(x,y,r)$ are differentiable of the n-th order with respect to x and*

$$\frac{\partial^n w(x,y)}{\partial x^n} = \left(\frac{\partial^n \underline{w}(x,y,r)}{\partial x^n}, \frac{\partial^n \overline{w}(x,y,r)}{\partial x^n}\right), \tag{1}$$

2. *if $w(x,y)$ is (ii)-differentiable of the n-th order with respect to x, then $\underline{w}(x,y,r)$ and $\overline{w}(x,y,r)$ are differentiable of the n-th order with respect to x and*

$$\frac{\partial^n w(x,y)}{\partial x^n} = \left(\frac{\partial^n \overline{w}(x,y,r)}{\partial x^n}, \frac{\partial^n \underline{w}(x,y,r)}{\partial x^n}\right). \tag{2}$$

3. Two-Dimensional Fuzzy Sumudu Transform

In this part, we give DFST definition and its inverse. We introduced the concept of double fuzzy convolution and give two new results of DFST for fuzzy partial derivative of the n-th order.

Definition 9 ([35]). *Let $w : \mathbb{R} \times \mathbb{R} \to E^1$ be a continuous fuzzy-valued function and the function $e^{-x-y} \odot w(ux, vy)$ is improper fuzzy Riemann-integrable on D, then*

$$(FR) \int_0^\infty (FR) \int_0^\infty e^{-x-y} \odot w(ux, vy) dxdy,$$

is called DFST and is denote by

$$W(u,v) = S[w(x,y)] = (FR) \int_0^\infty (FR) \int_0^\infty e^{-x-y} \odot w(ux, vy) dxdy, \tag{3}$$

for $u \in [-\tau_1, \tau_2]$ and $v \in [-\sigma_1, \sigma_2]$, where the variables u, v are used to factor the variables x, y in the argument of the fuzzy-valued function and $\tau_1,\ \tau_2,\ \sigma_1,\ \sigma_2 > 0$.

The parametric form of DFST is follows

$$S[w(x,y)] = (s[\underline{w}(x,y,r)], s[\overline{w}(x,y,r)]), \tag{4}$$

where

$$s[\underline{w}(x,y,r)] = \int_0^\infty \int_0^\infty e^{-x-y} \underline{w}(ux, vy, r) dxdy, \tag{5}$$

$$s[\overline{w}(x,y,r)] = \int_0^\infty \int_0^\infty e^{-x-y} \overline{w}(ux, vy, r) dxdy. \tag{6}$$

The equation (3) we can rewrite in the form

$$W(u,v) \ = S[w(x,y)] = \tfrac{1}{uv}(FR) \int_0^\infty (FR) \int_0^\infty e^{-(\frac{x}{u}+\frac{y}{v})} \odot w(x,y) dxdy. \tag{7}$$

Definition 10 ([35]). *Double fuzzy inverse Sumudu transform can be written as the formula*

$$S^{-1}\left[W(u,v)\right] = w(x,y) = \left(s^{-1}[\underline{W}(u,v,r)], s^{-1}[\overline{W}(u,v,r)]\right), \tag{8}$$

where

$$s^{-1}[\underline{W}(u,v,r)] = \frac{1}{2\pi\imath} \int_{\gamma-\imath\infty}^{\gamma+\imath\infty} e^{\frac{x}{u}} du \frac{1}{2\pi\imath} \int_{\delta-\imath\infty}^{\delta+\imath\infty} e^{\frac{y}{v}} \underline{W}(u,v,r) dv,$$

$$s^{-1}[\overline{W}(u,v,r)] = \frac{1}{2\pi\imath} \int_{\gamma-\imath\infty}^{\gamma+\imath\infty} e^{\frac{x}{u}} du \frac{1}{2\pi\imath} \int_{\delta-\imath\infty}^{\delta+\imath\infty} e^{\frac{y}{v}} \overline{W}(u,v,r) dv.$$

For all $r \in [0,1]$ the functions $\underline{W}(u,v,r)$ and $\overline{W}(u,v,r)$ must be analytic functions for all u and v in the region defined by the inequalities $Reu \geq \gamma$ and $Rev \geq \delta$, where γ and δ are real constants to be chosen suitably.

In [40] classical double Sumudu transform is applied on some special functions.

1. Let $g(x,y) = 1$ for $x > 0,\ y > 0$, then $s[g(x,y)] = 1$.
2. Let $g(x,y) = x^m y^n$, where $m,\ n$ are positive integers, then

$$s[g(x,y)] = (m!)(n!)u^m v^n. \tag{9}$$

3. Let $g(x,y) = e^{ax+by}$, where $a,\ b$ are any constants, then

$$s[g(x,y)] = \frac{1}{(1-au)(1-bv)}. \tag{10}$$

4.

$$s[cos(ax+by)] = \frac{(1-abuv)}{(1+a^2u^2)(1+b^2v^2)}, \tag{11}$$

$$s[sin(ax+by)] = \frac{(bv+au)}{(1+a^2u^2)(1+b^2v^2)}. \tag{12}$$

Theorem 3 ([35]). *Let $g(x,y)$ be a continuous fuzzy-valued function. If $G(u,v)$ is the double fuzzy Sumudu transform of $g(x,y)$ and $a,\ b$ are arbitrary constants, then*

$$S[e^{ax+by} \odot g(x,y)] = \frac{1}{(1-au)(1-bv)} \odot G\left(\frac{u}{1-au}, \frac{v}{1-bv}\right). \tag{13}$$

In [35] DFST theorems and properties generated by DFST are given.

Definition 11 ([35]). *If $k(x,y)$ and $w(x,y)$ are fuzzy Riemann integrable functions, then double fuzzy convolution of $k(x,y)$ and $w(x,y)$ is given by*

$$(k**w)(x,y) = (FR)\int_0^y (FR)\int_0^x k(x-\alpha, y-\beta) w(\alpha,\beta) d\alpha d\beta \tag{14}$$

and the symbol $**$ denotes the double convolution respect to x and y.

Theorem 4 ([35]). *Let $k : D \to \mathbb{R}$ and $w(x,y)$ be fuzzy functions. Then the DFST of the double fuzzy convolution k and w, is given by*

$$S[(k**w)(x,y)] = uvS[k(x,y)] \odot S[w(x,y)]. \tag{15}$$

We introduce results of DFST for fuzzy partial derivatives.

Theorem 5. *Let $w : \mathbb{R} \times \mathbb{R} \to E^1$ be a continuous fuzzy-valued function. The functions $e^{-x-y} \odot w(ux,vy)$, $e^{-x-y} \odot \frac{\partial^n w(ux,vy)}{\partial x^n}$ are improper fuzzy Riemann-integrable on D. Then*

$$S\left[\frac{\partial^n w(x,y)}{\partial x^n}\right] = \frac{\partial^n}{\partial x^n} S[w(x,y)], \tag{16}$$

where $S[w(x,y)]$ denotes the DFST of the function w and $n \in \mathbb{N}$.

Proof. Let the function $w(x,y)$ is (i)-differentiable. From definition of DFST, we have

$$S[\frac{\partial^n w(x,y)}{\partial x^n}] = (FR)\int_0^\infty (FR)\int_0^\infty e^{-x-y} \odot \frac{\partial^n w(ux,vy)}{\partial x^n} dxdy$$

$$= \left(\int_0^\infty\int_0^\infty e^{-x-y}\frac{\partial^n \underline{w}(ux,vy)}{\partial x^n} dxdy,\ \int_0^\infty\int_0^\infty e^{-x-y}\frac{\partial^n \overline{w}(ux,vy)}{\partial x^n} dxdy\right)$$

$$= \frac{\partial^n}{\partial x^n}\left[\int_0^\infty\int_0^\infty e^{-x-y}\underline{w}(ux,vy)dxdy,\ \int_0^\infty\int_0^\infty e^{-x-y}\overline{w}(ux,vy)dxdy\right] = \frac{\partial^n}{\partial x^n}S[w(x,y)].$$

□

Theorem 6. *Let $w : \mathbb{R}\times\mathbb{R} \to E^1$ be a fuzzy-valued function. The functions $e^{-x-y}\odot w(ux,vy)$, $e^{-x-y}\odot\frac{\partial^n w(ux,vy)}{\partial x^n}$ are improper fuzzy Riemann-integrable on D. For all $x>0$ and $n\in\mathbb{N}$ there exist to continuous partial H-derivatives to $(n-1)-th$ order with respect to x and there exists $\frac{\partial^n w(x,y)}{\partial x^n}$. Then*

1. *if the function $w(x,y)$ is (i)-differentiable then*

$$S\left[\frac{\partial^n w(x,y)}{\partial x^n}\right] = \left(s\left[\frac{\partial^n \underline{w}(x,y)}{\partial x^n}\right], s\left[\frac{\partial^n \overline{w}(x,y)}{\partial x^n}\right]\right),$$

2. *if the function $w(x,y)$ is (ii)-differentiable then*

$$S\left[\frac{\partial^n w(x,y)}{\partial x^n}\right] = \left(s\left[\frac{\partial^n \overline{w}(x,y)}{\partial x^n}\right], s\left[\frac{\partial^n \underline{w}(x,y)}{\partial x^n}\right]\right),$$

where

$$s\left[\frac{\partial^n \underline{w}(x,y,r)}{\partial x^n}\right] = \frac{1}{u^n}s\left[\underline{w}(x,y,r)\right] - \sum_{j=1}^{n}\frac{1}{u^j}s\left[\frac{\partial^{n-j}\underline{w}(0,y,r)}{\partial x^{n-j}}\right], \tag{17}$$

$$s\left[\frac{\partial^n \overline{w}(x,y,r)}{\partial x^n}\right] = \frac{1}{u^n}s\left[\overline{w}(x,y,r)\right] - \sum_{j=1}^{n}\frac{1}{u^j}s\left[\frac{\partial^{n-j}\overline{w}(0,y,r)}{\partial x^{n-j}}\right]. \tag{18}$$

Proof. Let the function $w(x,y)$ is (i)-differentiable. By induction we proof the equation (17). For $n=1$ from condition (4) we have

$$S\left[w'_x(x,y)\right] = (s[\underline{w}'_x(x,y,r)],\ s\left[\overline{w}'_x(x,y,r)\right]).$$

By us part integration on x and condition (4) we obtain

$$s[\underline{w}'_x(x,y,r)] = \int_0^\infty\int_0^\infty e^{-x-y}\underline{w}'_x(ux,vy,r)dxdy = \frac{1}{u}(s[\underline{w}(x,y,r)-s[\underline{w}(0,y,r)).$$

Let for $n=k$ the equation (17) holds. Then

$$s[\frac{\partial^k \underline{w}(x,y,r)}{\partial x^k}] = \frac{1}{u^k}s[\underline{w}(x,y,r)] - \sum_{j=1}^{k}\frac{1}{u^j}s[\frac{\partial^{k-j}\underline{w}(0,y,r)}{\partial x^{k-j}}].$$

Hence, for $n=k+1$ we get

$$s[\tfrac{\partial^{k+1}\underline{w}(x,y,r)}{\partial x^{k+1}}] = \tfrac{\partial}{\partial x}s[\tfrac{\partial^k \underline{w}(x,y,r)}{\partial x^k}] = \tfrac{\partial}{\partial x}(\tfrac{1}{u^k}s[\underline{w}(x,y,r)]) - \tfrac{\partial}{\partial x}(\sum_{j=1}^{k}\tfrac{1}{u^j}s[\tfrac{\partial^{k-j}\underline{w}(0,y,r)}{\partial x^{k-j}}])$$
$$= \tfrac{1}{u^k}s[\underline{w}'_x(x,y,r)] - \sum_{j=1}^{k}\tfrac{1}{u^j}s[\tfrac{\partial^{k+1-j}\underline{w}(0,y,r)}{\partial x^{k+1-j}}]$$
$$= \tfrac{1}{u^{k+1}}(s[\underline{w}(x,y,r)] - s[\underline{w}(0,y,r)]) - \sum_{j=1}^{k}\tfrac{1}{u^j}s[\tfrac{\partial^{k+1-j}\underline{w}(0,y,r)}{\partial x^{k+1-j}}]$$
$$= \tfrac{1}{u^{k+1}}s[\underline{w}(x,y,r)] - \sum_{j=1}^{k+1}\tfrac{1}{u^j}s[\tfrac{\partial^{k+1-j}\underline{w}(0,y,r)}{\partial x^{k+1-j}}].$$

□

4. DFST for Solving PVFIDE

In this section, we application of the DFST method for solving of PVFIDE. This equation is defined as

$$\sum_{i=1}^{m} a_i \odot \frac{\partial^i w(x,y)}{\partial x^i} \oplus \sum_{j=1}^{n} b_j \odot \frac{\partial^j w(x,y)}{\partial y^j} \oplus c \odot w(x,y) = g(x,y) \oplus (FR)\int_0^y (FR)\int_0^x k(x-\alpha, y-\beta) \odot w(\alpha,\beta) d\alpha d\beta, \tag{19}$$

with initial conditions

$$\frac{\partial^i w(0,y)}{\partial x^i} = \varphi_i(y), \quad i = 0,1,...,m-1 \tag{20}$$

$$\frac{\partial^i w(x,0)}{\partial y^j} = \psi_j(x), \quad i = 0,1,...,n-1, \tag{21}$$

where $k : [0,b] \times [0,d] \to \mathbb{R}$, is a continuous functions and $g,\ w : [0,b] \times [0,d] \to E^1$, $\varphi_i : [0,d] \to E^1$, $\psi_j : [0,b] \to E^1$ are continuous fuzzy functions and a_i, $i = 1,2,...m$, b_j, $j = 1,2,...n$, c, are constants.

Applying DFST on both side of it we get the following

$$S\left[\sum_{i=1}^{m} a_i \frac{\partial^i w(x,y)}{\partial x^i}\right] \oplus S\left[\sum_{j=1}^{n} b_j \frac{\partial^j w(x,y)}{\partial y^j}\right] \oplus S[c \odot w(x,y)] = S[g(x,y)] \oplus S\left[(FR)\int_0^y (FR)\int_0^x k(x-\alpha, y-\beta) \odot w(\alpha,\beta) d\alpha d\beta\right],$$

By using double fuzzy convolution (15) we obtain

$$\sum_{i=1}^{m} a_i \odot S\left[\frac{\partial^i w(x,y)}{\partial x^i}\right] \oplus \sum_{j=1}^{n} b_j \odot S\left[\frac{\partial^j w(x,y)}{\partial y^j}\right] \oplus c \odot S[w(x,y)] = S[g(x,y)] \oplus uvs[k(x,y)] \odot S[w(x,y)]$$

Let the constants a_i, $i = 1,...,m$, b_j, $j = 1,...n$, c be positive and the function $k(x,y) > 0$.

1. if $w(x,y)$ is (i)-differentiable, then

$$\sum_{i=1}^{m} a_i s\left[\frac{\partial^i \underline{w}(x,y,r)}{\partial x^i}\right] + \sum_{j=1}^{n} b_j s\left[\frac{\partial^j \underline{w}(x,y,r)}{\partial y^j}\right] + cs[\underline{w}(x,y,r)] = s[\underline{g}(x,y,r)] + uvs[k(x,y)]s[\underline{w}(x,y,r)]$$

and

$$\sum_{i=1}^{m} a_i s\left[\frac{\partial^i \overline{w}(x,y,r)}{\partial x^i}\right] + \sum_{j=1}^{n} b_j s\left[\frac{\partial^j \overline{w}(x,y,r)}{\partial y^j}\right] + cs[\overline{w}(x,y,r)] = s[\overline{g}(x,y,r)] + uvs[k(x,y)]s[\overline{w}(x,y,r)]$$

Then from (17) and (18) we have

$$\left(\sum_{i=1}^{m} \frac{a_i}{u^i} + \sum_{j=1}^{n} \frac{b_j}{v^j} + c - uvs[k(x,y)]\right) s[\underline{w}(x,y,r)] = s[\underline{g}(x,y,r)] + \sum_{i=1}^{m}\sum_{k=1}^{i} \frac{a_i}{u^k} s\left[\frac{\partial^{i-k} \underline{w}(0,y,r)}{\partial x^{i-k}}\right] + \sum_{j=1}^{n}\sum_{k=1}^{j} \frac{b_j}{v^k} s\left[\frac{\partial^{j-k} \underline{w}(x,0,r)}{\partial y^{j-k}}\right]$$

and

$$\left(\sum_{i=1}^{m} \frac{a_i}{u^i} + \sum_{j=1}^{n} \frac{b_j}{v^j} + c - uvs[k(x,y)]\right) s[\overline{w}(x,y,r)] = s[\overline{g}(x,y,r)] + \sum_{i=1}^{m}\sum_{k=1}^{i} \frac{a_i}{u^k} s\left[\frac{\partial^{i-k} \overline{w}(0,y,r)}{\partial x^{i-k}}\right] + \sum_{j=1}^{n}\sum_{k=1}^{j} \frac{b_j}{v^k} s\left[\frac{\partial^{j-k} \overline{w}(x,0,r)}{\partial y^{j-k}}\right]$$

Using the initial conditions (20) and (21) we get

$$\left(\sum_{i=1}^{m}\frac{a_i}{u^i}+\sum_{j=1}^{n}\frac{b_j}{v^j}+c-uvs[k(x,y)]\right)s[\underline{w}(x,y,r)]$$
$$=s[\underline{g}(x,y,r)]+\sum_{i=1}^{m}\sum_{k=1}^{i}\frac{a_i}{u^k}s\left[\underline{\varphi}_{i-k}^{i-k}(0,y,r)\right]+\sum_{j=1}^{n}\sum_{k=1}^{j}\frac{b_j}{v^k}s\left[\underline{\psi}_{j-k}^{j-k}(x,0,r)\right]$$

and

$$\left(\sum_{i=1}^{m}\frac{a_i}{u^i}+\sum_{j=1}^{n}\frac{b_j}{v^j}+c-uvs[k(x,y)]\right)s[\overline{w}(x,y,r)]$$
$$=s[\overline{g}(x,y,r)]+\sum_{i=1}^{m}\sum_{k=1}^{i}\frac{a_i}{u^k}s\left[\overline{\varphi}_{i-k}^{i-k}(0,y,r)\right]+\sum_{j=1}^{n}\sum_{k=1}^{j}\frac{b_j}{v^k}s\left[\overline{\psi}_{j-k}^{j-k}(x,0,r)\right]$$

Then

$$s[\underline{w}(x,y,r)]=\frac{s[\underline{g}(x,y,r)]+\sum_{i=1}^{m}\sum_{k=1}^{i}\frac{a_i}{u^k}s\left[\underline{\varphi}_{i-k}^{i-k}(0,y,r)\right]+\sum_{j=1}^{n}\sum_{k=1}^{j}\frac{b_j}{v^k}s\left[\underline{\psi}_{j-k}^{j-k}(x,0,r)\right]}{\sum_{i=1}^{m}\frac{a_i}{u^i}+\sum_{j=1}^{n}\frac{b_j}{v^j}+c-uvs[k(x,y)]} \tag{22}$$

and

$$s[\overline{w}(x,y,r)]=\frac{s[\overline{g}(x,y,r)]+\sum_{i=1}^{m}\sum_{k=1}^{i}\frac{a_i}{u^k}s\left[\overline{\varphi}_{i-k}^{i-k}(0,y,r)\right]+\sum_{j=1}^{n}\sum_{k=1}^{j}\frac{b_j}{v^k}s\left[\overline{\psi}_{j-k}^{j-k}(x,0,r)\right]}{\sum_{i=1}^{m}\frac{a_i}{u^i}+\sum_{j=1}^{n}\frac{b_j}{v^j}+c-uvs[k(x,y)]} \tag{23}$$

2. if $w(x,y)$ is (ii)-differentiable, then

$$s[\underline{w}(x,y,r)]=\frac{s[\underline{g}(x,y,r)]+\sum_{i=1}^{m}\sum_{k=1}^{i}\frac{a_i}{u^k}s\left[\overline{\varphi}_{i-k}^{i-k}(0,y,r)\right]+\sum_{j=1}^{n}\sum_{k=1}^{j}\frac{b_j}{v^k}s\left[\overline{\psi}_{j-k}^{j-k}(x,0,r)\right]}{\sum_{i=1}^{m}\frac{a_i}{u^i}+\sum_{j=1}^{n}\frac{b_j}{v^j}+c-uvs[k(x,y)]} \tag{24}$$

and

$$s[\overline{w}(x,y,r)]=\frac{s[\overline{g}(x,y,r)]+\sum_{i=1}^{m}\sum_{k=1}^{i}\frac{a_i}{u^k}s\left[\underline{\varphi}_{i-k}^{i-k}(0,y,r)\right]+\sum_{j=1}^{n}\sum_{k=1}^{j}\frac{b_j}{v^k}s\left[\underline{\psi}_{j-k}^{j-k}(x,0,r)\right]}{\sum_{i=1}^{m}\frac{a_i}{u^i}+\sum_{j=1}^{n}\frac{b_j}{v^j}+c-uvs[k(x,y)]} \tag{25}$$

By using the inverse of DFST we obtain $w(y,y)=(\underline{w}(x,y,r),\overline{w}(x,y,r)$.

5. Examples

In this section, we find the solution of partial convolution Volterra fuzzy integro-differential equation using DFST.

Example 1. *Consider the following PVFIDE*

$$w''_{xx}(x,y)\oplus w''_{yy}(x,y)\oplus w(x,y)=g(x,y)\oplus(FR)\int_0^y(FR)\int_0^x e^{x-\alpha+y-\beta}w(\alpha,\beta)d\alpha d\beta,$$
$$(x,y)\in[0,1]\times[0,1]\ \ r\in[0,1]$$

with initial conditions

$$w(x,0) = (e^x(2+r), e^x(4-r)), \ \ w'_y(x,0) = (e^x(2+r), e^x(4-r)),$$

$$w(0,y) = (e^y(2+r), e^y(4-r)), \ \ w'_x(0,y) = (e^y(2+r), e^y(4-r))$$

and

$$g(x,y) = (e^{x+y}(2+xy)(2+r), e^{x+y}(2+xy)(4-r)).$$

In this case $m = n = 2$, $a_1 = b_1 = 0$, $a_2 = b_2 = 1$, $c = 1$,
$k(x-\alpha, y-\beta) = e^{x-\alpha+y-\beta} > 0$ *for* $0 \le \alpha \le x \le 1$ *and* $0 \le \beta \le y \le 1$,

$$\psi_0(x,0) = (e^x(2+r), e^x(4-r)), \ \ \varphi_0(y) = (e^y(2+r), e^y(4-r)).$$

$$\psi_1(x,0) = (e^x(2+r), e^x(4-r)), \ \ \varphi_1(y) = (e^y(2+r), e^y(4-r)).$$

From (10), *we find*

$$s[k(x,y)] = s[e^{x+y}] = \frac{1}{(1-u)(1-v)},$$

$$S[\psi_0(x)] = S[\psi_1(x)] = (s[e^x(2+r)], s[e^x(4-r)]) = \left(\frac{1}{1-u}(2+r), \frac{1}{1-u}(4-r)\right),$$

$$S[\varphi_0(y)] = S[\varphi_1(y)] = (s[e^y(2+r)], s[e^y(4-r)]) = \left(\frac{1}{1-v}(2+r), \frac{1}{1-v}(4-r)\right)$$

From Theorem 3 we obtain

$$S[g(x,y)] = (s[(2+xy)e^{x+y}(2+r), s[(2+xy)e^{x+y}(4-r)])$$

$$= \left(\left(3 - \frac{uv}{(1-u)(1-v)}\right)\frac{(2+r)}{(1-u)(1-v)}, \left(3 - \frac{uv}{(1-u)(1-v)}\right)\frac{(4-r)}{(1-u)(1-v)}\right).$$

Then, of (22) *and* (23) *for the solution of the equation we have*

$$s[\underline{w}(x,y,r)] = \frac{1}{(1-u)(1-v)}(2+r).$$

and

$$s[\underline{w}(x,y,r)] = \frac{1}{(1-u)(1-v)}(4-r).$$

By inverse double Sumudu transform the solution of the equation is $w(x,y) = (e^{x+y}(2+r), e^{x+y}(4-r))$.

6. Conclusions

In this paper, the double fuzzy Sumudu transform method for solving partial convolution Volterra fuzzy integro-differential equations have been studied. The concept of double fuzzy convolution have been introduced. New results on DFST for fuzzy partial H-derivative of the n-th order have been proposed.

By using the parametric form of fuzzy functions we convert the investigated equation to a nonlinear system of partial Volterra integro-differential equations in a crisp case. Applying DFST method for this system we obtain system of algebraic equations. Hence we find the lower and upper functions of the solution. Finally, the examples to show that the investigation method is effective in solving the equations of considered kind.

Funding: A.G. is partially supported by the Bulgarian National Science Fund under Project KP-06-N32/7.

Conflicts of Interest: The author declares no conflict of interest.

References

1. Aslam, M.; Khan, N.; Ali Hussein, A.-M. Design of Variable Sampling Plan for Pareto Distribution Using Neutrosophic Statistical Interval Method. *Symmetry* **2019**, *11*, 80. [CrossRef]
2. Aslam, M. A New Sampling Plan Using Neutrosophic Process Loss Consideration. *Symmetry* **2018**, *10*, 132. [CrossRef]
3. Aslam, M.; Arif, O.H.; Khan Sherwan, A.R. New Diagnosis Test under the Neutrosophic Statistics: An Application to Diabetic Patients. *BioMed Res. Int.* **2020**, 2020, 2086185. [CrossRef]
4. Aslam, M. Attribute Control Chart Using the Repetitive Sampling Under Neutrosophic System. *IEEE Access* **2019**, *7*, 15367–15374. [CrossRef]
5. Chang, S.; Zadeh, L. On fuzzy mapping and control. *IEEE Trans. Syst. Man Cybern.* **1972**, *1*, 30–34. [CrossRef]
6. Seikkala, S. On the fuzzy initial value problem. *Fuzzy Sets Syst.* **1987**, *24*, 319–330. [CrossRef]
7. Dubois, D.; Prade, H. Towards fuzzy differential calculus I,II,III. *Fuzzy Sets Syst.* **1982**, *8*, 105–116, 225–233. [CrossRef]
8. Buckley, J.J.; Feuring, T. Introduction to fuzzy partial differential equations. *Fuzzy Sets Syst.* **1999**, *105*, 241–248. [CrossRef]
9. Allahveranloo, T. Difference methods for fuzzy partial differential equations. *Comput. Methods Appl. Math.* **2002**, *2*, 233–242. [CrossRef]
10. Allahviranloo, T.; Kermani, M. Numerical methods for fuzzy linear partial differential equations under new definition for derivative. *Iran. J. Fuzzy Syst.* **2010**, *7*, 33–50.
11. Chalco-Cano, Y.; Roman-Flores, H. On new solutions of fuzzy differential quations. *Chaos Solitons Fractals* **2008**, *38*, 112–119. [CrossRef]
12. Friedman, M.; Ma, M.; Kandel, A. Numerical solutions of fuzzy differential and integral equations. *Fuzzy Sets Syst.* **1999**, *106*, 35–48. [CrossRef]
13. Alidema, A.; Georgieva, A. Adomian decomposition method for solving two-dimensional nonlinear Volterra fuzzy integral equations. *AIP Conf. Proc.* **2018**, *2048*, 050009.
14. Georgieva, A. Solving two-dimensional nonlinear Volterra-Fredholm fuzzy integral equations by using Adomian decomposition method. *Dyn. Syst. Appl.* **2018**, *27*, 819–837.
15. Naydenova, I.; Georgieva, A. Approximate solution of nonlinear mixed Volterra-Fredholm fuzzy integral equations using the Adomian method. *AIP Conf. Proc.* **2019**, *2172*, 060005.
16. Georgieva, A.; Alidema, A. Convergence of homotopy perturbation method for solving of two-dimensional fuzzy Volterra functional integral equations. *Adv. Comput. Ind. Math. Stud. Computat. Intell.* **2019**, *793*, 129–145.
17. Mordeson, J.; Newman, W. Fuzzy integral equations. *Inf. Sci.* **1995**, *87*, 215–229. [CrossRef]
18. Abbasbandy, S.; Hashemi, M. A series solution of fuzzy integro-differential equations. *J. Fuzzy Set Valued Anal.* **2012**, *2012*, jfsva-00066. [CrossRef]
19. Hooshangian, L. Nonlinear Fuzzy Volterra Integro-differential Equation of N-th Order: Analytic Solution and Existence and Uniqueness of Solution. *Int. J. Ind. Math.* **2019**, *11*, 12.
20. Majid, Z.; Rabiei, F.; Hamid, F.; Ismail, F. Fuzzy Volterra Integro-Differential Equations Using General Linear Method. *Symmetry* **2019**, *11*, 381. [CrossRef]
21. Chalishajar, D.; Ramesh, R. Controllability for impulsive fuzzy neutral functional integrodifferential equations. *AIP Conf. Proc.* **2019**, *2159*, 030007.
22. Ahmad, J.; Nosher, H. Solution of Different Types of Fuzzy Integro-Differential Equations Via Laplace Homotopy Perturbation Method. *J. Sci. Arts* **2017**, *17*, 5.
23. Mikaeilv, N.; Khakrangin, S.; Allahviranloo, T. Solving fuzzy Volterra integro-differential equation by fuzzy differential transform method. In Proceedings of the 7th Conference of the European Society for Fuzzy Logic and Technology, Aix-Les-Bains, France, 18–22 July 2011; pp. 18–22.
24. Arnoldus, H.F. Application of the magnetic field integral equation to diffraction and reection by a conducting sheet. *Int. J. Theor. Phys. Group Theory Nonlinear Opt.* **2011**, *14*, 1–12.

25. Ma, S.Q.; Chem, F.C.; Zhao, Z.Q. Choquet type fuzzy complex-values integral and its application in classification. *Fuzzy Eng. Oper. Res.* **2012**, *147*, 229--237.
26. Watugala, G.K. Sumudu transform a new integral transform to solve differential equations and control engineering problems. *Internat. J. Math. Ed. Sci. Tech.* **1993**, *24*, 35–43. [CrossRef]
27. Watugala, G.K. The Sumudu transform for functions of two variables. *Math. Eng. Ind.* **2002**, *8*, 293–302.
28. Weerakoon, S. Application of Sumudu transform to partial differential equations. *Internat. J. Math. Ed. Sci. Tech.* **1994**, *25*, 277–283. [CrossRef]
29. Abdul Rahman, N.A.; Ahmad, M.Z. Fuzzy Sumudu transform for solving fuzzy partial differential equations. *J. Nonlinear Sci. Appl.* **2016**, *9*, 3226–3239. [CrossRef]
30. Belgacem, F.B.M. Sumudu transform applications to Bessel functions and equations. *Appl. Math. Sci.* **2010**, *4*, 3665–3686.
31. Abdul Rahman, N.A.; Ahmad, M.Z. Applications of the Fuzzy Sumudu Transform for the Solution of First Order Fuzzy Differential Equations. *Entropy* **2015**, *17*, 4582–4601. [CrossRef]
32. Abdul Rahman, N.A.; Ahmad, M.Z. Solving Fuzzy Volterra Integral Equations via Fuzzy Sumudu Transform. *Appl. Math. Comput. Intell.* **2017**, *6*, 19–28.
33. Min Kang, S.; Iqbal, Z.; Habib, M.; Nazeer, W. Sumudu Decomposition Method for Solving Fuzzy Integro-Differential Equations. *Axioms* **2019**, *8*, 74. [CrossRef]
34. Abdul Rahman, N.A.; Ahmad, M.Z. Solving Fuzzy Fractional Differential Equations using Fuzzy Sumudu Transform. *J. Nonlinear Sci. Appl.* **2017**, *6*, 19–28. [CrossRef]
35. Alidema, A.; Georgieva, A. Applications of the double fuzzy Sumudu transform for solving Volterra fuzzy integral equations. *AIP Conf. Proc.* **2019**, *2172*, 060006.
36. Mohand, M.; Mahgob A. Solution of Partial Integro-Differential Equations by Double Elzaki Transform Method. *Math. Theory Model.* **2015**, *5*, 61–65.
37. Goetschel, R.; Voxman, W. Elementary fuzzy calculus. *Fuzzy Sets Syst.* **1986**, *18*, 31–43. [CrossRef]
38. Kaufmann A.; Gupta M.M. *Introduction to Fuzzy Arithmetic: Theory and Applications;* Van Nostrand Reinhold Co.: New York, NY, USA, 1991.
39. Puri, M.L.; Ralescu, D.A. Differentials of fuzzy functions. *J. Math. Anal. Appl.* **1983**, *91*, 552–558. [CrossRef]
40. Kilicman, A.; Omran, M. On double Natural transform and its applications. *J. Nonlinear Sci. Appl.* **2017**, *10*, 1744–1754. [CrossRef]

mathematics

MDPI

Article

Evolution Inclusions in Banach Spaces under Dissipative Conditions

Tzanko Donchev [1,*], Shamas Bilal [2], Ovidiu Cârjă [3,4], Nasir Javaid [5] and Alina I. Lazu [6]

1 Department of Mathematics, University of Architecture, Civil Engineering and Geodesy, Sofia 1164, Bulgaria
2 Department of Mathematics, University of Sialkot, Sialkot 51040, Pakistan; Shams.bilal@uskt.edu.pk
3 Department of Mathematics, "Al. I. Cuza" University, Iaşi 700506, Romania; ocarja@uaic.ro
4 "Octav Mayer" Mathematics Institute, Romanian Academy, Iaşi 700505, Romania
5 Abdus Salam School of Mathematical Sciences, Lahore 54000, Pakistan; nasir.jav7000@gmail.com
6 Department of Mathematics, "Gh. Asachi" Technical University, Iaşi 700506, Romania; vieru_alina@yahoo.com
* Correspondence: tzankodd@gmail.com

Received: 23 March 2020; Accepted: 5 May 2020; Published: 9 May 2020

Abstract: We develop a new concept of a solution, called the limit solution, to fully nonlinear differential inclusions in Banach spaces. That enables us to study such kind of inclusions under relatively weak conditions. Namely we prove the existence of this type of solutions and some qualitative properties, replacing the commonly used compact or Lipschitz conditions by a dissipative one, i.e., one-sided Perron condition. Under some natural assumptions we prove that the set of limit solutions is the closure of the set of integral solutions.

Keywords: m-dissipative operators; limit solutions; integral solutions; one-sided Perron condition; Banach spaces

1. Introduction and Preliminaries

Let X be a real Banach space with the norm $|\cdot|$, $A : D(A) \subset X \rightrightarrows X$ an m–dissipative operator generating the semigroup $\{S(t) : \overline{D(A)} \to \overline{D(A)};\ t \geq 0\}$ and $F : I \times X \rightrightarrows X$ a multifunction with nonempty, closed and bounded values, where $I = [t_0, T]$.

In this paper, we study evolution inclusions of the form

$$\dot{x}(t) \in Ax(t) + F(t, x(t)),\ x(t_0) = x_0 \in \overline{D(A)}. \tag{1}$$

Notice that many parabolic systems can be written in the form (1). We refer the reader to [1–3] for the general theory of the system (1) when F is single valued. In the case when X^* is uniformly convex, the system (1) is comprehensively studied in [4]. We recall also the monograph [5], where (1) is studied in different settings.

An important problem regarding the system (1) is to find the closure of the set of integral solutions. This problem is not solved in the case of general Banach spaces.

We consider the associated Cauchy problem

$$\dot{x}(t) \in Ax(t) + f(t),\ x(t_0) = x_0 \in \overline{D(A)}, \tag{2}$$

where $f(\cdot)$ is a Bochner integrable function. We denote by $[\cdot,\cdot]_+$ the right directional derivative of the norm, i.e., $[x, y]_+ = \lim_{h \to 0^+} h^{-1}(|x + hy| - |x|)$ (see, e.g., ([6], Section 1.2) for definition and properties).

Following [7], we say that a continuous function $x : [t_0, T] \to \overline{D(A)}$ is an integral solution of (2) on $[t_0, T]$ if $x(t_0) = x_0$ and for every $u \in D(A)$, $v \in Au$ and $t_0 \leq \tau < t \leq T$ the following inequality holds

$$|x(t) - u| \leq |x(\tau) - u| + \int_\tau^t [x(s) - u, f(s) + v]_+ ds.$$

Definition 1. *The Bochner integrable function $g(\cdot)$ is said to be pseudoderivative of the continuous function $y(\cdot)$ (with respect to A) if $y(\cdot)$ is an integral solution of (2) on $[t_0, T]$ with $f(\cdot)$ replaced by $g(\cdot)$.*

Notice that the pseudoderivative $g(\cdot)$ (if it exists) depends on A and $y(\cdot)$. However, along this paper A is fixed and we assume without loss of generality that the pseudoderivative depends only on $y(\cdot)$. To stress this dependence on y, we will denote the pseudoderivative $g(\cdot)$ by $g_y(\cdot)$.

It is well known that for each $x_0 \in \overline{D(A)}$ the Cauchy problem (2) has a unique integral solution on $[t_0, T]$. Moreover, if $x(\cdot)$ and $y(\cdot)$ are integral solutions of (2) with $x(t_0) = x_0$ and $y(t_0) = y_0$ then

$$|x(t) - y(t)| \leq |x_0 - y_0| + \int_{t_0}^t [x(s) - y(s), f_x(s) - f_y(s)]_+ ds, \tag{3}$$

$$|x(t) - y(t)| \leq |x_0 - y_0| + \int_{t_0}^t |f_x(s) - f_y(s)| ds, \tag{4}$$

for every $t \in [t_0, T]$ (see, e.g., [7]).

We define now the notion of integral solution for the differential inclusion (1). Moreover, following [8], where the semilinear case was considered, we define the notions of ε-solution (called outer ε-solution in [8]) and limit solution for (1). In the following, $\mathbb{B}$ denotes the closed unit ball in X.

Definition 2. *The function $x : I \to \overline{D(A)}$ is said to be an integral solution of (1) on I if it is an integral solution of (2) such that its pseudoderivative $f_x(\cdot)$ satisfies $f_x(t) \in F(t, x(t))$ for a.a. $t \in I$.*

Consider the following system

$$\begin{cases} \dot{x}(t) \in Ax(t) + F(t, x(t) + \mathbb{B}) + \mathbb{B}, \\ x(t_0) = x_0. \end{cases} \tag{5}$$

Definition 3. *(i) Let $\varepsilon > 0$. The continuous function $x : I \to \overline{D(A)}$ is said to be an ε–solution of (1) on I if it is a solution of (5) and its pseudoderivative $f_x(\cdot)$ satisfies*

$$\int_I \text{dist}(f_x(t), F(t, x(t)))dt \leq \varepsilon.$$

(ii) The function $x(\cdot)$ is said to be a limit solution of (1) on I if $x(t) = \lim_{n\to\infty} x_n(t)$ uniformly on I for some sequence $(x_n(\cdot))$ of ε_n–solutions as $\varepsilon_n \downarrow 0^+$.

Recall that the distance between a point $u \in X$ and a subset C of X is given by $\text{dist}(u; C) = \inf\{\|u - c\|;\ c \in C\}$.

In the literature, we can find different definitions for ε–solutions. Maybe the most popular is when its pseudoderivative satisfies $f_x(t) \in F(t, x(t) + \varepsilon\mathbb{B})$ a.e. on I. However, our definition given above is more convenient for the study of the qualitative properties of the set of integral solutions of (1) in the case when X is an arbitrary Banach space.

For ordinary differential inclusions ($A = 0$), the limit solutions are usually called quasitrajectories (cf., [9]). We prefer the notion of limit solution because it is the original definition of the integral solution in the case of m–dissipative systems (cf. [6]). For ordinary differential inclusions in $\mathbb{R}^n$, the limit solutions are the integral solutions of the relaxed system. In our case, the relaxed system has the form

$$\dot{x}(t) \in Ax(t) + \overline{co}\, F(t, x(t)),\ x(t_0) = x_0, \tag{6}$$

where $\overline{co}\, F(t, x(t))$ stands for the closed convex hull of the set $F(t, x(t))$. In this general setting, the limit solutions are not integral solutions of the relaxed system (6).

It is well known that the set of integral solutions of (6) is not necessarily closed in $C(I, X)$ even if X is finite dimensional. For instance, in [10] the author constructed an example in which a sequence $(x_n(\cdot))$ of integral solutions of

$$\dot{x}(t) \in Ax(t) + f_n(t),\ x(t_0) = x_0,$$

converges uniformly on $[t_0, T]$ to a function $x(\cdot)$, $(f_n(\cdot))$ converges weakly in $L^1(t_0, T; X)$ to $f(\cdot)$, but $x(\cdot)$ is not an integral solution of

$$x'(t) \in Ax(t) + f(t),\ x(t_0) = x_0.$$

The main results of this paper are summarized as follows.

(I) We prove that the set of limit solutions of (1) is nonempty and closed in $C(I, X)$ when X is a general Banach space and $F(\cdot, \cdot)$ is almost continuous and satisfies a one-sided Perron condition.
(II) We prove that in the case when A generates a compact semigroup, the closure of the set of integral solutions of (1) is exactly the set of limit solutions, which in general does not coincide with the set of integral solutions of the relaxed system. The same result is proved also when $F(t, \cdot)$ is full Perron, but without any restrictions on the semigroup A.

The limit solutions in the case when A is linear were studied in [8]. It was shown there that the limit solutions of (1) and (6) coincide. It is not the case for the nonlinear problem.

Let us now define a few classes of multifunctions which will be used in the following.

We say that $F(\cdot, \cdot)$ is lower semicontinuous (LSC) at $(t_0, x_0) \in I \times X$ if for every $f_0 \in F(t_0, x_0)$, every $x_k \to x_0$ and every $t_k \to t_0$ there exists $f_k \in F(t_k, x_k)$ such that $f_k \to f_0$. This definition is equivalent to the following property of the graph: for every $\alpha \in F(t_0, x_0)$ and every $\varepsilon > 0$, there exists $\delta > 0$ such that $\alpha \in F(t, x) + \varepsilon\mathbb{B}$, when $|t - t_0| \leq \delta$ and $|x - x_0| \leq \delta$.

The multifunction $F(\cdot, \cdot)$ is called LSC if it is LSC at every $(t, x) \in I \times X$.

The multifunction $F(\cdot, \cdot)$ is called continuous if it is continuous with respect to the Hausdorff distance. We recall that the Hausdorff distance between the bounded sets B and C is defined by

$$D_H(B; C) = \max\{e(B; C), e(C; B)\},$$

where $e(B; C)$ is the excess of B to C, defined by $e(B; C) = \sup_{x \in B} \operatorname{dist}(x; C)$.

The multifunction $F(\cdot, \cdot)$ is called almost LSC (continuous) if for every $\varepsilon > 0$ there exists a compact set $I_\varepsilon \subset I$ with Lebesgue measure $\operatorname{meas}(I \setminus I_\varepsilon) \leq \varepsilon$ such that $F\,|_{I_\varepsilon \times X}$ is LSC (continuous).

Let $v : I \times \mathbb{R}_+ \to \mathbb{R}_+$ be Carathéodory and integrally bounded on the bounded sets. As is well known, the scalar differential equation

$$\dot{r}(t) = v(t, r(t)), \quad r(t_0) = r_0 \geq 0, \tag{7}$$

has maximal solutions $h(\cdot)$, i.e., $0 \leq r(t) \leq h(t)$ for every solution $r(\cdot)$ of (7) on the existence interval of $h(\cdot)$ (see, e.g., [6]).

We introduce now the standing hypotheses of this paper.

Hypothesis 1 (H1). *The multifunction $F(\cdot, \cdot)$ is almost continuous.*

Hypothesis 2 (H2). *There exists $\gamma > 0$ such that $\|F(t,x)\| \le \gamma(1+|x|)$ for a.a. $t \in I$ and every $x \in X$. We recall that $\|F(t,x)\| = \sup_{y \in F(t,x)} |y|$.*

Hypothesis 3 (H3). *(One-sided Perron condition) There exist a Perron function $w(\cdot,\cdot)$ and a null set $\mathcal{N} \subset I$ such that such for every x, $y \in X$, for every $\varepsilon > 0$ and for every $f \in F(t,x)$ there exists $g \in F(t,y)$ such that*

$$[x-y, f-g]_+ \le w(t, |x-y|) + \varepsilon$$

on $I \setminus \mathcal{N}$.

We recall that the Carathéodory function $w : I \times \mathbb{R}_+ \to \mathbb{R}_+$ is said to be *Perron function* if it is integrally bounded on bounded sets, $w(t,0) \equiv 0$, $w(t,\cdot)$ is nondecreasing for every $t \in I$ and the zero function is the only solution of the scalar differential equation $r'(t) = w(t, r(t))$, $r(t_0) = 0$, on I.

Notice that it is more popular to call such kind of functions Kamke functions. We refer the reader to [11], where Perron and Kamke functions are comprehensively studied. That paper is the main reason to use here the notion of Perron (not Kamke) function. In [12] some examples of the Perron (Kamke) functions different from the Lipshitz one are given (see, e.g., Corollary 1.13 and Corollary 1.15).

Remark 1. *Due to Gronwall's lemma, there exists a constant $M > 0$ such that $|x(t)| \le M$ for every $t \in I$ and every solution $x(\cdot)$ of (5). Let $N = 1 + \gamma(2+M)$. Then $\|F(t, x(t) + \mathbb{B}) + \mathbb{B}\| \le N$ for every solution $x(\cdot)$ of (5).*

Clearly, for every solution $x(\cdot)$ of (5), in particular for every ε–solution $x(\cdot)$ of (1), with the pseudoderivative $f_x(\cdot)$, we have that $\operatorname{dist}(f_x(t), F(t, x(t))) \le 2N$ on I, since $|f_x(t)| \le N$ and $\|F(t,x(t))\| \le N$ for every $t \in I$.

2. Main Results

The main results are given in three subsections. In the first one, we prove the existence of limit solutions. In the second subsection, we prove the most interesting results of this paper, namely, that the set of limit solutions of (1) is the closure of the set of integral solutions of (1) when A generates a compact semigroup or when $F(t,\cdot)$ is full Perron. An example and some applications are discussed in the last two subsections.

2.1. Existence of Limit Solutions

In this subsection we prove an existence result of ε-solutions of the Cauchy problem (1) on I and a variant of the well known lemma of Filippov–Pliś.

First, recall that $\bar{t}$ is said to be a right dense point of a closed subset $\mathcal{I} \subset I$ if for every $\tau > 0$ there exists a point $s \in (\bar{t}, \bar{t}+\tau) \bigcap \mathcal{I}$. Clearly, $\bar{t}$ is not a right dense point of $\mathcal{I}$ if there exists $\tau > 0$ such that $(\bar{t}, \bar{t}+\tau) \bigcap \mathcal{I} = \emptyset$.

Lemma 1. *Assume that $F(\cdot,\cdot)$ is almost LSC and satisfies* (H2). *Then for every $\varepsilon > 0$ there exists at least one ε–solution of* (1) *defined on the whole I.*

Proof. Let $\varepsilon > 0$. We take $\varepsilon' \le \dfrac{\varepsilon}{T - t_0 + 2N}$. There exists $I' \subset I$ a closed set with Lebesgue measure $\operatorname{meas}(I') \ge T - t_0 - \varepsilon'$ such that $F|_{I' \times X}$ is LSC on $I' \times X$.

We take $f_0 \in F(t_0, x_0)$ arbitrary but fixed and let $f_1(\cdot)$ be Bochner integrable with $f_1(t) \in F(t, x_0)$ on I. Two cases are possible.

Case 1. If t_0 is a right dense point of I'. Since $F|_{I'\times X}$ is LSC at (t_0, x_0), then there exists $\delta \in (0, 1/2)$ such that if $t \in I'$ with $t - t_0 \leq \delta$ and $|y - x_0| \leq \delta$ then $f_0 \in F(t, y) + \varepsilon'\mathbb{B}$. We pick

$$f_y(t) = \begin{cases} f_0, \ t \in I' \\ f_1(t), \ t \in I \setminus I'. \end{cases}$$

Let $y_1(\cdot)$ be the integral solution of the Cauchy problem

$$\dot{y}(t) \in Ay(t) + f_y(t), \ y(t_0) = x_0.$$

Since $\lim_{t\downarrow t_0} y_1(t) = x_0$, we deduce that there exists $\tau \in (t_0, t_0 + \delta)$ such that $|y_1(t) - x_0| \leq \delta$ whenever $t \in [t_0, \tau)$. Thus, $f_0 \in F(t, y_1(t)) + \varepsilon'\mathbb{B}$ for every $t \in [t_0, \tau) \bigcap I'$ and $f_1(t) \in F(t, y_1(t) + \mathbb{B})$ for $t \in [t_0, \tau) \bigcap (I \setminus I')$. Therefore, $f_y(t) \in F(t, y_1(t) + \mathbb{B}) + \mathbb{B}$ for every $t \in [t_0, \tau)$, i.e., $y_1(\cdot)$ is a solution of (5) on $[t_0, \tau)$.

We let $y(t) = y_1(t)$ for every $t \in [t_0, \tau)$. Thus, $\text{dist}(f_y(t), F(t, y(t))) \leq \varepsilon'$ for every $t \in [t_0, \tau) \bigcap I'$ and, due to Remark 1, $\text{dist}(f_y(t), F(t, y(t))) \leq 2N$ for every $t \in [t_0, \tau) \bigcap (I \setminus I')$.

Case 2. If t_0 is not a right dense point of I', let $y_1(\cdot)$ be the integral solution of the Cauchy problem

$$\dot{y}(t) \in Ay(t) + f_1(t), \ y(t_0) = x_0.$$

Then there exists $\tau > t_0$ such that $[t_0, \tau) \subset I \setminus I'$ and $|y_1(t) - x_0| < \varepsilon'$ for $t \in [t_0, \tau)$. Thus, $y_1(\cdot)$ is a solution of (5) on $[t_0, \tau)$.

We let $y(t) = y_1(t)$ and $f_y(t) = f_1(t)$ for every $t \in [t_0, \tau)$. Moreover, $\text{dist}(f_y(t), F(t, y(t))) \leq 2N$ for every $t \in [t_0, \tau)$.

In both cases we let $y_\tau = \lim_{t\uparrow\tau} y(t)$. We continue the above construction in a similar way by replacing t_0 by τ and x_0 by y_τ.

Let $[t_0, \bar{t})$ be the maximal interval of the existence of solution $y(\cdot)$ of (5), with the properties that $\text{dist}(f_y(t), F(t, y(t))) \leq \varepsilon'$ on $[t_0, \bar{t}) \bigcap I'$ and $\text{dist}(f_y(t), F(t, y(t))) \leq 2N$ on $[t_0, \bar{t}) \bigcap (I \setminus I')$, where $f_y(\cdot)$ is the pseudoderivative of $y(\cdot)$. Suppose that $\bar{t} < T$. Due to the growth condition $\lim_{t\uparrow\bar{t}} y(t)$ exists. Let $y_{\bar{t}} = \lim_{t\uparrow\bar{t}} y(t)$. Then, using a similar construction as above with $\bar{t}$ instead of t_0 and $y_{\bar{t}}$ instead of x_0, we can extend the solution $y(\cdot)$ on some interval $[t_0, \bar{t} + \theta)$, $\theta > 0$, such that $\text{dist}(f_y(t), F(t, y(t))) \leq \varepsilon'$ on $[t_0, \bar{t} + \theta) \bigcap I'$ and $\text{dist}(f_y(t), F(t, y(t))) \leq 2N$ on $[t_0, \bar{t} + \theta) \bigcap (I \setminus I')$, which contradicts the maximality of $[t_0, \bar{t})$. Hence $\bar{t} = T$.

It is clear that the pseudoderivative $f_y(\cdot)$ satisfies $\text{dist}(f_y(t), F(t, y(t))) = k_y(t)$ with $k_y(t) \leq \varepsilon'$ for every $t \in I'$ and $k_y(t) \leq 2N$ for every $t \in I \setminus I'$. One checks easily that $\int_I k_y(t)dt \leq \varepsilon$. Hence, $y(\cdot)$ is an ε–solution of (1) on I. □

The next lemma will play a crucial role in the sequel.

Lemma 2. *Assume* (H1)–(H3). *Let* $\varepsilon > 0$ *and let* $x(\cdot)$ *be an* ε*–solution of* (1) *on* I*. Then, there exist* $l(\cdot)$ *positive and bounded on* I *with* $\int_I l(t)dt \leq 2\varepsilon$ *and* $\eta > 0$ *such that for every* $y_0 \in \overline{D(A)}$ *with* $|x_0 - y_0| < \eta$ *we have that:*

(i) the maximal solution $\tilde{v}(\cdot)$ *of the scalar differential equation*

$$\dot{v}(t) = w(t, v(t)) + l(t), \ v(t_0) = |x_0 - y_0|,$$

exists on I *and*

(ii) for every $0 < \delta < \varepsilon$ there exists a δ–solution $y(\cdot)$ of (1) on I with x_0 replaced by y_0, satisfying

$$|x(t) - y(t)| \leq \tilde{v}(t),$$

for all $t \in I$.

Proof. The assertion (i) follows from ([13], Lemma 2.4) (see also Lemma 3 below).

Let $\varepsilon > 0$ be fixed and let $f_x(\cdot)$ be the pseudoderivative of $x(\cdot)$. Then, due to Definition 3, $f_x(t) \in F(t, x(t) + \mathbb{B}) + \mathbb{B}$ a.e. on I and $k_x(t) = \text{dist}(f_x(t), F(t, x(t)))$ satisfies $\int_I k_x(t)dt \leq \varepsilon$. Moreover, due to Remark 1, $k_x(t) \leq 2N$ for any $t \in I$.

We take $\varepsilon' \leq \dfrac{\varepsilon}{5(T - t_0 + N)}$. We can assume without loss of generality that there exists a compact set $I_\varepsilon \subset I$, with $\text{meas}(I \setminus I_\varepsilon) < \varepsilon'$, such that the functions $f_x|_{I_\varepsilon}$, $k_x|_{I_\varepsilon}$ and $w|_{I_\varepsilon \times \mathbb{R}}$ are continuous.

Let $\delta < \varepsilon$. We can assume that there exists a compact set $I_\delta \subset I$ such that $I_\varepsilon \subset I_\delta$, $\text{meas}(I \setminus I_\delta) < \delta'$, where $\delta' \leq \min\left\{\dfrac{\delta}{5(T - t_0 + N)}, \varepsilon'\right\}$, and $F|_{I_\delta \times X}$ is continuous.

We take $f_x \in F(t_0, x_0)$ such that $|f_x - f_x(t_0)| \leq k_x(t_0) + \varepsilon'$. Let $\eta \in (0,1)$ and $y_0 \in \overline{D(A)}$ with $|x_0 - y_0| < \eta$. By (H3), there exists $f_1 \in F(t_0, y_0)$ such that

$$[x_0 - y_0, f_x - f_1]_+ \leq w(t_0, |x_0 - y_0|) + \varepsilon'. \tag{8}$$

Hence,

$$[x_0 - y_0, f_x(t_0) - f_1]_+ \leq [x_0 - y_0, f_x - f_1]_+ + |f_x(t_0) - f_x| \leq w(t_0, |x_0 - y_0|) + 2\varepsilon' + k_x(t_0).$$

Let $f(\cdot)$ be a Bochner integrable function such that $f(t) \in F(t, y_0)$ for every $t \in I$.

We consider the following cases.

Case 1. t_0 is a right dense point of I_ε (hence it is a right dense point also for I_δ).

We pick

$$f_y(t) = \begin{cases} f_1, & \text{if } t \in I_\delta \\ f(t), & \text{if } t \in I \setminus I_\delta. \end{cases}$$

Let $y^1(\cdot)$ be the integral solution of

$$\dot{y}(t) \in Ay(t) + f_y(t),\ y(t_0) = y_0. \tag{9}$$

Then, by the continuity of $F|_{I_\delta \times X}$ and $y^1(\cdot)$, there exists $\tau > t_0$ such that $f_1 \in F(t, y^1(t)) + \delta'\mathbb{B}$ for every $t \in [t_0, \tau) \bigcap I_\delta$.

Due to the continuity of $y^1(\cdot)$, the upper semicontinuity of $[\cdot,\cdot]_+$ and the continuity of $w(\cdot,\cdot)$ at $(t_0, |x_0 - y_0|)$ and of $k_x(\cdot)$ at t_0, the number $\tau > t_0$ can be chosen such that $|y^1(t) - y_0| \leq \dfrac{1}{2}$ for every $t \in [t_0, \tau)$, and moreover,

$$\begin{aligned}
&[x(t) - y^1(t), f_x(t) - f_1]_+ \leq [x_0 - y_0, f_x(t_0) - f_1]_+ + \varepsilon' \\
&\leq w(t_0, |x_0 - y_0|) + 3\varepsilon' + k_x(t_0) \\
&\leq |w(t_0, |x_0 - y_0|) - w(t, |x(t) - y^1(t)|)| + w(t, |x(t) - y^1(t)|) + 4\varepsilon' + k_x(t) \\
&\leq w(t, |x(t) - y^1(t)|) + 5\varepsilon' + k_x(t),
\end{aligned}$$

for every $t \in [t_0, \tau) \bigcap I_\varepsilon$.

Clearly, due to our choice of τ, we have that $f_y(t) \in F(t, y^1(t) + \mathbb{B}) + \mathbb{B}$ for any $t \in [t_0, \tau)$, hence $y^1(\cdot)$ is a solution of (5) on $[t_0, \tau)$.

We set $y(t) = y^1(t)$ for any $t \in [t_0, \tau)$ and let $k_y(t) = \text{dist}(f_y(t), F(t, y(t)))$. Then $k_y(t) \le \delta'$ for $t \in [t_0, \tau) \bigcap I_\delta$ and $k_y(t) \le 2N$ for $t \in [t_0, \tau) \bigcap (I \setminus I_\delta)$.

Hence, for any $t \in [t_0, \tau) \bigcap I_\varepsilon$,

$$[x(t) - y(t), f_x(t) - f_y(t)]_+ \le w(t, |x(t) - y(t)|) + 5\varepsilon' + k_x(t).$$

On the other hand, for any $t \in [t_0, \tau) \bigcap (I \setminus I_\varepsilon)$ we have that

$$[x(t) - y(t), f_x(t) - f_y(t)]_+ \le |f_x(t) - f_y(t)| \le 2N \le 2N + w(t, |x(t) - y(t)|).$$

Case 2. t_0 is not a right dense point of I_ε but it is a right dense point of I_δ.

Let $y^1(\cdot)$ be the integral solution of (9), where $f_y(\cdot)$ is chosen as in Case 1. Then there exists $\tau > t_0$ such that $|y^1(t) - y_0| \le \frac{1}{2}$ for every $t \in [t_0, \tau)$, and moreover, $[t_0, \tau) \subset I \setminus I_\varepsilon$. Moreover, we can choose τ such that $f_1 \in F(t, y^1(t)) + \delta' \mathbb{B}$ for every $t \in [t_0, \tau) \bigcap I_\delta$.

We set, as in the previous case, $y(t) = y^1(t)$ for any $t \in [t_0, \tau)$. Hence $k_y(t) \le \delta'$ for $t \in [t_0, \tau) \bigcap I_\delta$ and $k_y(t) \le 2N$ for $t \in [t_0, \tau) \bigcap (I \setminus I_\delta)$.

Case 3. t_0 is not a right dense point of I_δ.

In this case, we let $y^1(\cdot)$ to be the integral solution of

$$\dot{y}(t) \in Ay(t) + f(t), \; y(t_0) = y_0.$$

Then there exists $\tau > t_0$ such that $|y^1(t) - y_0| \le \frac{1}{2}$ for every $t \in [t_0, \tau)$, and moreover, $[t_0, \tau) \subset I \setminus I_\delta \subset I \setminus I_\varepsilon$. We have that $y^1(\cdot)$ is a solution of (5) on $[t_0, \tau)$.

We let $y(t) = y^1(t)$ and $f_y(t) = f(t)$ for every $t \in [t_0, \tau)$ and hence $k_y(t) \le 2N$ on $[t_0, \tau)$.

Moreover, in both cases 2 and 3, for any $t \in [t_0, \tau)$ we have that

$$[x(t) - y(t), f_x(t) - f_y(t)]_+ \le 2N + w(t, |x(t) - y(t)|).$$

We continue the above construction in a similar way by replacing t_0 by τ and y_0 by $y_\tau = \lim_{t \uparrow \tau} y(t)$.

Finally, reasoning as in the proof of Lemma 1, we define $y(\cdot)$ on I, solution of (5). Its pseudoderivative $f_y(\cdot)$ satisfies $\text{dist}(f_y(t), F(t, y(t))) = k_y(t)$ with $k_y(t) \le \delta'$ for every $t \in I_\delta$ and $k_y(t) \le 2N$ for every $t \in I \setminus I_\delta$. One checks easily that $\int_I k_y(t)dt \le \delta$. Hence, $y(\cdot)$ is a δ–solution of (1) on I.

Moreover, for any $t \in I_\varepsilon$, we have that

$$[x(t) - y(t), f_x(t) - f_y(t)]_+ \le w(t, |x(t) - y(t)|) + 5\varepsilon' + k_x(t)$$

and, for any $t \in I \setminus I_\varepsilon$,

$$[x(t) - y(t), f_x(t) - f_y(t)]_+ \le 2N + w(t, |x(t) - y(t)|).$$

Furthermore, using (3), we have that

$$\begin{aligned} |x(t) - y(t)| &\le |x_0 - y_0| + \int_{t_0}^{t} [x(s) - y(s), f_x(s) - f_y(s)]_+ ds \\ &\le |x_0 - y_0| + \int_{[t_0, t] \bigcap I_\varepsilon} (w(s, |x(s) - y(s)|) + 5\varepsilon' + k_x(s))ds \\ &\quad + \int_{[t_0, t] \bigcap (I \setminus I_\varepsilon)} (w(s, |x(s) - y(s)|) + 2N)ds \end{aligned}$$

$$\leq |x_0 - y_0| + \int_{t_0}^{t} w(s, |x(s) - y(s)|)ds + \int_{[t_0,t]\bigcap I_\varepsilon} (5\varepsilon' + k_x(s))ds + \int_{[t_0,t]\bigcap(I\setminus I_\varepsilon)} 2Nds$$

for any $t \in I$. Let $l(t) = 5\varepsilon' + k_x(t)$ for $t \in I_\varepsilon$ and $l(t) = 2N$ for $t \in I \setminus I_\varepsilon$. Then, for any $t \in I$,

$$|x(t) - y(t)| \leq |x_0 - y_0| + \int_{t_0}^{t} w(s, |x(s) - y(s)|)ds + \int_{t_0}^{t} l(s)ds.$$

Hence, $|x(t) - y(t)| \leq \tilde{v}(t)$ for every $t \in I$, where $\tilde{v}(\cdot)$ is the maximal solution of the scalar differential equation

$$\dot{v}(t) = w(t, v(t)) + l(t),\ v(t_0) = |x_0 - y_0|$$

on I. Clearly, $l(\cdot)$ is bounded on I and

$$\int_I l(s)ds = \int_{I_\varepsilon} (5\varepsilon' + k_x(s))ds + \int_{I\setminus I_\varepsilon} 2Nds \leq 5\varepsilon'(T - t_0) + \varepsilon + 2N\varepsilon' \leq 2\varepsilon.$$

The proof is completed. □

The proof of the following result follows the same steps as the proof of ([13], Lemma 2.4) and it is omitted.

Lemma 3. *Let $\lambda \in L^1(I;\mathbb{R}_+)$ and let $v : I \times \mathbb{R}_+ \to \mathbb{R}_+$ be a Carathéodory function, integrally bounded on the bounded sets, with $v(t,\cdot)$ nondecreasing for every $t \in I$. If the maximal solution $h(\cdot)$ of (7) exists on I, then for every $\varepsilon > 0$ there exists $\delta > 0$ such that the maximal solution $\bar{r}(\cdot)$ of*

$$\dot{r}(t) = v(t, r(t)) + \mu(t),\ r(t_0) = \bar{r}_0 \in [r_0, r_0 + \delta],$$

exists on I and $\bar{r}(t) \leq h(t) + \varepsilon$ on I, for every function $\mu(\cdot)$ such that $0 \leq \mu(t) \leq \lambda(t)$ for $t \in I$ and $\int_I \mu(t)dt \leq \delta$.

Now, by using the previous lemmas, we will prove the following existence result of a limit solution for the Cauchy problem (1).

Theorem 1. *Assume* (H1)–(H3). *Let $\varepsilon > 0$ and let $x(\cdot)$ be an ε–solution of (1). Then, there exist a positive and bounded function $l(\cdot)$ with $\int_I l(t)dt \leq 2\varepsilon$ and $\eta > 0$ such that for every $y_0 \in \overline{D(A)}$ with $|x_0 - y_0| < \eta$ we have that:*

(i) the maximal solution $\tilde{v}(\cdot)$ of the scalar differential equation

$$\dot{v}(t) = w(t, v(t)) + l(t),\ v(t_0) = |x_0 - y_0|, \tag{10}$$

exists on I and

(ii) there exists a limit solution $y(\cdot)$ of (1) on I with $y(t_0) = y_0$ such that

$$|x(t) - y(t)| \leq \tilde{v}(t) + \varepsilon,$$

for every $t \in I$.

Proof. Let $\delta > 0$ be given by Lemma 3, corresponding to $\varepsilon/2$. Take $\varepsilon_1 \leq \min\{\varepsilon/2, \delta/2\}$. By Lemma 2 there exist $l_1(\cdot)$ a positive and bounded function with $\int_I l_1(t)dt \leq 2\varepsilon$ and $\eta > 0$ such that for any $y_0 \in \overline{D(A)}$ with $|x_0 - y_0| < \eta$ there exists $y_1(\cdot)$ an ε_1–solution of (1) with $y_1(t_0) = y_0$ satisfying

$$|x(t) - y_1(t)| \leq v_1(t),$$

where $v_1(\cdot)$ is the maximal solution of

$$\dot{v}(t) = w(t, v(t)) + l_1(t),\ v(t_0) = |x_0 - y_0|, \tag{11}$$

on I.

Let $\delta_1 > 0$ be given by Lemma 3 corresponding to $\varepsilon_1/2$. Take $\varepsilon_2 \leq \min\{\varepsilon_1/2, \delta_1/2\}$. By Lemma 2 there exists an ε_2–solution $y_2(\cdot)$ of (1) on I with $y_2(t_0) = y_0$ such that

$$|y_2(t) - y_1(t)| \leq v_2(t),$$

for every $t \in I$. Here $v_2(\cdot)$ is the maximal solution of

$$\dot{v}(t) = w(t, v(t)) + l_2(t),\ v(t_0) = 0,$$

where $l_2(\cdot)$ is positive and bounded on I and $\int_I l_2(t)dt \leq 2\varepsilon_1 \leq \delta$. Then, by Lemma 3, $v_2(t) \leq \varepsilon/2$ for any $t \in I$.

We construct by induction a sequence of ε_n–solutions $(y_n(\cdot))$ of (1) on I, where $\varepsilon_n \leq \min\{\varepsilon_{n-1}/2, \delta_{n-1}/2\}$, for any $n = 2, 3, \ldots$, such that

$$|y_{n+1}(t) - y_n(t)| \leq v_{n+1}(t),$$

for every $t \in I$. Here $v_{n+1}(\cdot)$ is the maximal solution of

$$\dot{v}(t) = w(t, v(t)) + l_{n+1}(t),\ v(t_0) = 0,$$

where $l_{n+1}(\cdot)$ is positive and bounded on I and satisfies $\int_I l_{n+1}(t)dt \leq 2\varepsilon_n \leq \delta_{n-1}$. Moreover, $v_{n+1}(t) \leq \varepsilon_{n-1}/2$ for every $t \in I$ and every $n = 2, 3, \ldots$. Therefore,

$$|y_{n+1}(t) - y_n(t)| \leq \varepsilon_{n-1}$$

for every $t \in I$ and every $n = 2, 3, \ldots$. Taking into account that $\sum_{n=1}^{\infty} \varepsilon_n \leq \varepsilon$, we conclude that $(y_n(\cdot))$ is a Cauchy sequence in $C(I; X)$. Thus, there exists a continuous function $y : I \to X$ such that $\lim_{n\to\infty} y_n(t) = y(t)$ uniformly on I. Furthermore, $|x(t) - y(t)| \leq v_1(t) + \varepsilon$, where $v_1(\cdot)$ is the maximal solution of (11). □

The next theorem is a variant of the well known lemma of Filippov–Pliś. This lemma has numerous applications in optimal control theory and had been proved on different variants by different authors. In the next theorem, we extend this result to the case when the integral solutions do not necessarily exist. Variants of this lemma have been proved in [14,15] for the case of uniformly convex dual space and in [16] for the case when A generates a compact semigroup.

Theorem 2. *Assume* (H1)–(H3). *Let $x(\cdot)$ be an integral solution of the differential inclusion*

$$\begin{cases} \dot{x}(t) \in Ax(t) + F(t, x(t)) + g(t)\mathbb{B}, \\ x(t_0) = x_0 \in \overline{D(A)}, \end{cases} \tag{12}$$

on I, where $g \in L^1(I; \mathbb{R}_+)$. Then for every $\varepsilon > 0$ and every $y_0 \in \overline{D(A)}$ for which the maximal solution $v(\cdot)$ of the scalar differential equation

$$\dot{v}(t) = w(t, v(t)) + g(t),\ v(t_0) = |x_0 - y_0|, \tag{13}$$

exists on I, there exists a limit solution $z(\cdot)$ of (1) on I with $z(t_0) = y_0$, satisfying

$$|x(t) - z(t)| \le v(t) + \varepsilon,$$

for all $t \in I$.

Proof. Let $f_x(\cdot)$ be the pseudoderivative of $x(\cdot)$. Then $f_x(t) \in F(t, x(t)) + g(t)\mathbb{B}$ for every $t \in I$. Furthermore, for every $\varepsilon > 0$ there exists a compact $I_\varepsilon \subset I$ with Lebesgue measure $\text{meas}(I \setminus I_\varepsilon) < \varepsilon$ such that $f_x|_{I_\varepsilon}$, $g|_{I_\varepsilon}$, $F|_{I_\varepsilon \times X}$ and $w|_{I_\varepsilon \times \mathbb{R}_+}$ are continuous. We fix $\nu > 0$ and define the multifunction

$$G(t,u) = \overline{\{v \in F(t,u);\ [x(t) - u, f_x(t) - v]_+ < w(t, |x(t) - u|) + g(t) + \nu\}}.$$

It follows from (H3) that $G(\cdot,\cdot)$ has nonempty closed values. Moreover, $G(\cdot,\cdot)$ is almost LSC (the proof follows, with obvious modifications, the same lines as the proof of ([16], Theorem 2). Due to Lemma 1, for every $\mu > 0$ the evolution inclusion

$$\begin{cases} \dot{x}(t) \in Ax(t) + G(t, x(t)), \\ x(t_0) = y_0 \end{cases}$$

has a μ–solution $y(\cdot)$ defined on the whole I. Then, its pseudoderivative $f_y(\cdot)$ satisfies $f_y(t) \in G(t, y(t)) + h_y(t)\mathbb{B}$ for any $t \in I$, where $h_y(t) \le 2N$ on I and $\int_I h_y(s)ds \le \mu$. It follows from the properties of $[\cdot,\cdot]_+$ that

$$[x(t) - y(t), f_x(t) - f_y(t)]_+ \le w(t, |x(t) - y(t)|) + g(t) + \nu + h_y(t).$$

Thus, $|x(t) - y(t)| \le r(t)$, where $r(\cdot)$ is the maximal solution of the inequality $\dot{r}(t) \le w(t, r(t)) + g(t) + \nu + h_y(t)$ with $r(t_0) = |x_0 - y_0|$.

Due to Lemma 3, $r(\cdot)$ exists on the whole I for sufficiently small ν and μ and moreover, for every $\varepsilon > 0$ there exists $\kappa > 0$ such that $r(t) \le v(t) + \varepsilon$ for $\mu, \nu < \kappa$.

Clearly, $y(\cdot)$ is a μ–solution also of (1). It follows from Theorem 1 that there exists a limit solution $z(\cdot)$ of (1) such that $|z(t) - y(t)| \le \varepsilon$. The proof is therefore complete thanks to the triangle inequality. □

Remark 2. *In fact, Theorem 2 says that the solution set of (1) depends continuously on small perturbations of the initial condition and the right-hand side.*

2.2. Limit and Integral Solutions

We start this subsection by giving a simple example to illustrate the notion of limit solutions.

Example 1. *Let $A \equiv 0$. We consider the ordinary differential inclusion:*

$$\dot{x}(t) \in \mathbb{B},\ t \in (0,1),\ x(0) = 0. \tag{14}$$

Here $\mathbb{B}$ denotes the unit ball in $L^1(0,1;\mathbb{R}^n)$. Clearly, the limit solutions of (14) are all Lipschitz functions (of Lipschitz constant 1). However, there exists such kind of functions nowhere differentiable, i.e., which are not integral solutions.

First, we will prove that the set of limit solutions is the closure of the set of integral solutions of (1) when $F(\cdot,\cdot)$ satisfies the following stronger assumption than (H3).

Hypothesis 3′ (H3′). *(Full Perron condition) There exists a Perron function $w(\cdot,\cdot)$ such that*

$$D_H(F(t,x),F(t,y)) \le w(t,|x-y|)$$

for every $x, y \in X$ and every $t \in I$.

Theorem 3. *Assume* (H1), (H2) *and* (H3′). *Then* (1) *has integral solutions. Furthermore, the set of integral solutions of* (1) *is dense in the set of limit solutions of* (1).

Proof. Let $\varepsilon > 0$ and let $y(\cdot)$ be an ε-solution (1) with the pseudoderivative $f_y(\cdot)$. Then $f_y(t) \in F(t,y(t)) + h_y(t)\mathbb{B}$ for any $t \in I$, where $h_y(t) \le 2N$ on I and $\int_I h_y(t)dt \le \varepsilon$.

Let $0 < \delta < \varepsilon$. Since the function $w(\cdot,\cdot)$ is Perron, there exists $0 < \mu < \varepsilon$ such that $\int_I w(t,\mu)dt < \delta$. Furthermore, there exists $t_1 > t_0$ such that $|y(t) - x_0| < \mu$ for $t \in [t_0,t_1]$. Let $z(t) := y(t)$ on $[t_0,t_1)$ and denote $z^1 = z(t_1)$. By (H1) and (H3′), there exists a strongly measurable function $f_1(\cdot)$ such that $f_1(t) \in F(t,z^1)$ and

$$|f_y(t) - f_1(t)| \le w(t,|y(t) - z^1|) + h_y(t) + \mu$$

a.e. on $[t_1,T]$. Consider the problem

$$\begin{cases} \dot z(t) \in Az(t) + f_1(t) \\ z(t_1) = z^1 \end{cases} \tag{15}$$

and let $z_1(\cdot)$ be a solution of (15) on $[t_1,T]$. There exists $t_2 > t_1$ such that $|z_1(t) - z^1| < \mu$ for any $t \in [t_1,t_2]$. Then, on $[t_1,t_2]$,

$$|f_y(t) - f_1(t)| \le w(t,|y(t) - z_1(t)|) + |w(t,|y(t) - z_1(t)| + \mu) - w(t,|y(t) - z_1(t)|) + h_y(t) + \mu.$$

Denote $M_w(\mu) := \sup\limits_{|x|\le 2N} |w(t,|x|+\mu) - w(t,|x|)|$ and let $z(t) := z_1(t)$ on $[t_1,t_2]$. Then, $z(\cdot)$ is a solution of $\dot z(t) \in Az(t) + F(t,z(t)) + w(t,\mu)\mathbb{B}$ and

$$|f_y(t) - f_z(t)| \le w(t,|y(t) - z(t)|) + M_w(\mu) + h_y(t) + \mu$$

on $[t_1,t_2]$.

Using the same method as above, as in the proof of Lemma 1, we can extend $z(\cdot)$ on the whole interval I, such that $\dot z(t) \in Az(t) + F(t,z(t)) + w(t,\mu)\mathbb{B}$ and

$$|f_y(t) - f_z(t)| \le w(t,|y(t) - z(t)|) + M_w(\mu) + h_y(t) + \mu$$

for any $t \in I$. Moreover, $\int_I \operatorname{dist}(f_z(t),F(t,z(t)))dt \le \int_I w(t,\mu) < \delta$ on I. Hence, $z(\cdot)$ is a δ-solution of (1). Using (4), we get that $|y(t) - z(t)| \le r(t)$, where $r(\cdot)$ is the maximal solution of

$$\dot r(t) \le w(t,r(t)) + M_w(\varepsilon) + h_y(t) + \varepsilon,\ r(t_0) = 0.$$

Now, let $\varepsilon_n \downarrow 0$ and let $(x_n(\cdot))$ be a sequence of ε_n-solutions of (1), constructed as above, with $(f_n(\cdot))$ the corresponding sequence of pseudoderivatives. Then

$$|x_n(t) - x_{n+1}(t)| \le r_n(t)$$

and

$$|f_n(t) - f_{n+1}(t)| \le w(t,r_n(t)) + M_w(\varepsilon_n) + h_{n+1}(t) + \varepsilon_n,$$

where $r_n(\cdot)$ is the maximal solution of

$$\dot{r}_n(t) \leq w(t, r_n(t)) + M_w(\varepsilon_n) + h_{n+1}(t),\ r_n(t_0) = 0.$$

Due to the definitions of $M_w(\varepsilon_n)$ and since $w(\cdot,\cdot)$ is Perron, one can choose (ε_n) such that $\sum_{n=1}^{\infty} |x_n(t) - x_{n+1}(t)|$ converges uniformly to 0 and $(f_n(\cdot))$ converges L^1-strongly. Therefore, $\lim_{n\to\infty} x_n(t) = x(t)$ and $\lim_{n\to\infty} f_n(t) = f(t)$. Then $f(t) \in F(t, x(t))$ since $F(\cdot,\cdot)$ is almost continuous and $\dot{x}(t) \in Ax(t) + f(t)$ with $x(t_0) = x_0$. Therefore, $x(\cdot)$ is an integral solution of (1).

To prove the second part of the theorem, let $\delta > 0$. Let $z(\cdot)$ be a limit solution of (1). Therefore, for any $\varepsilon > 0$ there exists an ε-solution $z_\varepsilon(\cdot)$ such that $|z(t) - z_\varepsilon(t)| < \varepsilon$ for $t \in I$. As in the first part of the proof starting from $z_\varepsilon(\cdot)$, we can choose $\varepsilon_n \downarrow 0$ with $\varepsilon_1 = \varepsilon$ such that there exists an integral solution $x(\cdot)$ of (1) with $|x(t) - z_\varepsilon(t)| < \delta$ on I. Hence, $|z(t) - x(t)| < \varepsilon + \delta$ for any $t \in I$. The proof is completed. □

We refer the reader to ([4], pp. 25–27), where the author gives one example of nonexistence of solutions even when $X = \mathbb{R}^n$. In this case, the set of limit solutions is nonempty and closed.

In [4] it is also studied another example where the solution set of

$$\dot{x}(t) \in Ax(t) + K,\ x(t_0) = x_0 \in \overline{D(A)},$$

with K convex compact, is not closed. In this case, since the multivalued term is constant, due to Theorem 3, the set of integral solutions is nonempty and dense in the set of limit solutions.

Remark 3. *Consider the relaxed problem (6). The solution set of this problem is not closed, in general. We are not able to prove that it is contained in the set of limit solutions of (1), even if $F(t,\cdot)$ is Lipshitz continuous. Nevertheless, if the solution set of (1) is dense in the solution set of (6), then every relaxed solution is also a limit solution. We refer the reader to [16,17], where this type of relaxation theorems are proved in Banach spaces with some additional properties. In our opinion, the limit solution set is more adequate, because it is compact and, under mild assumptions, it is the closure of the solution set of (1).*

Definition 4. *(see, e.g., [18]) The m-dissipative operator A is said to be of complete continuous type if for every $a < b$ and every $(f_n(\cdot))$ in $L^1(a, b; X)$ and $(x_n(\cdot))$ in $C([a, b], X)$, with $x_n(\cdot)$ a solution on $[a, b]$ of $\dot{x}_n(t) \in Ax_n(t) + f_n(t)$, $n = 1, 2, \ldots$, $\lim_{n\to\infty} f_n = f$ weakly in $L^1(a, b; X)$ and $\lim_{n\to\infty} x_n = x$ uniformly in $C([a, b], X)$, it follows that x is a solution on $[a, b]$ of*

$$\dot{x}(t) \in Ax(t) + f(t).$$

We need the following assumption:

Hypothesis 4 (H4). *$F(\cdot,\cdot)$ has nonempty convex weakly compact values.*

We give now sufficient conditions that the limit solutions to be integral ones.

Theorem 4. *Let A be of complete continuous type. If (H1)–(H4) hold, then every limit solution of (1) is also an integral solution of (1).*

Proof. Let $(x_n(\cdot))$ be a sequence of ε_n–solutions of (1) with $\varepsilon_n \downarrow 0$ such that $\lim_{n\to\infty} x_n(t) = x(t)$ uniformly on I. Consequently, the set $\mathcal{M} = \bigcup_{t\in I} \bigcup_{n=1}^{\infty} \{x_n(t)\}$ is compact. Denote by $(f_n(\cdot))$ the corresponding sequence of pseudoderivatives, hence $\int_I \text{dist}(f_n(t), F(t, x_n(t)))dt \leq \varepsilon_n$ for any natural n. Let $\bar{f}_n(\cdot) \in$

$L^1(I;X)$ be such that $\bar{f}_n(t) \in F(t, x_n(t))$ and $|f_n(t) - \bar{f}_n(t)| \leq \frac{3}{2}\text{dist}(f_n(t), F(t, x_n(t)))$ for a.a. $t \in I$. Take $y_n(\cdot)$ the solutions of

$$\dot{y}_n(t) \in Ay_n(t) + \bar{f}_n(t),\ y_n(t_0) = x_0.$$

Due to (4), $|x_n(t) - y_n(t)| \leq \int_{t_0}^{t} |f_n(t) - \bar{f}_n(t)| dt \leq \frac{3}{2}\varepsilon_n$. Consequently, $(y_n(\cdot))$ converges uniformly to $x(\cdot)$.

On the other hand, since $F(\cdot,\cdot)$ is almost continuous, for any $\varepsilon > 0$ there exists a compact set $I_\varepsilon \subset I$ with $\text{meas}(I \setminus I_\varepsilon) \leq \varepsilon$ such that $F|_{I_\varepsilon \times X}$ is continuous. Therefore, $F : I_\varepsilon \times X \rightrightarrows X_w$ is also continuous. Here X_w is X endowed with the weak topology. Due to (H4), the set $K_\varepsilon := \overline{co}(\bigcup_{t \in I_\varepsilon} \bigcup_{n=1}^{\infty} F(t, x_n(t)))$ is weakly compact. We have that $\bar{f}_n(t) \in K_\varepsilon$ on I_ε. Moreover, since $(\bar{f}_n(\cdot))$ is uniformly integrable, it is relatively weakly compact. Then, passing to subsequences, $\bar{f}_n(\cdot) \to f(\cdot)$ weakly in $L^1(I;X)$. Moreover, as $F(\cdot,\cdot)$ is almost continuous, $f(t) \in F(t, x(t))$ a.e. on I.

Finally, since A is of complete continuous type, we get that $x(\cdot)$ is the solution of

$$\dot{x}(t) \in Ax(t) + f(t),\ x(t_0) = x_0.$$

The proof is therefore complete. □

2.3. *m–Dissipative Inclusions with Compact Semigroup*

In this section, we will study the differential inclusion (1) under the following additional assumption on A.

(A) The semigroup $\{S(\cdot);\ t \geq 0\}$ is compact, i.e., $S(t)$ is a compact operator for every $t > 0$.

Since $\|F(t, x(T))\| \leq N$ for every solution $x(\cdot)$ of (5) the following result is a consequence of ([4], Lemma 3.1).

Lemma 4. *Under hypotheses* (H1)–(H3) *and* (A)*, the set of integral solutions of* (1) *is* $C(I, X)$ *precompact (if nonempty).*

Notice also the following theorem which is proved in [19].

Theorem 5. *Let $F(\cdot,\cdot)$ be almost LSC with closed bounded values and let X be a separable Banach space. Under hypotheses* (H2) *and* (A)*, the set of integral solutions of* (1) *is nonempty.*

As a corollary, one can prove the following variant of Filippov–Pliś Lemma (see ([16], Theorem 3) for the separable case).

Proposition 1. *Assume* (H1)–(H3) *and* (A)*. Let $x(\cdot)$ be an integral solution of the Cauchy problem*

$$\dot{x}(t) \in Ax(t) + f_x(t),\ x(t_0) = x_0 \in \overline{D(A)},$$

on I, where $\text{dist}(f_x(t); F(t, x(t))) \leq g(t)$ *for all $t \in I$ and $g \in L^1(t_0, T; \mathbb{R}^+)$. Then for any $\varepsilon > 0$ and any $y_0 \in \overline{D(A)}$, there exists a solution $y(\cdot)$ of the Cauchy problem* (1) *on I with x_0 replaced by y_0 such that*

$$|x(t) - y(t)| \leq v(t) + \varepsilon,$$

for all $t \in I$, where $v(\cdot)$ is the maximal solution of the scalar differential equation $\dot{v}(t) = w(t, v(t)) + g(t)$, $v(t_0) = |x_0 - y_0|$, on I.

We are ready to prove the following interesting result.

Theorem 6. *Under hypotheses* (H1)–(H3) *and* (A)*, the set of integral solutions of* (1) *is dense in the set of limit solutions of* (1).

Proof. Let $x(\cdot)$ be a limit solution of (1) on I. Then there exists a sequence $(x_n(\cdot))$ of ε_n–solutions of (1) with $\varepsilon_n \downarrow 0$ such that $\lim_{n\to\infty} |x_n(t) - x(t)| = 0$ uniformly on I. Then, for any natural n, $x_n(\cdot)$ is a solution of $\dot{x}_n(t) \in Ax_n(t) + f_n(t)$, where $\text{dist}(f_n(t); F(t, x_n(t)) = g_n(t)$ with $0 < g_n(t) \leq 2N$ on I and $\int_I g_n(t)dt \leq \varepsilon_n$. Due to Proposition 1, to every n there exists a solution $y_n(\cdot)$ of (1) such that

$$|x_n(t) - y_n(t)| \leq v_n(t) + \frac{\varepsilon}{2^n},$$

where $v_n(\cdot)$ is the maximal solution of the scalar differential equation $\dot{v}(t) = w(t, v(t)) + g_n(t)$, $v(t_0) = 0$, on I. From Lemma 1, we have that $\lim_{n\to\infty} v_n(t) = 0$ uniformly on I. Consequently, $\lim_{n\to\infty} |x_n(t) - y_n(t)| = 0$ uniformly on I, i.e., $x(t) = \lim_{n\to\infty} y_n(t)$ uniformly on I. □

2.4. Example

The following example is a modification of ([20], Example) and ([16], Example 1).

Let $\Omega \subset \mathbb{R}^n$ with $n \geq 4$ be a domain with smooth boundary $\partial\Omega$. Define $\varphi(r) = |r|^{\gamma-1}r$ for $r \neq 0$ and $0 < \gamma < \frac{n-2}{n}$. We consider the following system:

$$\begin{cases} u_t \in \Delta\varphi(u) + G(t, y, u) \\ -\dfrac{\partial\varphi(u)}{\partial\nu} \in \beta(u) \text{ on } (0, T) \times \partial\Omega \\ u(0, y) = u_0(y). \end{cases}$$

Here, $u \in \mathbb{R}$, $\frac{\partial\varphi(u)}{\partial\nu}$ is the outward normal derivative on $\partial\Omega$ and $\beta(\cdot)$ is a maximal monotone graph in $\mathbb{R}$ with $\beta(0) \ni 0$. The multifunction G has nonempty compact values, is measurable on all variables and continuous on the third one.

Define the operator B in $L^1(\Omega)$ by

$$Bu = \Delta\varphi(u), \text{ for } u \in D(B), \text{ where}$$

$$D(B) = \{u \in L^1(\Omega);\ \varphi(u) \in W^{1,1}(\Omega),\ \Delta\varphi(u) \in L^1(\Omega),\ -\frac{\partial\varphi(u)}{\partial\nu} \in \beta(u) \text{ on } \partial\Omega\}.$$

The derivatives here are understood in the sense of distributions.

As it is shown in ([4], p. 97), the operator B defined above is m-dissipative in $L^1(\Omega)$ and generates a noncompact semigroup. Notice that in [4] the author works with m-accretive operators A; however A is m-dissipative iff $-A$ is m-accretive.

Let

$$F(t, x) = \{f \in L^1(\Omega);\ f(y) \in G(t, y, x(t, y)) \text{ a.e. in } \Omega\},$$

which is jointly measurable and continuous on x. We assume also that there exists $h \in L^1([0, T])$ such that $\|F(t, x)\| \leq h(t)(1 + |x|)$. Let $x_0 = u(\cdot) \in D(B)$. Therefore (H1), (H2) hold true.

Suppose also that there exists a Perron function $w(\cdot, \cdot)$ such that for every $x, z \in \Omega$ and every $f \in F(t, x)$ there exists $g \in F(t, z)$ such that

$$\int_{\Omega^+(x\to z)} (f(y) - g(y))dy - \int_{\Omega^-(x\to z)} (f(y) - g(y))dy$$
$$\pm \int_{\Omega^0(x\to z)} (f(y) - g(y))dy \leq w\left(t, \int_\Omega |f(y) - g(y)|\right) dy.$$

Here, $\Omega^{+(-,0)}_{x\to y} = \{y \in \Omega;\ f(y) > g(y)(<,=)\}$. It follows from the characterization of $[\cdot,\cdot]_+$ (see, e.g., [21], Example 1.4.3) that (H3) also hold true.

In the case when $\gamma > \dfrac{n-2}{n}$ the operator B generates a compact semigroup and it is of complete continuous type.

2.5. Applications to Optimal Control

Our results can be applied to the following optimal control problem:

$$\min\left\{g(x(T)) + \int_{t_0}^{T} f(t,x(t))dt\right\}, \tag{16}$$

where $x(\cdot)$ is a solution of (1). Here, $f(\cdot,\cdot)$ is Carathéodory and integrally bounded on the bounded sets and the function $g : X \to \mathbb{R}$ is assumed to be lower semicontinuous.

Assume (H1)–(H3) and (A). In this case, the limit solution set of (1) is compact and moreover, the set of integral solutions of (1) is dense in the set of limit solutions (see Theorem 6 and Lemma 4).

Clearly, in general, the problem (16) has no optimal solution.

Theorem 7. *Under the above conditions, the problem (16) admits an optimal limit solution.*

Proof. The functional $x(\cdot) \to \int_{t_0}^{T} f(t,x(t))dt$ is continuous from $C(I,X)$ into $\mathbb{R}$. Furthermore, $x(\cdot) \to g(x(T))$ is lower semicontinuous. Consequently, the functional $J(x(\cdot)) = g(x(T)) + \int_{t_0}^{T} f(t,x(t))dt$ is lower semicontinuous from $C(I,X)$ into $\mathbb{R}$. The proof follows from the facts that the limit solution set is $C(I,X)$ compact and every lower semicontinuous real valued function attains its minimum on a compact set. □

3. Conclusions

As we pointed out, the theory of parabolic differential equations and inclusions written in the abstract operator form is growing rapidly. We refer the reader to [1–3] for the theory of PDE and their investigations as abstract equations. Especially the multivalued evolution equations are comprehensively studied in [4,5,18]. In the book by [5], the authors study differential inclusions in evolution (Gelfand) triple. The authors provide many interesting results and examples. In that case, the compactness assumptions are crucially used. In [17], the author prove relaxation theorem in that case.

In [4], the author restricted the study to Banach spaces with uniformly convex duals and A generating a compact semigroup, or he used compactness-type assumptions regarding the Kuratowski (or Hausdorff) measure of noncompactness. In that case, every limit solution is also an integral one. That implies that our existence results extend the existence result there. Notice also [19] where lower semicontinuous perturbations of m-dissipative operators are considered. The existence theorem there is used in the proof of Theorem 6 in this paper. We recall also the book by [18], devoted to nonlocal problems of evolution inclusions with time lag. The main assumptions there are that A is completely continuous and generates a compact semigroup. We mention also [22] where functional evolution inclusions are studied.

In [12], the author uses full Perron condition in the case of ordinary differential inclusions in Banach spaces. The author assumes that the multifunction F has strongly compact values.

The one-sided Perron condition as used here was introduced in [23]. Using integral representation of the solutions the author defined the so-called weak solutions (which are developed in [8]). Here the integral representation of the solution does not hold when A is nonlinear and we use limit solutions. The case of a Banach space with uniformly convex dual was studied in [13] where it was shown that if F has compact values, then the solution set of (1) is compact R_δ and a relaxation theorem has been proved. No other compactness conditions were used. The paper [14] was devoted to Lemma of Filippov–Pliś. The papers [15,16] study the problem (1) in the case when the Banach space has uniformly convex dual.

In the present paper we introduce the so-called limit solutions for the fully nonlinear evolution inclusion (1) and we study their properties. In general, the limit solutions of (1) are not solutions of the relaxed system (6).

(a) The set of limit solutions is nonempty and always $C(I, X)$ closed when the right hand side F is almost continuous with closed bounded values and one-sided Perron in the state variable. Furthermore, every integral solution is also a limit solution.
(b) The set of limit solutions is the closure of the set of integral solutions when $F(t, \cdot)$ is full Perron or A generates a compact semigroup. In the last case every control problem admits an optimal limit solution. We extend the existence and relaxation results of [4,5,15,16].
(c) The existence of limit solutions can be also shown for a large class of evolution inclusions.

It appears that the notion of limit solutions is meaningful and it deserves further investigations.

Author Contributions: Conceptualization, methodology, investigation, writing–original draft preparation, writing–review and editing, T.D., S.B., O.C., N.J. and A.I.L. All authors contributed equally in writing this article. All authors have read and agreed to the published version of the manuscript.

Funding: The work of A. I. Lazu was supported by a grant of Romanian Ministry of Research and Innovation, CNCS - UEFISCDI, project number PN-III-P1-1.1-TE-2016-0868, within PNCDI III. The work of the other authors was supported by the Bulgarian National Science Fund under Project KP-06-N32/7.

Acknowledgments: The authors thank the reviewers for their valuable comments and suggestions which improved the paper.

Conflicts of Interest: The authors declare no conflict of interest.

References

1. Barbu, V. *Nonlinear Differential Equations of Monotone Types in Banach Spaces*; Springer: New York, NY, USA, 2010.
2. Ito, K.; Kappel, F. *Evolution Equations and Approximations*; World Scientific: Singapore, 2002.
3. Roubicek, T. *Nonlinear Partial Differential Equations with Applications*; Birkhauser: Basel, Switzerland, 2005.
4. Bothe, D. *Nonlinear Evolutions in Banach Spaces*; Habilitationsschritt: Paderborn, Germany, 1999.
5. Hu, S.; Papageorgiou, N. *Handbook of Multivalued Analysis Volume II Applications*; Kluwer: Dordrecht, The Netherlands, 2000.
6. Lakshmikantham, V.; Leela, S. *Nonlinear Differential Equations in Abstract Spaces*; Pergamon Press: Oxford, UK, 1981.
7. Benilan, P. Solutions intégrales d'équations d'évolution dans un espace de Banach. *C. R. Acad. Sci. Paris Ser.* **1972**, *A-B 274*, A47–A50.
8. Cârjă, O.; Donchev, T.; Lazu, A.I. Generalized solutions of semilinear evolution inclusions. *SIAM J. Optim.* **2016**, *26*, 891–1409. [CrossRef]
9. Ważewski, T. On an optimal control problem. In *Differential Equations and Their Applications*; Springer: New York, NY, USA, 1963; pp. 229–242.
10. Bothe, D. Multivalued perturbations of m-accretive differential inclusions. *Israel J. Math.* **1998**, *108*, 109–138. [CrossRef]
11. Augustynowicz, A. Some remarks on comparison functions. *Ann. Polon. Math.* **2009**, *96*, 97–106. [CrossRef]
12. Tolstonogov, A. *Differential Inclusions in a Banach Space*; Springer: New York, NY, USA, 2000.
13. Bilal, S.; Cârjă, O.; Donchev, T.; Lazu, A.I. Nonlocal problems for evolution inclusions with one-sided Perron nonlinearities. *RACSAM* **2019**, *113*, 1917–1933. [CrossRef]
14. Donchev, T.; Farkhi, E. On the theorem of Filippov-Pliss and some applications. *Control Cybernet.* **2009**, *38*, 1251–1271.
15. Cârjă, O.; Donchev, T.; Postolache, V. Nonlinear evolution inclusions with one-sided Perron right-hand side. *J. Dyn. Control Syst.* **2013**, *19*, 439–456. [CrossRef]
16. Cârjă, O.; Donchev, T.; Postolache, V. Relaxation results for nonlinear evolution inclusions with one-sided Perron right-hand side. *Set-Valued Var. Anal.* **2014**, *22*, 657–671. [CrossRef]

17. Migórski, S. Existence and relaxation results for nonlinear evolution inclusions revisited. *J. Appl. Math. Stoch. Anal.* **1995**, *8*, 143–149. [CrossRef]
18. Burlică, M.; Necula, M.; Roşu, D.; Vrabie, I.I. *Delay Differential Evolutions Subjected to Nonlocal Initial Conditions*; Monographs and Research Notes in Mathematics; CRC Press: New York, NY, USA, 2016.
19. Avgerinos, E.; Papageorgiou, N. Nonconvex perturbations of evolution equations with m-dissipative operators in Banach spaces. *Comment. Math. Univ. Carolin.* **1989**, *30*, 657–664.
20. Ahmed, R.; Donchev, T.; Lazu, A.I. Nonlocal m-dissipative evolution inclusions in general Banach spaces. *Mediterr. J. Math.* **2017**, *14*, 215. [CrossRef]
21. Vrabie, I. *Compactness Methods for Nonlinear Evolutions*; Wiley: Harlow, UK, 1995.
22. Ke, T.D. Cauchy problems for functional evolution inclusions involving m-accretive operators. *Electron. J. Qual. Theory Differ. Equ.* **2013**, *75*, 1–13. [CrossRef]
23. Donchev, T. Functional differential inclusion with monotone right-hand side. *Nonlinear Anal.* **1991**, *16*, 533–542. [CrossRef]

mathematics

MDPI

Article

Generalized Fixed Point Results with Application to Nonlinear Fractional Differential Equations

Hanadi Zahed [1], Hoda A. Fouad [1,2], Snezhana Hristova [3] and Jamshaid Ahmad [4,*]

1 Department of Mathematics, College of Science, Taibah University, Al Madina Al Munawara 41411, Saudi Arabia; hzahed@taibahu.edu.sa (H.Z.); Htarad@taibahu.edu.sa (H.A.F.)
2 Department of Mathematics and Computer Science, Faculty of Science, Alexandria University, Alexandria 21500, Egypt
3 Department of Applied Mathematics and Modeling, University of Plovdiv "Paisii Hilendarski", 4000 Plovdiv, Bulgaria; snehri@gmail.com
4 Department of Mathematics, University of Jeddah, P.O. Box 80327, Jeddah 21589, Saudi Arabia
* Correspondence: jamshaid_jasim@yahoo.com

Received: 13 June 2020; Accepted: 11 July 2020; Published: 16 July 2020

Abstract: The main objective of this paper is to introduce the (α,β)-type ϑ-contraction, (α,β)-type rational ϑ-contraction, and cyclic (α-ϑ) contraction. Based on these definitions we prove fixed point theorems in the complete metric spaces. These results extend and improve some known results in the literature. As an application of the proved fixed point Theorems, we study the existence of solutions of an integral boundary value problem for scalar nonlinear Caputo fractional differential equations with a fractional order in $(1,2)$.

Keywords: fixed point; complete metric space; fractional differential equations

1. Introduction

Fixed point theorems are useful tools in nonlinear analysis, the theory of differential equations, and many other related areas of mathematics. One of the most applicable method for various investigations is Banach's contraction principle [1]. Many researchers generalized and extended this theorem to different directions. For example, Boyd and Wong [2] elongated the main result of Banach and they replaced the constant in the contractive condition by an appropriate function. Recently, Samet et al. in [3] defined α-admissible and α-ψ-contractive type mappings and studied some of their properties in the framework of complete metric spaces. Later on, Salimi et al. in [4] introduced and investigated the twisted (α,β)-admissible mappings. Many extensions of the notion of α-ψ-contractive type mappings have been developed, see, for example, [5–9] and the references therein.

In 2012, Wardowski ([10]) defined ϑ-contraction in the setting of metric space. Wardowski et al. [11] also presented the concept of ϑ-weak contraction and generalized the conception of ϑ-contraction. Kaddouri et al. [12] extended the notion of ϑ-contraction and gave applications of their results to integral inclusions. Arshad et al. in [13] instigated the rational ϑ-contraction and obtained some fixed points results in a metric space. Concerning ϑ-contractions, we mention the researchers in [14–22].

In all these investigations, the underlying space was complete metric space. There were some open problems for fixed point theorems in ordered metric spaces and cyclic representations of ϑ-contraction. To solve the first problem, we define (α,β)-type ϑ-contraction with the help of control functions α and β. With this new notion, we not only generalize the main theorem of Wardowski [10] but also derive the results for ordered metric spaces by these control functions. We also introduce (α,β)-type rational ϑ-contraction which extend the notion of ϑ-contraction. Moreover, a cyclic (α-ϑ) contraction and cyclic ordered (α-ϑ) contraction are also introduced to solve the second problem.

To illustrate some of the applications of the fixed point theorems studied in this paper, we use the Caputo fractional differential equation. Note that nonlinear fractional differential equations play a very useful role in modeling in various fields of science, such as physics, engineering, bio-physics, fluid mechanics, chemistry, and biology [23,24]. In this paper, based on the proved fixed point theorems, we provide some new sufficient conditions for the existence of the solutions of an integral boundary value problem for a scalar nonlinear Caputo fractional differential equations with fractional order in $(1, 2)$. We also compare the obtained existence results with known ones in the literature.

2. Preliminaries

Let (Ω, ω) (Ω, for short) and $C_L(\Omega)$ be the complete metric space Ω with a metric ω and the set of all non-empty closed subsets of Ω, respectively.

To be more precise and to be easier for readers to see the novelty of the results in this paper, we will initially give some that are known in the literature definitions.

In 2012, Samet et al. ([3]) defined α-admissibility of mapping in the following way:

Definition 1. *([3]) Let the function $\alpha : \Omega \times \Omega \to [0, +\infty)$. The mapping $\mathcal{J} \colon \Omega \to \Omega$ is α-admissible if:*

$$\alpha(\iota, \kappa) \geq 1 \quad \textit{implies} \quad \alpha(\mathcal{J}(\iota), \mathcal{J}(\kappa)) \geq 1 \quad \textit{for } \iota, \kappa \in \Omega.$$

Later, Salimi et al. ([4]) defined twisted (α,β)-admissible mappings in the following way:

Definition 2. *([4]). Let the functions $\alpha, \beta : \Omega \times \Omega \to [0, +\infty)$. The mapping $\mathcal{J} \colon \Omega \to \Omega$ is twisted (α,β) -admissible if:*

$$\begin{cases} \alpha(\iota, \kappa) \geq 1 \\ \beta(\iota, \kappa) \geq 1 \end{cases} \implies \begin{cases} \alpha(\mathcal{J}(\iota), \mathcal{J}(\kappa)) \geq 1 \\ \beta(\mathcal{J}(\iota), \mathcal{J}(\kappa)) \geq 1 \quad \textit{for } \iota, \kappa \in \Omega. \end{cases}$$

Wardowski ([10]) presented a new family of mappings named Wardowski-contractions.

Definition 3. *([10]) The mapping $\mathcal{J} : \Omega \to \Omega$ is ϑ-contraction if there exists a number $\pi > 0$ such that:*

$$\omega(\mathcal{J}(\iota), \mathcal{J}(\kappa)) > 0 \Longrightarrow \pi + \vartheta(\omega(\mathcal{J}(\iota), \mathcal{J}(\kappa))) \leq \vartheta(\omega(\iota, \kappa)), \quad \iota, \kappa \in \Omega \tag{1}$$

where $\vartheta : (0, +\infty) \to \mathbb{R}$ is a function satisfying the assertions:

(F_1) for all $0 < x < y$ the inequality $\vartheta(x) < \vartheta(y)$ holds;
(F_2) for $\{x_j\}_{j=1}^{\infty} \subseteq (0, +\infty)$ the equality $\lim_{j\to\infty} x_j = 0$ holds if $\lim_{j\to\infty} \vartheta(x_j) = -\infty$;
(F_3) $\exists\, 0 < k < 1$ such that $\lim_{x\to 0^+} x^k \vartheta(x) = 0$.

Let Δ be the set of all mappings $\vartheta : (0, +\infty) \to \mathbb{R}$ satisfying the assertions (F_1)–(F_3).

Theorem 1. *([10]) Let $\vartheta \in \Delta$ and $\mathcal{J} \colon \Omega \to \Omega$ is ϑ-contraction, then the mapping $\mathcal{J}$ has a fixed point in Ω, i.e., there exists a point $\iota^* \in \Omega$ such that $\mathcal{J}(\iota^*) = \iota^*$.*

We will give some examples of functions from the set Δ which will be used later.

Example 1. *([10]) Let the function $\vartheta(\iota) = \ln(\iota)$, $\iota > 0$. Then ϑ satisfies conditions (F_1)-(F_3), i.e., $\vartheta \in \Delta$.*
Any function $\mathcal{J} \colon \Omega \to \Omega$ satisfying (1) is a ϑ-contraction because:

$$\omega(\mathcal{J}(\iota), \mathcal{J}(\kappa)) \leq e^{-\pi} \omega(\iota, \kappa)$$

$\forall\, \iota, \kappa \in \Omega$ with $\omega(\mathcal{J}(\iota), \mathcal{J}(\kappa)) > 0$ and $\pi > 0$. Note that $e^{-\pi} \in (0, 1)$, and therefore, the above condition is also the contractive condition of Banach ([1]).

Example 2. *Let the function $\vartheta(\iota) = \iota - \frac{1}{\sqrt{\iota}}$, $\iota > 0$. Then ϑ satisfies conditions (F_1)-(F_3) with $k \in (\frac{1}{2}, 1)$, i.e., $\vartheta \in \Delta$.*

Any function $\mathcal{J}\colon \Omega \to \Omega$ satisfying (1) is a ϑ-contraction because:

$$\pi - \frac{1}{\sqrt{\omega(\mathcal{J}(\iota), \mathcal{J}(\kappa))}} + \omega(\mathcal{J}(\iota), \mathcal{J}(\kappa)) \leq -\frac{1}{\sqrt{\omega(\iota, \kappa)}} + \omega(\iota, \kappa)$$

$\forall\, \iota, \kappa \in \Omega$ with $\omega(\mathcal{J}(\iota), \mathcal{J}(\kappa)) > 0$ and $\pi > 0$.

3. Fixed Point Results

We will introduce a new type of contraction mapping.

Definition 4. *Let the functions $\vartheta \in \Delta$ and $\alpha, \beta : \Omega \times \Omega \to \{-\infty\} \cup (0, \infty)$. The mapping $\mathcal{J}\colon \Omega \to \Omega$ is (α,β)-type ϑ-contraction if for all $\iota, \kappa \in \Omega :\ \omega(\mathcal{J}(\iota), \mathcal{J}(\kappa)) > 0$ the inequality:*

$$\pi + \alpha(\iota, \kappa)\beta(\iota,\ \kappa)\vartheta\left(\omega(\mathcal{J}(\iota), \mathcal{J}(\kappa))\right) \leq \vartheta\left(\omega(\iota, \kappa)\right) \tag{2}$$

holds where $\pi > 0$ is a real number.

Definition 5. *Let the functions $\vartheta \in \Delta$ and $\alpha, \beta : \Omega \times \Omega \to \{-\infty\} \cup (0, \infty)$. The mapping $\mathcal{J}\colon \Omega \to \Omega$ is (α,β)-type rational ϑ-contraction if for all $\iota, \kappa \in \Omega :\ \omega(\mathcal{J}(\iota), \mathcal{J}(\kappa)) > 0$ the inequality:*

$$\pi + \alpha(\iota, \kappa)\beta(\iota,\ \kappa)\vartheta\left(\omega(\mathcal{J}(\iota), \mathcal{J}(\kappa))\right) \leq \vartheta\left(\mathcal{R}(\iota, \kappa)\right) \tag{3}$$

holds, where $\pi > 0$ is a real number and

$$\mathcal{R}(\iota, \kappa) = \max\left\{\omega(\iota, \kappa), \frac{\omega(\iota, \mathcal{J}(\iota))\omega(\kappa, \mathcal{J}(\kappa))}{1 + \omega(\iota, \kappa)}\right\}. \tag{4}$$

Remark 1. *Note that the (α,β)-type ϑ-contraction defined in Definition 4 is a generalization of ϑ-contraction given in [10] with $\alpha(\iota, \kappa) = \beta(\iota, \kappa) = 1$ (see Definition 3).*

We will obtain some new fixed point results applying the introduced above types of mappings.

Theorem 2. *Let the functions $\vartheta \in \Delta$ and $\alpha, \beta : \Omega \times \Omega \to \{-\infty\} \cup (0, \infty)$ and $\mathcal{J}\colon \Omega \to \Omega$ be (α,β)-type ϑ-contraction and the following conditions be satisfied:*

(a) The mapping $\mathcal{J}$ is twisted (α,β) -admissible;
(b) There exists an element $\iota_0 \in \Omega$ such that $\alpha(\iota_0, \mathcal{J}(\iota_0)) \geq 1$ and $\beta(\iota_0, \mathcal{J}(\iota_0)) \geq 1$;
(c) The mapping $\mathcal{J}$ is continuous.

Then the mapping $\mathcal{J}$ has a fixed point in Ω, i.e., there exists a point $\iota^* \in \Omega$ such that $\mathcal{J}(\iota^*) = \iota^*$.

Proof. Let $\iota_0 \in \Omega$ be the element from condition (b). Define the sequence $\{\iota_j\}_{j=0}^{\infty}$ in Ω by $\iota_{j+1} = \mathcal{J}(\iota_j)$ for $j = 0, 1, 2, \ldots$. If $\iota_{j+1} = \iota_j$ for some $j = 0, 1, 2, \ldots$, then $\iota^* = \iota_j$ is the fixed point of the mapping $\mathcal{J}$. Assume $\iota_{j+1} \neq \iota_j$ for all $j = 0, 1, 2, \ldots$. Then from condition (a) and the choice of ι_0 it follows that $\alpha(\iota_1, \iota_2) = \alpha(\mathcal{J}(\iota_0), \mathcal{J}(\iota_1)) \geq 1$ and $\beta(\iota_1, \iota_2) = \beta(\mathcal{J}(\iota_0), \mathcal{J}(\iota_1)) \geq 1$. By induction we get $\alpha(\iota_j, \iota_{j+1}) \geq 1$ and $\beta(\iota_j, \iota_{j+1}) \geq 1$ for $j \in \mathbb{N}$. Now by inequality (2) with $\iota = \iota_{j-1}$ and $\kappa = \iota_j$, we have:

$$\begin{aligned} \pi + \vartheta\left(\omega(\iota_j, \iota_{j+1})\right) &= \pi + \vartheta\left(\omega(\mathcal{J}(\iota_{j-1}), \mathcal{J}(\iota_j))\right) \\ &\leq \pi + \alpha(\iota_{j-1}, \iota_j)\beta(\iota_{j-1}, \iota_j)\vartheta\left(\omega(\mathcal{J}(\iota_{j-1}), \mathcal{J}(\iota_j))\right) \\ &\leq \vartheta(\omega(\iota_{j-1}, \iota_j)). \end{aligned} \tag{5}$$

From inequality (5) it follows that:

$$\vartheta\left(\omega(\iota_j,\iota_{j+1})\right) \leq \vartheta(\omega(\iota_{j-1},\iota_j)) - \pi. \tag{6}$$

Therefore, applying inequality (6) step by step we obtain:

$$\begin{aligned}\vartheta\left(\omega(\iota_j,\iota_{j+1})\right) &\leq \vartheta(\omega(\iota_{j-1},\iota_j)) - \pi \leq \vartheta(\omega(\iota_{j-2},\iota_{j-1})) - 2\pi \\ &\leq \ldots \leq \vartheta(\omega(\iota_0,\iota_1)) - j\pi.\end{aligned} \tag{7}$$

Since $\vartheta \in \Delta$, so letting $j \to \infty$ in (7), we get:

$$\lim_{j\to\infty} \vartheta\left(\omega(\iota_j,\iota_{j+1})\right) = -\infty \Longleftrightarrow \lim_{j\to\infty} \omega(\iota_j,\iota_{j+1}) = 0. \tag{8}$$

From condition (F_3), $\exists\, 0 < k < 1$ such that:

$$\lim_{j\to\infty} \omega(\iota_j,\iota_{j+1})^k \vartheta\left(\omega(\iota_j,\iota_{j+1})\right) = 0. \tag{9}$$

From Equation (7) we get:

$$\begin{aligned}&\left(\omega(\iota_j,\iota_{j+1})\right)^k \vartheta\left(\omega(\iota_j,\iota_{j+1})\right) - \left(\omega(\iota_j,\iota_{j+1})\right)^k \vartheta\left(\omega(\iota_0,\iota_1)\right) \\ \leq\ & \left(\omega(\iota_j,\iota_{j+1})\right)^k \left(\vartheta(\omega(\iota_0,\iota_1)) - j\pi\right) - \left(\omega(\iota_j,\iota_{j+1})\right)^k \vartheta\left(\omega(\iota_0,\iota_1)\right) \\ \leq\ & -\left(\omega(\iota_j,\iota_{j+1})\right)^k j\pi \leq 0, \quad j \in \mathbb{N}.\end{aligned} \tag{10}$$

From inequality (10) for $j \to \infty$ and (8), (9) we obtain:

$$\lim_{j\to\infty} \left(j\left(\omega(\iota_j,\iota_{j+1})\right)^k\right) = 0. \tag{11}$$

Thus there exists $j_1 \in \mathbb{N}$ such that $j\left(\omega(\iota_j,\iota_{j+1})\right)^k \leq 1$ for $j \geq j_1$, or:

$$\omega(\iota_j,\iota_{j+1}) \leq \frac{1}{j^{\frac{1}{k}}}, \quad j \geq j_1. \tag{12}$$

Then for $m, j \in \mathbb{N}$ with $m > j \geq j_1$, we have:

$$\begin{aligned}&\omega(\iota_j,\iota_m) \\ &\leq \omega(\iota_j,\iota_{j+1}) + \omega(\iota_{j+1},\iota_{j+2}) + \omega(\iota_{j+2},\iota_{j+3}) + \ldots + \omega(\iota_{m-1},\iota_m) \\ &= \sum_{i=j}^{m-1} \omega(\iota_i,\iota_{i+1}) \leq \sum_{i=j}^{\infty} \omega(\iota_i,\iota_{i+1}) \leq \sum_{i=j}^{\infty} \frac{1}{i^{\frac{1}{k}}} < \infty.\end{aligned} \tag{13}$$

Hence $\{\iota_j\}$ is a Cauchy sequence in Ω. From completeness of Ω there exists an element $\iota^* \in \Omega L \ \lim_{j\to\infty} \iota_{j+1} = \iota^*$. As $\mathcal{J}$ is continuous, we have $\mathcal{J}(\iota^*) = \lim_{j\to\infty} \mathcal{J}(\iota_j) = \lim_{j\to\infty} \iota_{j+1} = \iota^*$. It proves the claim. □

In the partial case of α-admissible mapping we get the following result:

Corollary 1. *Let the assumptions be satisfied:*

1. *The functions* $\vartheta \in \Delta$ *and* $\alpha : \Omega \times \Omega \to \{-\infty\} \cup (0,\infty)$, *the mapping* $\mathcal{J} : \Omega \to \Omega$ *is* α*-admissible mapping and for* $\iota, \kappa \in \Omega$ *and* $\omega(\mathcal{J}(\iota), \mathcal{J}(\kappa)) > 0$ *the inequality:*

$$\pi + \alpha(\iota,\kappa)\vartheta\left(\omega(\mathcal{J}(\iota),\mathcal{J}(\kappa))\right) \leq \vartheta\left(\omega(\iota,\kappa)\right),$$

holds.

2. *There exists an element $\iota_0 \in \Omega$ such that $\alpha(\iota_0, \mathcal{J}(\iota_0)) \geq 1$.*
3. *The mapping $\mathcal{J}$ is continuous.*

Then the mapping $\mathcal{J}$ has a fixed point in Ω.

Proof. The claim follows from Theorem 2 with $\beta(\iota, \kappa) \equiv 1$ for $\iota, \kappa \in \Omega$. □

In the case when the mapping $\mathcal{J}$ is not continuous we get the following result:

Theorem 3. *Let $\mathcal{J} : \Omega \to \Omega$ be an (α,β)-type rational ϑ-contraction and the following condition be satisfied:*

(a) The mapping $\mathcal{J}$ is twisted (α,β) -admissible;
(b) There exists $\iota_0 \in \Omega$ such that $\alpha(\iota_0, \mathcal{J}(\iota_0)) \geq 1$ and $\beta(\iota_0, \mathcal{J}(\iota_0)) \geq 1$;
(c) If the sequence $\{\iota_j\}_{j=0}^{\infty}$: $\iota_{j+1} = \mathcal{J}(\iota_j) \in \Omega$ for $j = 0, 1, 2, \ldots$ with ι_0 from condition (b), is convergent to $\iota^* \in \Omega$, i.e., $\lim_{j\to\infty} \omega(\iota_j, \iota^*) = 0$ and $\alpha(\iota_j, \iota_{j+1}) \geq 1$ and $\beta(\iota_j, \iota_{j+1}) \geq 1$, then the inequalities $\alpha(\iota_j, \iota^*) \geq 1$ and $\beta(\iota_j, \iota^*) \geq 1$, $j \in \mathbb{N}$, hold.

Then the point ι^* from condition (c) is a fixed point of the mapping $\mathcal{J}$.

Proof. As in the proof of Theorem 2 we construct the sequence $\{\iota_j\}_{j=0}^{\infty}$ and obtain the inequalities $\alpha(\iota_j, \iota_{j+1}) \geq 1$, $\beta(\iota_j, \iota_{j+1}) \geq 1$. The sequence $\{\iota_j\}_{j=0}^{\infty}$ is a Cauchy sequence in Ω and $\lim_{j\to\infty} \omega(\iota_j, \iota^*) = 0$ with $\iota^* \in \Omega$.

Therefore by condition (c) of Theorem 3, we have $\alpha(\iota_j, \iota^*) \geq 1$ and $\beta(\iota_j, \iota^*) \geq 1$ for all $j \in \mathbb{N}$. We will prove that $\mathcal{J}(\iota^*) = \iota^*$. Assuming the contrary that $\mathcal{J}(\iota^*) \neq \iota^*$. Then there exists a number $j_0 \in \mathbb{N}$ such that $\iota_{j+1} \neq \mathcal{J}(\iota^*)$, for all $j \geq j_0$. Therefore, $\omega(\mathcal{J}(\iota_j), \mathcal{J}(\iota^*)) > 0$, for $j \geq j_0$. By (2), we have:

$$\begin{aligned} \pi + \vartheta(\omega(\iota_{j+1}, \mathcal{J}(\iota^*))) &= \pi + \vartheta(\omega(\mathcal{J}(\iota_j), \mathcal{J}(\iota^*))) \\ &\leq \pi + \alpha(\iota_j, \iota^*)\beta(\iota_j, \iota^*)\vartheta(\omega(\mathcal{J}(\iota_j), \mathcal{J}(\iota^*))) \\ &\leq \vartheta(\omega(\iota_j, \iota^*)). \end{aligned} \tag{14}$$

This implies that:

$$\begin{aligned} \vartheta(\omega(\iota_{j+1}, \mathcal{J}(\iota^*))) &\leq \vartheta(\omega(\iota_j, \iota^*)) - \pi \\ &< \vartheta(\omega(\iota_j, \iota^*)). \end{aligned}$$

By (F_1), we have:

$$\omega(\iota_{j+1}, \mathcal{J}(\iota^*)) < \omega(\iota_j, \iota^*).$$

Letting $j \to \infty$ and using the fact that $\lim_{j\to\infty} \omega(\iota_j, \iota^*) = 0$ and $\lim_{j\to\infty} \omega(\iota_j, \iota_{j+1}) = 0$ we get $\omega(\iota^*, \mathcal{J}(\iota^*)) \leq 0$ which is a contradiction. Therefore $\omega(\iota^*, \mathcal{J}(\iota^*)) = 0$, i.e., $\mathcal{J}(\iota^*) = \iota^*$. □

In the partial case of α-admissible mapping we obtain the result:

Corollary 2. *Let the assumptions be fulfilled:*

1. *The functions $\vartheta \in \Delta$ and $\alpha : \Omega \times \Omega \to \{-\infty\} \cup (0, \infty)$ and the mapping $\mathcal{J} : \Omega \to \Omega$ is α-admissible mapping such that for $\iota, \kappa \in \Omega$ and $\omega(\mathcal{J}(\iota), \mathcal{J}(\kappa)) > 0$ the inequality:*

$$\pi + \alpha(\iota, \kappa)\vartheta\left(\omega(\mathcal{J}(\iota), \mathcal{J}(\kappa))\right) \leq \vartheta\left(\omega(\iota, \kappa)\right)$$

holds.

2. *The conditions (b) and (c) of Theorem 3 are fulfilled.*

Then the point $\mathfrak{l}^*$ from condition (c) is a fixed point of the mapping $\mathcal{J}$.

Proof. The claim follows from Theorem 3 with $\beta(\mathfrak{l},\kappa)\equiv 1$ for $\mathfrak{l},\kappa\in\Omega$. □

We state the following property.
(P) $\alpha(\mathfrak{l},\kappa)\geq 1$ and $\beta(\mathfrak{l},\kappa)\geq 1$ for all fixed points $\mathfrak{l},\kappa\in\Omega$.

Theorem 4. *Suppose that the assertions of Theorem 2 are satisfied and the property (P) holds, then the fixed point of the mapping $\mathcal{J}$ is unique.*

Proof. Let $\mathfrak{l}^*,\widehat{\mathfrak{l}}\in\Omega$ be such that $\mathcal{J}(\mathfrak{l}^*)=\mathfrak{l}^*$ and $\mathcal{J}(\widehat{\mathfrak{l}})=\widehat{\mathfrak{l}}$ but $\mathfrak{l}^*\neq\widehat{\mathfrak{l}}$. Then by (P), $\alpha(\mathfrak{l}^*,\widehat{\mathfrak{l}})\geq 1$ and $\beta(\mathfrak{l}^*,\widehat{\mathfrak{l}})\geq 1$. Thus by (2), we have:

$$\begin{aligned}\pi+\vartheta(\omega(\mathfrak{l}^*,\widehat{\mathfrak{l}})) &= \pi+\vartheta(\omega(\mathcal{J}(\mathfrak{l}^*),\mathcal{J}(\widehat{\mathfrak{l}})))\\ &\leq \pi+\vartheta(\alpha(\mathfrak{l}^*,\widehat{\mathfrak{l}})\beta(\mathfrak{l}^*,\widehat{\mathfrak{l}})\omega(\mathcal{J}(\mathfrak{l}^*),\mathcal{J}(\widehat{\mathfrak{l}})))\\ &\leq \vartheta(\omega(\mathfrak{l}^*,\widehat{\mathfrak{l}})).\end{aligned}$$

The above inequality is a contradiction because $\pi>0$. Hence, $\mathfrak{l}^*$ is unique. □

The fixed point result in Theorem 4 generalize the known in the literature result.

Corollary 3. *([10]). Let $\mathcal{J}:\Omega\to\Omega$ be ϑ -contraction . Then the mapping $\mathcal{J}$ has a fixed point in Ω.*

Proof. The claim follows from the proof of Theorem 4 with $\alpha(\mathfrak{l},\kappa)$=$\beta(\mathfrak{l},\kappa)\equiv 1$ for all $\mathfrak{l},\kappa\in\Omega$. □

Example 3. *Consider the set $\Omega=\{\mathfrak{l}_j : j\in\mathbb{N}\}$ where the natural numbers:*

$$\mathfrak{l}_j=1\times 2+3\times 4+\ldots+(2j-1)(2j)=\frac{j(j+1)(4j-1)}{3},\quad \text{for } j=1,2,\ldots.$$

Let $\omega(\mathfrak{l},\kappa)=|\mathfrak{l}-\kappa|$ for any $\mathfrak{l},\kappa\in\Omega$. Define the mapping $\mathcal{J}:\Omega\to\Omega$ by,

$$\mathcal{J}(\mathfrak{l}_1)=\mathfrak{l}_1,\quad \mathcal{J}(\mathfrak{l}_j)=\mathfrak{l}_{j-1},\quad \text{for all } j\geq 2.$$

Let the functions $\alpha:\Omega\times\Omega\to\{-\infty\}\cup(0,\infty)$ be defined by $\alpha(\mathfrak{l},\kappa)=\beta(\mathfrak{l},\kappa)\equiv 1$ for all $\mathfrak{l},\kappa\in\Omega$ and $\vartheta:(0,+\infty)\to\mathbb{R}$ be defined by $\vartheta(\mathfrak{l})=\mathfrak{l}-\frac{1}{\sqrt{\mathfrak{l}}}$, $\mathfrak{l}>0$. According to Example 2 the function $\Theta\in\Delta$.

Then the mapping $\mathcal{J}$ is (α,β)-type ϑ-contraction, with $\pi=12$. or it is ϑ-contraction (see Remark 1). Consider the following three possible cases:

***Case 1.** Let $1=j<\iota$. Then,*

$$|\mathcal{J}(\mathfrak{l}_\iota)-\mathcal{J}(\mathfrak{l}_1)|=|\mathfrak{l}_{\iota-1}-\mathfrak{l}_1|=3\times 4+5\times 6+\ldots+(2\iota-3)(2\iota-2) \tag{15}$$

and

$$\omega(\mathfrak{l}_\iota,\mathfrak{l}_1)=|\mathfrak{l}_\iota-\mathfrak{l}_1|=3\times 4+5\times 6+\ldots+(2\iota-1)(2\iota). \tag{16}$$

As $\iota>1$, so we get,

$$\frac{-1}{\sqrt{3\times 4+\ldots+(2\iota-3)(2\iota-2)}}<\frac{-1}{\sqrt{3\times 4+\ldots+(2\iota-1)(2\iota)}}. \tag{17}$$

From (17), we have,

$$\begin{aligned} & 12-\frac{-1}{\sqrt{3\times 4+\ldots+(2\iota-3)(2\iota-2)}}+3\times 4++5\times 6+\ldots+(2\iota-3)(2\iota-2) \\ < \; & 12-\frac{-1}{\sqrt{3\times 4+\ldots+(2\iota-1)(2\iota)}}+[3\times 4++5\times 6+\ldots+(2\iota-3)(2\iota-2)] \\ \leq \; & -\frac{-1}{\sqrt{3\times 4+\ldots+(2\iota-1)(2\iota)}}+[3\times 4++5\times 6+\ldots+(2\iota-3)(2\iota-2)]+(2\iota-1)(2\iota). \end{aligned}$$

By (15) and (16), we have,

$$12-\frac{1}{\sqrt{|\mathcal{J}(\mathfrak{l}_\iota),\mathcal{J}(\mathfrak{l}_1)|}}+|\mathcal{J}(\mathfrak{l}_\iota),\mathcal{J}(\mathfrak{l}_1)|<-\frac{1}{\sqrt{|\mathfrak{l}_\iota-\mathfrak{l}_1|}}+|\mathfrak{l}_\iota-\mathfrak{l}_1|. \tag{18}$$

Case 2. *Let $1=\iota<j$ This case is similar to Case 1 and therefore we omit it.*
Case 3. *Let $\iota>j>1$. Then we have,*

$$|\mathcal{J}(\mathfrak{l}_\iota)-\mathcal{J}(\mathfrak{l}_j)|=(2j-1)(2j)+(2j+1)(2j+2)+\ldots+(2\iota-3)(2\iota-2) \tag{19}$$

and

$$|\mathfrak{l}_\iota-\mathfrak{l}_j|=(2j+1)(2j+2)+(2j+3)(2j+4)+\ldots+(2\iota-1)(2\iota). \tag{20}$$

As $\iota>j>1$, we get:

$$(2\iota-1)(2\iota)\geq(2j+2)(2j+1)>(2j+2)(2j+2)=2j(2j+2)+2(2j+2)\geq 2j(2j+2)+12.$$

We know that,

$$\frac{-1}{\sqrt{(2j-1)(2j)+\ldots+(2\iota-3)(2\iota-2)}}<\frac{-1}{\sqrt{(2j+1)(2j+2)+\ldots+(2\iota-1)(2\iota)}}. \tag{21}$$

By (21), we have:

$$\begin{aligned} & 12-\frac{1}{\sqrt{(2j-1)(2j)+(2j+1)(2j+2)+\ldots+(2\iota-3)(2\iota-2)}} \\ & +(2j-1)(2j)+(2j+1)(2j+2)+\ldots+(2\iota-3)(2\iota-2) \\ < \; & 12-\frac{1}{\sqrt{(2j+1)(2j+2)+(2j+3)(2j+4)+\ldots+(2\iota-1)(2\iota)}} \\ & +(2j-1)(2j)+(2j+1)(2j+2)+\ldots+(2\iota-3)(2\iota-2) \\ < \; & -\frac{1}{\sqrt{(2j+1)(2j+2)+(2j+3)(2j+4)+\ldots+(2\iota-1)(2\iota)}} \\ & +(2j-1)(2j)+(2j+1)(2j+2)+\ldots+(2\iota-3)(2\iota-2) \\ & +(2\iota-1)(2\iota) \\ = \; & -\frac{1}{\sqrt{(2j+1)(2j+2)+(2j+3)(2j+4)+\ldots+(2\iota-1)(2\iota)}} \\ & +(2j-1)(2j)+(2j+1)(2j+2)+\ldots+(2\iota-1)(2\iota) \end{aligned}$$

By (19) and (20), we have:

$$12-\frac{1}{\sqrt{|\mathcal{J}(\mathfrak{l}_\iota)-\mathcal{J}(\mathfrak{l}_j)|}}+|\mathcal{J}(\mathfrak{l}_\iota)-\mathcal{J}(\mathfrak{l}_j)|<-\frac{1}{\sqrt{|\mathfrak{l}_\iota-\mathfrak{l}_j|}}+|\mathfrak{l}_\iota-\mathfrak{l}_j|.$$

Thus all the hypotheses of Theorem 3 hold and therefore, the mapping $\mathcal{J}$ has a unique fixed point $\mathfrak{l}_1$.

Now we provide some fixed point theorems for (α,β)-type rational ϑ-contraction.

Theorem 5. *Let the functions $\vartheta \in \Delta$ and $\alpha : \Omega \times \Omega \to \{-\infty\} \cup (0, \infty)$ and $\mathcal{J}$: $\Omega \to \Omega$ be (α,β)-type ϑ-contraction and:*

(a) *The mapping $\mathcal{J}$ is twisted (α,β) -admissible;*
(b) *$\exists\, \mathfrak{l}_0 \in \Omega$ such that $\alpha(\mathfrak{l}_0, \mathcal{J}(\mathfrak{l}_0)) \geq 1$ and $\beta(\mathfrak{l}_0, \mathcal{J}(\mathfrak{l}_0)) \geq 1'$*
(c) *The mapping $\mathcal{J}$ is continuous.*

Then the mapping $\mathcal{J}$ has a fixed point in Ω, i.e., there exists a point $\mathfrak{l}^* \in \Omega$ such that $\mathcal{J}(\mathfrak{l}^*) = \mathfrak{l}^*$.

Proof. As in the proof of Theorem 2 we construct the sequence $\{\mathfrak{l}_j\}_{j=0}^{\infty}$ in Ω. Assume that $\mathfrak{l}_{j+1} \neq \mathfrak{l}_j$ for all $j = 0, 1, 2, \ldots$. Then from condition (a) and the choice of $\mathfrak{l}_0$ it follows that $\alpha(\mathfrak{l}_1, \mathfrak{l}_2) = \alpha(\mathcal{J}(\mathfrak{l}_0), \mathcal{J}(\mathfrak{l}_1)) \geq 1$ and $\beta(\mathfrak{l}_1, \mathfrak{l}_2) = \beta(\mathcal{J}(\mathfrak{l}_0), \mathcal{J}(\mathfrak{l}_1)) \geq 1$. By induction we get $\alpha(\mathfrak{l}_j, \mathfrak{l}_{j+1}) \geq 1$ and $\beta(\mathfrak{l}_j, \mathfrak{l}_{j+1}) \geq 1$ for $j \in \mathbb{N}$. Now by inequality (3) with $\mathfrak{l} = \mathfrak{l}_{j-1}$ and $\kappa = \mathfrak{l}_j$, we have:

$$\begin{aligned} \pi + \vartheta\left(\omega(\mathfrak{l}_j, \mathfrak{l}_{j+1})\right) &= \pi + \vartheta\left(\omega(\mathcal{J}(\mathfrak{l}_{j-1}), \mathcal{J}(\mathfrak{l}_j))\right) \\ &\leq \pi + \alpha(\mathfrak{l}_{j-1}, \mathfrak{l}_j)\beta(\mathfrak{l}_{j-1}, \mathfrak{l}_j)\vartheta\left(\omega(\mathcal{J}(\mathfrak{l}_{j-1}), \mathcal{J}(\mathfrak{l}_j))\right) \\ &\leq \vartheta(\mathcal{R}(\mathfrak{l}_{j-1}, \mathfrak{l}_j)) \end{aligned} \tag{22}$$

where

$$\begin{aligned} \mathcal{R}(\mathfrak{l}_{j-1}, \mathfrak{l}_j) &= \max\left\{\omega(\mathfrak{l}_{j-1}, \mathfrak{l}_j), \frac{\omega(\mathfrak{l}_{j-1}, \mathcal{J}(\mathfrak{l}_{j-1}))\omega(\mathfrak{l}_j, \mathcal{J}(\mathfrak{l}_j))}{1 + \omega(\mathfrak{l}_{j-1}, \mathfrak{l}_j)}\right\} \\ &= \max\left\{\omega(\mathfrak{l}_{j-1}, \mathfrak{l}_j), \frac{\omega(\mathfrak{l}_{j-1}, \mathfrak{l}_j)\omega(\mathfrak{l}_j, \mathfrak{l}_{j+1})}{1 + \omega(\mathfrak{l}_{j-1}, \mathfrak{l}_j)}\right\}. \end{aligned} \tag{23}$$

If we assume $\max\left\{\omega(\mathfrak{l}_{j-1}, \mathfrak{l}_j), \frac{\omega(\mathfrak{l}_{j-1}, \mathfrak{l}_j)\omega(\mathfrak{l}_j, \mathfrak{l}_{j+1})}{1+\omega(\mathfrak{l}_{j-1}, \mathfrak{l}_j)}\right\} = \frac{\omega(\mathfrak{l}_{j-1}, \mathfrak{l}_j)\omega(\mathfrak{l}_j, \mathfrak{l}_{j+1})}{1+\omega(\mathfrak{l}_{j-1}, \mathfrak{l}_j)}$, then from (22) we obtain:

$$\pi + \vartheta\left(\omega(\mathfrak{l}_j, \mathfrak{l}_{j+1})\right) \leq \vartheta\left(\frac{\omega(\mathfrak{l}_{j-1}, \mathfrak{l}_j)\omega(\mathfrak{l}_j, \mathfrak{l}_{j+1})}{1 + \omega(\mathfrak{l}_{j-1}, \mathfrak{l}_j)}\right) < \vartheta\left(\omega(\mathfrak{l}_j, \mathfrak{l}_{j+1})\right).$$

The above inequality is a contradiction because $\pi > 0$. Hence,

$$\max\left\{\omega(\mathfrak{l}_{j-1}, \mathfrak{l}_j), \frac{\omega(\mathfrak{l}_{j-1}, \mathfrak{l}_j)\omega(\mathfrak{l}_j, \mathfrak{l}_{j+1})}{1 + \omega(\mathfrak{l}_{j-1}, \mathfrak{l}_j)}\right\} = \omega(\mathfrak{l}_{j-1}, \mathfrak{l}_j).$$

Therefore the inequality (22) is reduced to:

$$\pi + \vartheta\left(\omega(\mathfrak{l}_j, \mathfrak{l}_{j+1})\right) \leq \vartheta(\omega(\mathfrak{l}_{j-1}, \mathfrak{l}_j)). \tag{24}$$

Following the same procedure as we did in Theorem 2, we get $\mathfrak{l}^* \in \Omega$ such that $\mathcal{J}(\mathfrak{l}^*) = \mathfrak{l}^*$. Thus $\mathfrak{l}^*$ is a fixed point of $\mathcal{J}$ □

In the partial case of α-admissible mapping we obtain the result:

Corollary 4. *Let the following assumptions be satisfied:*

1. *The functions $\vartheta \in \Delta$ and $\alpha : \Omega \times \Omega \to \{-\infty\} \cup (0, \infty)$ and the mapping $\mathcal{J} : \Omega \to \Omega$ is α-admissible mapping such that for $\mathfrak{l}, \kappa \in \Omega$ and $\omega(\mathcal{J}(\mathfrak{l}), \mathcal{J}(\kappa)) > 0$ the inequality*

$$\pi + \alpha(\mathfrak{l}, \kappa)\vartheta\left(\omega(\mathcal{J}(\mathfrak{l}), \mathcal{J}(\kappa))\right) \leq \vartheta\left(\mathcal{R}(\mathfrak{l}, \kappa)\right),$$

holds where

$$\mathcal{R}(\mathfrak{l},\kappa) = \max\left\{\omega(\mathfrak{l},\kappa), \frac{\omega(\mathfrak{l},\mathcal{J}(\mathfrak{l}))\omega(\kappa,\mathcal{J}(\kappa))}{1+\omega(\mathfrak{l},\kappa)}\right\};$$

2. $\exists\, \mathfrak{l}_0 \in \Omega$ *such that* $\alpha(\mathfrak{l}_0, \mathcal{J}(\mathfrak{l}_0)) \geq 1$;
3. *The mapping* $\mathcal{J}$ *is continuous.*

Then the mapping $\mathcal{J}$ *has a fixed point in* Ω.

Proof. The claim follows from Theorem 5 with $\beta(\mathfrak{l},\kappa) \equiv 1$ for $\mathfrak{l},\kappa \in \Omega$. □

Now we prove a result for (α,β)-type rational ϑ-contraction when the mapping $\mathcal{J}$ is not continuous.

Theorem 6. *Let the functions* $\vartheta \in \Delta$ *and* $\alpha, \beta : \Omega \times \Omega \to \{-\infty\} \cup (0,\infty)$ *and* $\mathcal{J} \colon \Omega \to \Omega$ *be an* (α,β)*-type rational* ϑ*-contraction and the following condition be satisfied:*

(a) *The mapping* $\mathcal{J}$ *is twisted* (α,β) *-admissible;*
(b) *there exists a point* $\mathfrak{l}_0 \in \Omega$ *such that the inequalities* $\alpha(\mathfrak{l}_0, \mathcal{J}(\mathfrak{l}_0)) \geq 1$ *and* $\beta(\mathfrak{l}_0, \mathcal{J}(\mathfrak{l}_0)) \geq 1$ *hold;*
(c) *If the sequence* $\{\mathfrak{l}_j\}_{j=0}^{\infty}: \ \mathfrak{l}_{j+1} = \mathcal{J}(\mathfrak{l}_j) \in \Omega$ *for* $j = 0,1,2,...$ *with* $\mathfrak{l}_0$ *from condition (b), is convergent to* $\mathfrak{l}^* \in \Omega$, *i.e.,* $\lim_{j\to\infty} \omega(\mathfrak{l}_j, \mathfrak{l}^*) = 0$ *and* $\alpha(\mathfrak{l}_j, \mathfrak{l}_{j+1}) \geq 1$ *and* $\beta(\mathfrak{l}_j, \mathfrak{l}_{j+1}) \geq 1$, *then the inequalities* $\alpha(\mathfrak{l}_j, \mathfrak{l}^*) \geq 1$ *and* $\beta(\mathfrak{l}_j, \mathfrak{l}^*) \geq 1$, $j \in \mathbb{N}$, *hold.*

Then the point $\mathfrak{l}^*$ *from condition (c) is a fixed point of the mapping* $\mathcal{J}$ *in* Ω.

Proof. As in the proof of Theorem 2 we construct the sequence $\{\mathfrak{l}_j\}_{j=0}^{\infty}$ in Ω. Similarly to the proof of Theorem 5 we obtain the inequalities $\alpha(\mathfrak{l}_j, \mathfrak{l}_{j+1}) \geq 1$, $\beta(\mathfrak{l}_j, \mathfrak{l}_{j+1}) \geq 1$ and $\{\mathfrak{l}_j\}_{j=0}^{\infty}$ is a Cauchy sequence in Ω which converges to $\mathfrak{l}^*$, i.e., $\lim_{j\to\infty} \omega(\mathfrak{l}_j, \mathfrak{l}^*) = 0$.

Therefore by condition (c) of Theorem 6, we have $\alpha(\mathfrak{l}_j, \mathfrak{l}^*) \geq 1$ and $\beta(\mathfrak{l}_j, \mathfrak{l}^*) \geq 1$ for all $j \in \mathbb{N}$. We will prove that $\mathcal{J}(\mathfrak{l}^*) = \mathfrak{l}^*$. Assume the contrary that $\mathcal{J}(\mathfrak{l}^*) \neq \mathfrak{l}^*$. Then there exists $j_0 \in \mathbb{N}$ such that $\mathfrak{l}_{j+1} \neq \mathcal{J}(\mathfrak{l}^*)$, for all $j \geq j_0$. Therefore, $\omega(\mathcal{J}(\mathfrak{l}_j), \mathcal{J}(\mathfrak{l}^*)) > 0$, for $j \geq j_0$. By (3), we have:

$$\begin{aligned}
\pi + \vartheta(\omega(\mathfrak{l}_{j+1}, \mathcal{J}(\mathfrak{l}^*))) &= \pi + \vartheta(\omega(\mathcal{J}(\mathfrak{l}_j), \mathcal{J}(\mathfrak{l}^*))) \\
&\leq \pi + \alpha(\mathfrak{l}_j, \mathfrak{l}^*)\beta(\mathfrak{l}_j, \mathfrak{l}^*)\vartheta(\omega(\mathcal{J}(\mathfrak{l}_j), \mathcal{J}(\mathfrak{l}^*))) \\
&\leq \vartheta(\max\{\omega(\mathfrak{l}_j, \mathfrak{l}^*), \tfrac{\omega(\mathfrak{l}_j, \mathcal{J}(\mathfrak{l}_j))\omega(\mathfrak{l}^*, \mathcal{J}(\mathfrak{l}^*))}{1+\omega(\mathfrak{l}_j, \mathfrak{l}^*)}\}) \\
&= \vartheta(\max\{\omega(\mathfrak{l}_j, \mathfrak{l}^*), \tfrac{\omega(\mathfrak{l}_j, \mathcal{J}(\mathfrak{l}_j))\omega(\mathfrak{l}^*, \mathcal{J}(\mathfrak{l}^*))}{1+\omega(\mathfrak{l}_j, \mathfrak{l}^*)}\}) \\
&= \vartheta(\max\{\omega(\mathfrak{l}_j, \mathfrak{l}^*), \tfrac{\omega(\mathfrak{l}_j, \mathfrak{l}_{j+1})\omega(\mathfrak{l}^*, \mathcal{J}(\mathfrak{l}^*))}{1+\omega(\mathfrak{l}_j, \mathfrak{l}^*)}\})
\end{aligned} \tag{25}$$

which implies:

$$\begin{aligned}
\vartheta(\omega(\mathfrak{l}_{j+1}, \mathcal{J}(\mathfrak{l}^*))) &\leq \vartheta(\max\{\omega(\mathfrak{l}_j, \mathfrak{l}^*), \frac{\omega(\mathfrak{l}_j, \mathfrak{l}_{j+1})\omega(\mathfrak{l}^*, \mathcal{J}(\mathfrak{l}^*))}{1+\omega(\mathfrak{l}_j, \mathfrak{l}^*)}\}) - \pi \\
&< \vartheta(\max\{\omega(\mathfrak{l}_j, \mathfrak{l}^*), \frac{\omega(\mathfrak{l}_j, \mathfrak{l}_{j+1})\omega(\mathfrak{l}^*, \mathcal{J}(\mathfrak{l}^*))}{1+\omega(\mathfrak{l}_j, \mathfrak{l}^*)}\}).
\end{aligned}$$

By (F_1), we have:

$$\omega(\mathfrak{l}_{j+1}, \mathcal{J}(\mathfrak{l}^*)) < \max\{\omega(\mathfrak{l}_j, \mathfrak{l}^*), \frac{\omega(\mathfrak{l}_j, \mathfrak{l}_{j+1})\omega(\mathfrak{l}^*, \mathcal{J}(\mathfrak{l}^*))}{1+\omega(\mathfrak{l}_j, \mathfrak{l}^*)}\}$$

Letting $j \to \infty$ and using the fact that $\lim_{j\to\infty} \omega(\mathfrak{l}_j, \mathfrak{l}^*) = 0$ and $\lim_{j\to\infty} \omega(\mathfrak{l}_j, \mathfrak{l}_{j+1}) = 0$ we get $\omega(\mathfrak{l}^*, \mathcal{J}(\mathfrak{l}^*)) \leq 0$ which is a contradiction. Therefore $\omega(\mathfrak{l}^*, \mathcal{J}(\mathfrak{l}^*)) = 0$, i.e., $\mathcal{J}(\mathfrak{l}^*) = \mathfrak{l}^*$. □

Example 4. Let $\Omega = \{0\} \cup [\frac{9}{4}, 5]$ and $\omega(\mathfrak{l}, \kappa) = |\mathfrak{l} - \kappa|$ for $\mathfrak{l}, \kappa \in \Omega$. Clearly (Ω, ω) is a complete metric space. Consider the function $\vartheta(\mathfrak{l}) = \frac{-1}{\sqrt{\mathfrak{l}}} + \mathfrak{l} \in \Delta$ for $\mathfrak{l} \in \Omega$ (see Example 2) and $\pi \in \left(0, \frac{112 - 3\sqrt{5}}{15}\right)$. Define $\mathcal{J} : \Omega \to \Omega$ and $\alpha, \beta : \Omega \to \{-\infty\} \cup (0, \infty)$ by:

$$\mathcal{J}(\mathfrak{l}) = \begin{cases} \frac{9}{4}, & \mathfrak{l} \in \{0\} \cup [\frac{9}{4}, 5) \\ 0, & \mathfrak{l} = 5. \end{cases}$$

and

$$\alpha(\mathfrak{l}, \kappa) = \beta(\mathfrak{l}, \kappa) = 1$$

We prove that $\mathcal{J}$ is (α, β)-type rational ϑ-contraction. Consider these possible cases:

Case I. For $\mathfrak{l} = 0$ and $\kappa = 5$, we have

$$\omega(\mathcal{J}(0), \mathcal{J}(5)) = \omega\left(\{\frac{9}{4}\}, 0\right) = \frac{9}{4} > 0$$

and

$$\mathcal{R}(0, 5) = 5 = \max\left\{\omega(0, 5), \frac{\omega(0, \mathcal{J}(0)) \cdot \omega(5, \mathcal{J}(5))}{1 + \omega(0, 5)}\right\}.$$

Since,

$$\omega(\mathcal{J}(0), \mathcal{J}(5)) = \frac{9}{4} < 5 = \omega(0, 5) \leq \mathcal{R}(0, 5).$$

So, we have

$$-\frac{1}{\sqrt{\omega(\mathcal{J}(0), \mathcal{J}(5))}} < -\frac{1}{\sqrt{\mathcal{R}(0, 5)}},$$

which further implies:

$$-\frac{1}{\sqrt{\omega(\mathcal{J}0, \mathcal{J}5)}} + \omega(\mathcal{J}(0), \mathcal{J}(5)) < -\frac{1}{\sqrt{\mathcal{R}(0, 5)}} + \mathcal{R}(0, 5).$$

Thus we obtain:

$$\pi + \alpha(0, 5)\beta(0, 5)\vartheta(\omega(\mathcal{J}(0), \mathcal{J}(5))) = \pi + \vartheta(\omega(\mathcal{J}(0), \mathcal{J}(5)))$$

$$= \pi - \frac{1}{\sqrt{\omega(\mathcal{J}(0), \mathcal{J}(5))}} + \omega(\mathcal{J}(0), \mathcal{J}(5)) = \pi - \sqrt{\frac{4}{9}} + \frac{9}{5}$$

$$\leq -\sqrt{\frac{1}{5}} + 5 \leq -\frac{1}{\sqrt{\mathcal{R}(0, 5)}} + \mathcal{R}(0, 5) = \vartheta(\mathcal{R}(0, 5)).$$

Hence,

$$\pi + \alpha(0, 5)\beta(0, 5)\vartheta(\omega(\mathcal{J}(0), \mathcal{J}(5))) \leq \vartheta(\mathcal{R}(5, \kappa)).$$

Case II.
For $\mathfrak{l} \in [\frac{9}{4}, 5), \kappa = 0$

$$\omega(\mathcal{J}(\mathfrak{l}), \mathcal{J}(0)) = \omega\left(\{\frac{9}{4}\}, \{\frac{9}{4}\}\right) = 0.$$

Case III.
For $\mathfrak{l} = 5, \kappa \in [\frac{9}{4}, 5)$, we have:

$$\omega(\mathcal{J}(5), \mathcal{J}(\kappa)) = \omega\left(\{0\}, \frac{9}{4}\right) = \frac{9}{4} > 0.$$

Therefore,

$$\omega\left(\mathcal{J}(5), \mathcal{J}(\kappa)\right) < \max\left\{\omega\left(5, \kappa\right), \frac{\omega\left(5, \mathcal{J}(5)\right) \cdot \omega\left(\leq, \mathcal{J}(\kappa)\right)}{1 + \omega\left(5, \kappa\right)}\right\} = \mathcal{R}\left(5, \kappa\right).$$

Similarly to case I, we get:

$$\pi + \alpha\left(5, \kappa\right)\beta\left(5, \kappa\right)\vartheta\left(\omega\left(\mathcal{J}(5), \mathcal{J}(\kappa)\right)\right) \leq \vartheta\left(\mathcal{R}\left(5, \kappa\right)\right)$$

Thus $\mathcal{J}$ is (α, β)-type rational ϑ-contraction. Moreover all the assumptions of Theorem 6 are satisfied and $\frac{9}{4}$ is a fixed point of $\mathcal{J}$.

Corollary 5. Let:

1. The functions $\vartheta \in \Delta$ and $\alpha : \Omega \times \Omega \to \{-\infty\} \cup (0, \infty)$ and the mapping $\mathcal{J} : \Omega \to \Omega$ is α-admissible mapping such that for $\mathfrak{l}, \kappa \in \Omega$ and $\omega(\mathcal{J}(\mathfrak{l}), \mathcal{J}(\kappa)) > 0$ the inequality:

$$\pi + \alpha(\mathfrak{l}, \kappa)\vartheta\left(\omega(\mathcal{J}(\mathfrak{l}), \mathcal{J}(\kappa))\right) \leq \vartheta\left(\mathcal{R}(\mathfrak{l}, \kappa)\right)$$

holds where

$$\mathcal{R}(\mathfrak{l}, \kappa) = \max\left\{\omega(\mathfrak{l}, \kappa), \frac{\omega(\mathfrak{l}, \mathcal{J}(\mathfrak{l}))\omega(\kappa, \mathcal{J}(\kappa))}{1 + \omega(\mathfrak{l}, \kappa)}\right\}.$$

2. The conditions (b) and (c) of Theorem 6 are fulfilled.

Then the point $\mathfrak{l}^*$ from condition (c) is a fixed point of the mapping $\mathcal{J}$.

Proof. The claim follows from Theorem 6 with $\beta(\mathfrak{l}, \kappa) \equiv 1$ for $\mathfrak{l}, \kappa \in \Omega$. □

Theorem 7. Suppose that the assertions of Theorem 5 are satisfied and the property (P) holds. Then the fixed point of the mapping $\mathcal{J}$ is unique.

Proof. Let $\mathfrak{l}^*, \hat{\mathfrak{l}} \in \Omega$ be such that $\mathcal{J}(\mathfrak{l}^*) = \mathfrak{l}^*$ and $\mathcal{J}(\hat{\mathfrak{l}}) = \hat{\mathfrak{l}}$ but $\mathfrak{l}^* \neq \hat{\mathfrak{l}}$. Then by (P), $\alpha(\mathfrak{l}^*, \hat{\mathfrak{l}}) \geq 1$ and $\beta(\mathfrak{l}^*, \hat{\mathfrak{l}}) \geq 1$. Thus,

$$\pi + \vartheta\left(\omega(\mathfrak{l}^*, \hat{\mathfrak{l}})\right) = \pi + \vartheta\left(\omega(\mathcal{J}(\mathfrak{l}^*), \mathcal{J}(\hat{\mathfrak{l}}))\right) \leq \pi + \vartheta\left(\alpha(\mathfrak{l}^*, \hat{\mathfrak{l}})\beta(\mathfrak{l}^*, \hat{\mathfrak{l}})\omega(\mathcal{J}(\mathfrak{l}^*), \mathcal{J}(\hat{\mathfrak{l}}))\right)$$
$$\leq \vartheta(\max\{\omega(\mathfrak{l}^*, \hat{\mathfrak{l}}), \frac{\omega(\mathfrak{l}^*, \mathcal{J}(\mathfrak{l}^*))\omega(\hat{\mathfrak{l}}, \mathcal{J}(\hat{\mathfrak{l}}))}{1 + \omega(\mathfrak{l}^*, \hat{\mathfrak{l}})}\}) = \vartheta(\omega(\mathfrak{l}^*, \hat{\mathfrak{l}})).$$

The above inequality is a contradiction because $\pi > 0$. Hence, $\mathfrak{l}^*$ is unique. □

Now we define cyclic (α-ϑ) contraction and derive some results from our main theorems.

Definition 6. Let the functions $\alpha : \Omega \times \Omega \to \{-\infty\} \cup (0, \infty)$, $\vartheta \in \Delta$, the sets $S_1, S_2 \in C_L(\Omega)$, and $\mathcal{J} : S_1 \cup S_2 \to S_1 \cup S_2$ with $\mathcal{J}S_1 \subseteq S_2$ and $\mathcal{J}S_2 \subseteq S_1$. The mapping $\mathcal{J}$ is cyclic (α-ϑ) contraction if there exists a number $\pi > 0$ such that:

$$\omega(\mathcal{J}(\mathfrak{l}), \mathcal{J}(\kappa)) > 0 \Longrightarrow \pi + \alpha(\mathfrak{l}, \kappa)\vartheta(\omega(\mathcal{J}(\mathfrak{l}), \mathcal{J}(\kappa))) \leq \vartheta(\omega(\mathfrak{l}, \kappa))$$

holds for all $\mathfrak{l} \in S_1$ and $\kappa \in S_2$.

Theorem 8. Let the functions $\alpha : \Omega \times \Omega \to \{-\infty\} \cup (0, \infty)$, $\vartheta \in \Delta$, the mapping $\mathcal{J} : S_1 \cup S_2 \to S_1 \cup S_2$ is a cyclic (α-ϑ) contraction and the following conditions be satisfied:

(a) The mapping $\mathcal{J}$ is α- admissible;
(b) There exists $\mathfrak{l}_0 \in \Omega$ such that $\alpha(\mathfrak{l}_0, \mathcal{J}(\mathfrak{l}_0)) \geq 1$;

(c) *The mapping $\mathcal{J}$ is continuous.*

Then the mapping $\mathcal{J}$ has a fixed point in $S_1 \cap S_2$.

Proof. We take $\Omega = S_1 \cup S_2$. Then (Ω, ω) is a complete metric space. Define $\beta : \Omega \times \Omega \to \{-\infty\} \cup (0, \infty)$ by:

$$\beta(\mathfrak{l}, \kappa) = \begin{cases} 1, \text{ if } \mathfrak{l} \in S_1 \text{ and } \kappa \in S_2 \\ 0, \text{ otherwise.} \end{cases}$$

Then $\mathcal{J}$ is twisted (α,β)-admissible. Let the point $\mathfrak{l}_0 \in \Omega$ be defined in condition (b). Then $\beta(\mathfrak{l}_0, \mathcal{J}(\mathfrak{l}_0)) \geq 1$ holds. Therefore, the assumptions of Theorem 2 are fulfilled and there exists a point $\mathfrak{l}^* \in S_1 \cup S_2$ such that $\mathcal{J}(\mathfrak{l}^*) = \mathfrak{l}^*$. If $\mathfrak{l}^* \in S_1$, then $\mathfrak{l}^* = \mathcal{J}(\mathfrak{l}^*) \in S_2$ because $\mathcal{J}S_1 \subseteq S_2$. Thus $\exists$ $\mathfrak{l}^* \in S_1 \cap S_2$ such that $\mathcal{J}(\mathfrak{l}^*) = \mathfrak{l}^*$. Similarly, if $\mathfrak{l}^* \in S_2$, then $\mathfrak{l}^* = \mathcal{J}(\mathfrak{l}^*) \in S_1$ because $\mathcal{J}S_2 \subseteq S_1$. Thus $\exists$ $\mathfrak{l}^* \in S_1 \cap S_2$ such that $\mathcal{J}(\mathfrak{l}^*) = \mathfrak{l}^*$. □

Theorem 9. *Let the functions $\alpha : \Omega \times \Omega \to \{-\infty\} \cup (0, \infty)$, $\vartheta \in \Delta$, the mapping $\mathcal{J} : S_1 \cup S_2 \to S_1 \cup S_2$ is a cyclic (α-ϑ) contraction and the following conditions be satisfied:*

(a) *The mapping $\mathcal{J}$ is α- admissible;*
(b) *There exists a point $\mathfrak{l}_0 \in \Omega$ such that $\alpha(\mathfrak{l}_0, \mathcal{J}(\mathfrak{l}_0)) \geq 1$;*
(c) *If $\{\mathfrak{l}_j\} \subseteq \Omega$ such that $\alpha(\mathfrak{l}_j, \mathfrak{l}_{j+1}) \geq 1$ for all j and $\mathfrak{l}_j \to \mathfrak{l}^* \in \Omega$ as $j \to \infty$, then $\alpha(\mathfrak{l}_j, \mathfrak{l}^*) \geq 1$ for all $j \in \mathbb{N} \cup \{0\}$.*

Then the mapping $\mathcal{J}$ has a fixed point in $S_1 \cap S_2$.

Proof. We take $\Omega = S_1 \cup S_2$. As in the proof of Theorem 8 we define the function $\beta : \Omega \times \Omega \to \{-\infty\} \cup (0, \infty)$. Then $\mathcal{J}$ is twisted (α,β)-admissible. Let the point $\mathfrak{l}_0 \in \Omega$ be defined in condition (b). Then $\beta(\mathfrak{l}_0, \mathcal{J}(\mathfrak{l}_0)) \geq 1$ holds. Let $\{\mathfrak{l}_j\} \subseteq \Omega$ such that $\alpha(\mathfrak{l}_j, \mathfrak{l}_{j+1}) \geq 1$ and $\beta(\mathfrak{l}_j, \mathfrak{l}_{j+1}) \geq 1$ for all $j \in \mathbb{N} \cup \{0\}$ and $\mathfrak{l}_j \to \mathfrak{l}^*$ as $j \to +\infty$. Then $\mathfrak{l}_j \in S_1$ and $\mathfrak{l}_{j+1} \in S_2$. Now as S_2 is closed, so $\mathfrak{l}^* \in S_2$ and hence $\alpha(\mathfrak{l}_j, \mathfrak{l}^*) \geq 1$ and $\beta(\mathfrak{l}_j, \mathfrak{l}^*) \geq 1$. Therefore, the assumptions of Theorem 3 are fulfilled and $\exists$ $\mathfrak{l}^* \in S_1 \cup S_2$ such that $\mathcal{J}(\mathfrak{l}^*) = \mathfrak{l}^*$. If $\mathfrak{l}^* \in S_1$, then $\mathfrak{l}^* = \mathcal{J}(\mathfrak{l}^*) \in S_2$ because $\mathcal{J}S_1 \subseteq S_2$. Thus $\exists$ $\mathfrak{l}^* \in S_1 \cap S_2$ such that $\mathcal{J}(\mathfrak{l}^*) = \mathfrak{l}^*$. Similarly, if $\mathfrak{l}^* \in S_2$, then $\mathfrak{l}^* = \mathcal{J}(\mathfrak{l}^*) \in S_1$ because $\mathcal{J}S_2 \subseteq S_1$. Thus $\exists$ $\mathfrak{l}^* \in S_1 \cap S_2$ such that $\mathcal{J}(\mathfrak{l}^*) = \mathfrak{l}^*$. □

Corollary 6. *Let the function $\vartheta \in \Delta$, the sets $S_1, S_2 \in C_L(\Omega)$, and $\mathcal{J} : S_1 \cup S_2 \to S_1 \cup S_2$ with $\mathcal{J}S_1 \subseteq S_2$ and $\mathcal{J}S_2 \subseteq S_1$ is continuous and the inequality:*

$$\omega(\mathcal{J}(\mathfrak{l}), \mathcal{J}(\kappa)) > 0 \Longrightarrow \pi + \vartheta(\omega(\mathcal{J}(\mathfrak{l}), \mathcal{J}(\kappa))) \leq \vartheta(\omega(\mathfrak{l}, \kappa))$$

holds for all $\mathfrak{l} \in S_1$ and $\kappa \in S_2$.
Then the mapping $\mathcal{J}$ has a fixed point in $S_1 \cap S_2$.

Proof. The claim follows from Theorem 8 with $\alpha(\mathfrak{l}, \kappa) = 1$ for all $\mathfrak{l} \in S_1$ and $\kappa \in S_2$. □

Now we define cyclic ordered (α-ϑ) contraction and derive some results from our main theorems.

Definition 7. *Let $(\Omega, \omega, \preceq)$ be an ordered metric space and $S_1, S_2 \in C_L(\Omega)$, and $\mathcal{J} : S_1 \cup S_2 \to S_1 \cup S_2$ with $\mathcal{J}S_1 \subseteq S_2$ and $\mathcal{J}S_2 \subseteq S_1$. The mapping $\mathcal{J}$ is a cyclic ordered (α-ϑ) contraction if there exists a number $\pi > 0$ and $\alpha : \Omega \times \Omega \to \{-\infty\} \cup (0, \infty)$ such that:*

$$\omega(\mathcal{J}(\mathfrak{l}), \mathcal{J}(\kappa)) > 0 \Longrightarrow \pi + \alpha(\mathfrak{l}, \kappa)\vartheta(\omega(\mathcal{J}(\mathfrak{l}), \mathcal{J}(\kappa))) \leq \vartheta(\omega(\mathfrak{l}, \kappa))$$

holds for all $\mathfrak{l} \in S_1$ and $\kappa \in S_2$ with $\mathfrak{l} \preceq$, where $\vartheta \in \Delta$.

Theorem 10. *Let the functions $\alpha : \Omega \times \Omega \to \{-\infty\} \cup (0, \infty)$, $\vartheta \in \Delta$, the mapping $\mathcal{J} : S_1 \cup S_2 \to S_1 \cup S_2$ is decreasing continuous cyclic ordered (α-ϑ) contraction and the following conditions be satisfied:*

(a) The mapping $\mathcal{J}$ is α- admissible;
(b) There exists a point $\mathfrak{l}_0 \in \Omega$ such that $\alpha(\mathfrak{l}_0, \mathcal{J}(\mathfrak{l}_0)) \geq 1$ and $\mathfrak{l}_0 \preceq \mathcal{J}(\mathfrak{l}_0)$.

Then $\exists \mathfrak{l}^* \in S_1 \cap S_2$ such that $\mathfrak{l}^* = \mathcal{J}(\mathfrak{l}^*)$.

Proof. We take $\Omega = S_1 \cup S_2$. Then (Ω, ω) is a complete metric space. Define $\beta : \Omega \times \Omega \to \{-\infty\} \cup (0, \infty)$ by:

$$\beta(\mathfrak{l}, \kappa) = \begin{cases} 1, \text{ if } \mathfrak{l} \in S_1 \text{ and } \kappa \in S_2, \text{ with } \mathfrak{l} \preceq \kappa \\ 0, \text{ otherwise.} \end{cases}$$

Let $\beta(\mathfrak{l}, \kappa) \geq 1$, for all $\mathfrak{l}, \kappa \in \Omega$, then $\mathfrak{l} \in S_1$ and $\kappa \in S_2$ with $\mathfrak{l} \preceq \kappa$. It follows that $\mathcal{J}(\mathfrak{l}) \in S_2$ and $\mathcal{J}(\kappa) \in S_1$ with $\mathcal{J}(\kappa) \preceq \mathcal{J}(\mathfrak{l})$, since $\mathcal{J}$ is decreasing. Therefore $\beta(\mathcal{J}(\kappa), \mathcal{J}(\mathfrak{l})) \geq 1$, that is, $\mathcal{J}$ is twisted (α,β)-admissible. Now, let $\alpha(\mathfrak{l}_0, \mathcal{J}(\mathfrak{l}_0)) \geq 1$, with $\mathfrak{l}_0 \in S_1$ and $\mathfrak{l}_0 \preceq \mathcal{J}(\mathfrak{l}_0)$. From $\mathfrak{l}_0 \in S_1$, we have $\mathcal{J}(\mathfrak{l}_0) \in S_2$ with $\mathfrak{l}_0 \preceq \mathcal{J}(\mathfrak{l}_0)$, that is $\beta(\mathfrak{l}_0, \mathcal{J}(\mathfrak{l}_0)) \geq 1$. Then all assumptions of Theorem 2 are satisfied and $\mathcal{J}$ has a fixed point $\mathfrak{l}^*$ in $S_1 \cup S_2$. The remaining proof is identical to the proof of Theorem 9. □

Theorem 11. *Let the functions $\alpha : \Omega \times \Omega \to \{-\infty\} \cup (0, \infty)$, $\vartheta \in \Delta$, the mapping $\mathcal{J} : S_1 \cup S_2 \to S_1 \cup S_2$ is a cyclic ordered (α-ϑ) contraction and the following conditions be satisfied:*

(a) The mapping $\mathcal{J}$ is α- admissible;
(b) There exists a point $\mathfrak{l}_0 \in \Omega$ such that $\alpha(\mathfrak{l}_0, \mathcal{J}(\mathfrak{l}_0)) \geq 1$ and $\mathfrak{l}_0 \preceq \mathcal{J}(\mathfrak{l}_0)$;
(c) If $\{\mathfrak{l}_j\} \subseteq \Omega$ such that $\alpha(\mathfrak{l}_j, \mathfrak{l}_{j+1}) \geq 1$ for all j and $\mathfrak{l}_j \to \mathfrak{l}^* \in \Omega$ as $j \to \infty$, then $\alpha(\mathfrak{l}_j, \mathfrak{l}^*) \geq 1$ for all $j \in \mathbb{N} \cup \{0\}$;
(d) If $\{\mathfrak{l}_j\} \subseteq \Omega$ such that $\mathfrak{l}_j \preceq \mathfrak{l}_{j+1})$ for all j and $\mathfrak{l}_j \to \mathfrak{l}^* \in \Omega$ as $j \to \infty$, then $\mathfrak{l}_j \preceq \mathfrak{l}^*$ for all $j \in \mathbb{N} \cup \{0\}$.

Then $\exists \mathfrak{l}^* \in S_1 \cap S_2$ such that $\mathfrak{l}^* = \mathcal{J}(\mathfrak{l}^*)$.

Proof. We take $\Omega = S_1 \cup S_2$. As in the proof of Theorem 10 we define the function $\beta : \Omega \times \Omega \to [0, +\infty)$. Then $\mathcal{J}$ is twisted (α,β)-admissible. Let $\{\mathfrak{l}_j\} \subseteq \Omega$ such that $\alpha(\mathfrak{l}_j, \mathfrak{l}_{j+1}) \geq 1$ and $\beta(\mathfrak{l}_j, \mathfrak{l}_{j+1}) \geq 1$ for all $j \in \mathbb{N} \cup \{0\}$ and $\mathfrak{l}_j \to \mathfrak{l}^*$ as $j \to +\infty$. Then $\mathfrak{l}_j \in S_1$ and $\mathfrak{l}_{j+1} \in S_2$. Now as S_2 is closed, so $\mathfrak{l}^* \in S_2$ and hence $\mathfrak{l}_j \preceq \mathfrak{l}^*$ and $\beta(\mathfrak{l}_j, \mathfrak{l}^*) \geq 1$. Therefore, the assumptions of Theorem 3 are fulfilled and $\exists\, \mathfrak{l}^* \in S_1 \cup S_2$ such that $\mathcal{J}(\mathfrak{l}^*) = \mathfrak{l}^*$. The remaining proof is identical to the proof of Theorem 6. □

4. Applications to Caputo Fractional Differential Equations

Recently, many researchers have studied the existence of solutions of varies types of fractional differential equations. In this paper we will emphasize our study of Caputo fractional differential equations of the fractional order in $(1, 2)$ and the integral boundary condition. Note that similar problems are studied in [25–27] but the main condition is connected with enough small Lipschitz constant of the right hand side part of the equation. Based on the obtained fixed points theorems we can use weaker conditions for the right hand side part of the equation (see Example 5).

We will apply some of the proved above Theorems to investigate the existence of the solutions of the nonlinear Caputo fractional differential equation:

$$ {}_a^C D_t^q(x(t)) = f(t, x(t)) \quad \text{for } t \in (a, b) \tag{26}$$

with the integral boundary condition:

$$x(a) = 0,\ x(b) = \int_a^\lambda x(s)ds \quad (a < \lambda < b) \tag{27}$$

where $x \in \mathbb{R}$, $q \in (1,2)$, ${}_a^C D_t^q x(t) = \frac{1}{\Gamma(2-q)} \int\limits_a^t (t-s)^{1-q} x''(s)ds$ represents the Caputo fractional derivative, and $a, b : 0 \le a < b$ are given real numbers.

Let $\Omega = C([a,b], \mathbb{R})$ with a norm $\|x\|_{[a,b]} = \sup_{s \in [a,b]} |x(s)|$. For any $x, y \in \Omega$ we define $\omega(x,y) = \|x - y\|_{[a,b]}$.

Consider the linear fractional differential equation:

$$ {}_a^C D_t^q (x(t)) = g(t) \qquad \text{for } t \in (a,b) \tag{28} $$

with the integral boundary condition (27) where $g \in \Omega$.

Lemma 1. *Let $g \in \Omega$. Then the boundary value problem (28), (27) has a solution:*

$$ \begin{aligned} x(t) = & \frac{1}{\Gamma(q)} \int_a^t (t-s)^{q-1} g(s) ds \\ & + \frac{2(t-a)}{((\lambda-a)^2 - 2(b-a))\Gamma(q)} \int_a^b (b-s)^{q-1} g(s) ds \\ & - \frac{2(t-a)}{((\lambda-a)^2 - 2(b-a))\Gamma(q)} \int\limits_a^\lambda \int_a^s (s-\xi)^{q-1} g(\xi) d\xi ds. \end{aligned} \tag{29} $$

The proof of Lemma 1 is based on the presentation of the solution $x(t) = \frac{1}{\Gamma(q)} \int_a^t (t-s)^{q-1} g(s) ds - d_1 - d_2(t-a)$ given in [28].

Based on the presentation (29) we will define a mild solution of (26) and (27).

Definition 8. *The function $x \in \Omega$ is a mild solution of the boundary value problem (26) and (27) if it satisfies:*

$$ \begin{aligned} x(t) = & \frac{1}{\Gamma(q)} \int_a^t (t-s)^{q-1} f(s, x(s)) ds \\ & + \frac{2(t-a)}{((\lambda-a)^2 - 2(b-a))\Gamma(q)} \int_a^b (b-s)^{q-1} f(s, x(s)) ds \\ & - \frac{2(t-a)}{((\lambda-a)^2 - 2(b-a))\Gamma(q)} \int_a^\lambda \int_a^s (s-\xi)^{q-1} f(\xi, x(\xi)) d\xi ds, \quad t \in [a,b]. \end{aligned} $$

For any function $u \in \Omega$, we define the mapping $\mathcal{J} : \Omega \to \Omega$ by:

$$ \begin{aligned} \mathcal{J}(u)(t) = & \frac{1}{\Gamma(q)} \int_a^t (t-s)^{q-1} f(s, u(s)) ds \\ & + \frac{2(t-a)}{((\lambda-a)^2 - 2(b-a))\Gamma(q)} \int_a^b (b-s)^{q-1} f(s, u(s)) ds \\ & - \frac{2(t-a)}{((\lambda-a)^2 - 2(b-a))\Gamma(q)} \int_a^\lambda \int_a^s (s-\xi)^{q-1} f(\xi, u(\xi)) d\xi ds, \\ & \text{for } \ t \in [a,b]. \end{aligned} \tag{30} $$

Now, we establish the existence result as follows.

Theorem 12. *Suppose that:*

(i) The function $f \in C([a,b] \times \mathbb{R}, \mathbb{R})$ and there exists a constant K such that:

$$ \frac{K(b-a)^q}{\Gamma(1+q)} \left(1 + \frac{2K(b-a)}{(2(b-a) - (\lambda-a)^2)} \left(1 + \frac{\lambda - a}{1+q} \right) \right) \in (0,1) \tag{31} $$

and a number $p \in (0,1]$ such that:

$$|f(t,x) - f(t,y)| \leq K|x-y|^p, \quad x,y \in \mathbb{R}, \ t \in [a,b];$$

(ii) There exists a function $x_0 \in \Omega$ such that $\omega(x_0, \mathcal{J}(x_0)) > 0$, where the operator $\mathcal{J}$ is defined by (30);

(iii) For any two functions $x, y \in \Omega$ such that $\omega(x,y) > 0$ the inequality $\omega(\mathcal{J}(x), \mathcal{J}(y)) > 0$ holds.

Then the boundary value problem (26),(27) has a mild solution.

Proof. Note that any fixed point of the mapping $\mathcal{J}$ is a mild solution of the boundary value problem (26) and (27).

Now, let $x, y \in \Omega$ be such that $\omega(x,y) > 0$. By condition (i) of Theorem 12 we obtain:

$$\begin{aligned}
|\mathcal{J}(x)(t) - \mathcal{J}(y)(t)| \leq & \frac{1}{\Gamma(q)} \int_a^t (t-s)^{q-1} \left| f(s,x(s)) - f(s,y(s)) \right| \omega s \\
& + \frac{2(t-a)}{(2(b-a)-(\lambda-a)^2)\Gamma(q)} \int_a^b (1-s)^{q-1} \left| f(s,x(s)) - f(s,y(s)) \right| \omega s \\
& + \frac{2(t-a)}{(2(b-a)-(\lambda-a)^2)\Gamma(q)} \int_a^\lambda \left(\int_a^s (s-t)^{q-1} \left| f(t,x(t)) - f(t,y(t)) \right| \omega t \right) \omega s \\
\leq & \frac{K}{\Gamma(q)} \int_a^t (t-s)^{q-1} |x(s) - y(s)|^p ds \\
& + \frac{2K(t-a)}{(2(b-a)-(\lambda-a)^2)\Gamma(q)} \int_a^b (b-s)^{q-1} |x(s) - y(s)|^p ds \\
& + \frac{2K(t-a)}{(2(b-a)-(\lambda-a)^2)\Gamma(q)} \int_a^\lambda \left(\int_a^s (s-\xi)^{q-1} |x(\xi) - y(\xi)|^p d\xi \right) ds
\end{aligned}$$

$$\begin{aligned}
\leq & \left(\frac{K(t-a)^q}{q\Gamma(q)} + \frac{2K(t-a)}{(2(b-a)-(\lambda-a)^2)\Gamma(q)} \left(\frac{(b-a)^q}{q} + \frac{(\lambda-a)^{1+q}}{q(1+q)} \right) \right) ||x-y||_\infty^p \\
\leq & \frac{K(b-a)^q}{\Gamma(1+q)} \left(1 + \frac{2K(b-a)}{(2(b-a)-(\lambda-a)^2)} \left(1 + \frac{\lambda-a}{1+q} \right) \right) ||x-y||_\infty^p \\
= & \Lambda ||x-y||_\infty^p, \quad t \in [a,b]
\end{aligned}$$

with $\Lambda = \frac{K(b-a)^q}{\Gamma(1+q)} \left(1 + \frac{2K(b-a)}{(2(b-a)-(\lambda-a)^2)} \left(1 + \frac{\lambda-a}{1+q} \right) \right) \in (0,1)$ (see (31)).

Therefore,

$$||\mathcal{J}(x) - \mathcal{J}(y)||_\infty \leq \Lambda ||x-y||_\infty^p$$

or

$$\omega(\mathcal{J}(x), \mathcal{J}(y)) \leq \Lambda (\omega(x,y))^p. \tag{32}$$

From (32) applying condition (ii) we get:

$$\ln\left(\omega(\mathcal{J}(x), \mathcal{J}(y))\right) \leq \ln(\Lambda) + p \ln\left(\omega(x,y)\right).$$

Thus,

$$\ln\left(\frac{1}{\Lambda}\right)^{\frac{1}{p}} + \frac{1}{p} \ln\left(\omega(\mathcal{J}(x), \mathcal{J}(y))\right) \leq \ln(\omega(x,y)).$$

Therefore, the operator $\mathcal{J}$ is (α, β)-type ϑ-contraction with $\vartheta(u) = \ln u \in \Delta$ (see Example 1), $\pi = \ln\left(\frac{1}{\Lambda}\right)^{\frac{1}{p}} > 0$, and the mappings $\alpha, \beta : \Omega \times \Omega \to \{-\infty\} \cup (0, +\infty)$ are defined by:

$$\alpha(x,y) = \begin{cases} \frac{1}{p} & \text{if } \omega(x,y) > 0, \\ 0.1 & \text{otherwise} \end{cases} \qquad \beta(x,y) = \begin{cases} 1 & \text{if } \omega(x,y) > 0, \\ -\infty & \text{otherwise.} \end{cases}$$

Therefore, the assumption (i) of Theorem 3 is satisfied.

The operator $\mathcal{J}$ is twisted (α,β)-admissible because for any $x, y \in \Omega$ if $\alpha(x,y) \geq 1$ and $\beta(x,y) \geq 1$ then from definitions of α, β it follows that $\omega(x,y) > 0$ and from condition (iii) of Theorem 12 the inequality $\omega(\mathcal{J}(x)(t), \mathcal{J}(y)(t)) > 0$ holds. Thus, $\alpha(\mathcal{J}(x), \mathcal{J}(y)) \geq 1$ and $\beta(\mathcal{J}(x), \mathcal{J}(y)) \geq 1$. Therefore, the condition (a) of Theorem 2 is satisfied.

From condition (ii) of Theorem 12, there exists a point $x_0 \in \Omega$ such that $\omega(x_0, \mathcal{J}(x_0)) > 0$ and therefore, $\alpha(x_0, \mathcal{J}(x_0)) = \frac{1}{p} \geq 1$ and $\beta(x_0, \mathcal{J}(x_0)) = 1 \geq 1$. Thus condition (b) of Theorem 2 is satisfied.

According to Theorem 3 the operator $\mathcal{J}$ has a fixed point in Ω, i.e., there exists a function $x^* \in C([a,b], \mathbb{R})$ such that $x^* = \mathcal{J}(x^*)$. This function x^* is a mild solution of the boundary value problem for (26) and (27). □

Remark 2. *Note that the condition (i) of Theorem 12 for the function $f(t,x)$ is less restrictive than the Lipschitz condition used in many existence results (see, for example [25]).*

Now we will provide an example to demonstrate the existence result.

Example 5. *Consider the nonlinear Caputo fractional differential equation:*

$${}_2^C D_t^{1.75}(x(t)) = \frac{1}{\sqrt{t+14}} \arctan(\sqrt{|x(t)|} + e^t \cos t) + \sin t \quad \text{for } t \in (2,3) \tag{33}$$

with the integral boundary condition:

$$x(2) = 0 \ x(3) = \int_0^{2.5} x(s)ds. \tag{34}$$

In this case $f(t,u) = \frac{1}{\sqrt{t+14}} \arctan(\sqrt{|u|} + e^t \cos t) + \sin t$ *and* $|f(t,x) - f(t,u)| \leq 0.25\sqrt{|x-y|}$. *The condition (31) is reduced to:*

$$\begin{aligned} &\frac{K(b-a)^q}{\Gamma(1+q)}\left(1 + \frac{2K(b-a)}{(2(b-a) - (\lambda - a)^2)}\left(1 + \frac{\lambda - a}{1+q}\right)\right) \\ = \ &\frac{K}{\Gamma(2.75)}\left(1 + \frac{2K}{1.75}\frac{3.25}{2.75}\right) = 0.215998 \in (0,1) \end{aligned}$$

with $K = 0.25$.

According to Theorem 12 the boundary value problem (33) and (34) has a solution.

Remark 3. *Note that the boundary value problem (33) and (34) is studied in [25], but the absolute value is missing under the square root. Also, the function $f(t,x)$ is assumed as Lipschitz, but it is not (see Figure 1 for the particular value $t = 2.2 \in (2,3)$). At the same time the function $f(t,x)$ satisfies the condition 1 with $k = 0.25$ (see Figure 2 for the particular value $t = 2.2 \in (2,3)$), and by one of the fixed point theorems proved in this paper the existence of the solution follows.*

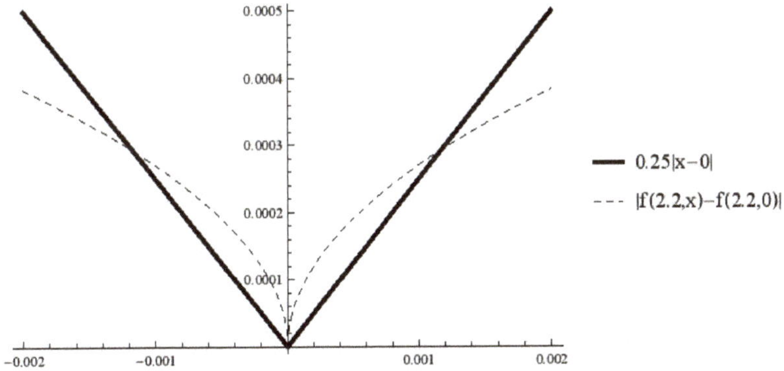

Figure 1. Graphs of $0.25|x - 0|$ and $f(2.2, x) - f(2.2, 0)|$.

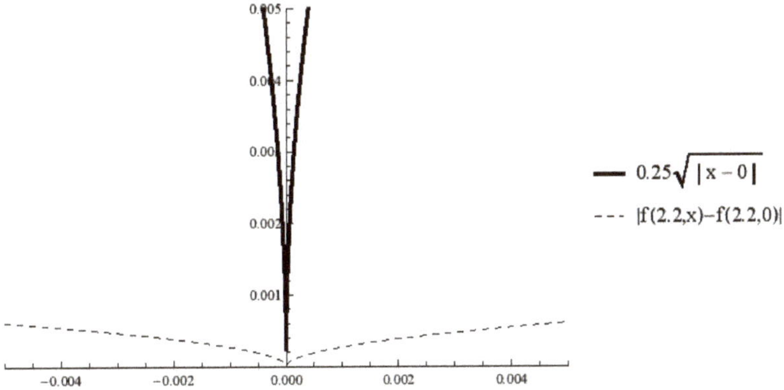

Figure 2. Graphs of $0.25\sqrt{|x - 0|}$ and $f(2.2, x) - f(2.2, 0)|$.

5. Discussion

In fixed point theory, the contractive inequality and underlying space play a significant role. A pioneer result in this theory is a Banach contraction principle that consists of compete metric space (Ω, ω) as underlying space and the following contractive inequality:

$$\omega(\mathcal{J}(\mathfrak{l}), \mathcal{J}(\kappa)) \leq \pi\omega(\mathfrak{l}, \kappa) \tag{35}$$

in which $\mathcal{J}$ is a self mapping and $\pi \in [0, 1)$. Over the years, many mathematicians have generalized and extended above contractive inequality in different ways.

In 2012, Wardowski ([10]) initiated the application of a mapping $\mathcal{J} : (\Omega, \omega) \to (\Omega, \omega)$ and $\pi > 0$ such that:

$$\omega(\mathcal{J}(\mathfrak{l}), \mathcal{J}(\kappa)) > 0 \Longrightarrow \pi + \vartheta(\omega(\mathcal{J}(\mathfrak{l}), \mathcal{J}(\kappa))) \leq \vartheta(\omega(\mathfrak{l}, \kappa)) \tag{36}$$

for $\mathfrak{l}, \kappa \in \Omega$, where $\vartheta : (0, +\infty) \to \mathbb{R}$ satisfies the following conditions:

- $\vartheta(\mathfrak{l}) < \vartheta(\kappa)$ for $0 < \mathfrak{l} < \kappa$;
- For $\{\mathfrak{l}_j\} \subseteq (0, +\infty)$, $\lim_{j\to\infty} \mathfrak{l}_j = 0$ iff $\lim_{j\to\infty} \vartheta(\mathfrak{l}_j) = -\infty$;
- There exists $0 < k < 1$ such that $\lim_{\mathfrak{l}\to 0^+} \mathfrak{l}^k \vartheta(\mathfrak{l}) = 0$.

As it is pointed out in [10] the introduced mapping and inequality (36) are a generalization of Banach contraction (35) with $\vartheta(\mathfrak{l}) = \ln(\mathfrak{l})$, for $\mathfrak{l} > 0$.

In this paper, we generalized the mapping used in [10] by introducing two new notions (α, β)-type ϑ-contraction and (α, β)-type rational ϑ-contraction.

As a partial case of some of our results, we obtained known results in the literature. For example, if $\alpha(\mathfrak{l},\kappa) = \beta(\mathfrak{l},\kappa) = 1$ in Theorem 2 then we obtain Theorem 1 ([10]) by which one can derive the result of [1].

6. Conclusions

In the present paper, we introduced two new types of contractions: (α,β)-type ϑ-contraction and (α,β)-type rational ϑ-contraction. Based on their applications we proved new fixed points theorems. These results generalized some known ones from fixed point theory. To support our results, we provided two non trivial examples. The obtained results are noteworthy contributions to the current results of literature in the theory of fixed points. In this field, one can establish (α,β)-type ϑ-contraction and (α,β)-type rational ϑ-contraction for the multivalued mappings in the perspective of complete metric spaces and generalized metric spaces. To illustrate the application of the new fixed point theorems, we considered an integral boundary value problem for a Caputo fractional scalar equation of order from the interval (1,2) and proved the existence of the solution.

Author Contributions: Conceptualization, H.Z., H.A.F., and J.A.; Methodology, H.Z., H.A.F., S.H., and J.A.; Validation, H.Z., H.A.F., S.H., and J.A.; Formal Analysis, H.Z., H.A.F., S.H., and J.A.; Writing—Original Draft Preparation, H.Z., H.A.F., S.H., and J.A.; Writing—Review and Editing, H.Z., H.A.F., S.H., and J.A.; Funding Acquisition, J.A. All authors have read and agreed to the published version of the manuscript.

Funding: The authors received no direct funding for this work.

Acknowledgments: The authors are thankful to the editor and the referees for their useful comments and suggestions.

Conflicts of Interest: The authors declare no conflict of interest.

References

1. Banach, S. Sur les opérations dans les ensembles abstraits et leur applications aux équations intégrales. *Fund. Math.* **1922**, *3*, 133–181 [CrossRef]
2. Boyd, D.W.; Wong, J.S.W. On nonlinear contractions. *Proc. Am. Math. Soc.* **1969**, *20*, 458–464 [CrossRef]
3. Samet, B.; Vetro, C.; Vetro, P. Fixed point theorem for $\alpha - \psi$ contractive type mappings. *Nonlinear Anal.* **2012**, *75*, 2154–2165. [CrossRef]
4. Salimi, P.; Vetro, C.; Vetro, P. Fixed point theorems for twisted (α, β)-ψ-contractive type mappings and applications. *Filomat* **2013**, *27*, 605–615. [CrossRef]
5. Ahmad, J.; Al-Rawashdeh, A.; Azam, A. Fixed point results for $\{\alpha, \xi\}$-expansive locally contractive mappings, *J. Ineq. Appl.* **2014**, *2014*, 364. [CrossRef]
6. Asl, J.H.; Rezapour, S.; Shahzad, N. On fixed points of $\alpha - \psi$ contractive multifunctions. *Fixed Point Theory Appl.* **2012**, *2012*, 212. [CrossRef]
7. Hussain, H.; Ahmad, J.; Azam, A. Generalized fixed point theorems for multi-valued $\alpha - \psi$-contractive mappings. *J. Ineq. Appl.* **2014**, *2014*, 348. [CrossRef]
8. Kutbi, M.A.; Ahmad, J.; Azam, A. On fixed points of $\alpha - \psi$-contractive multi-valued mappings in cone metric spaces. In *Abstract and Applied Analysis*; Hindawi: Warsaw, Poland, 2013; p. 313782.
9. Salimi, P.; Latif, A.; Hussain, N. Modified $\alpha - \psi$-contractive mappings with applications. *Fixed Point Theory Appl.* **2013**, *2013*, 151. [CrossRef]
10. Wardowski, D. Fixed points of a new type of contractive mappings in complete metric spaces. *Fixed Point Theory Appl.* **2012**, *2012*, 94. [CrossRef]
11. Wardowski, D.; Van Dung, N. Fixed points of F-weak contractions on complete metric spaces. *Demonstr. Math.* **2014**, *47*, 146–155. [CrossRef]
12. Kaddouri, H.; Huseyin, I.; Beloul, S. On new extensions of F-contraction with application to integral inclusions. *UPB Sci. Bull. Ser. A* **2019**, *81*, 31–42.
13. Arshad, M.; Khan, S.; Ahmad, J. Fixed Point Results for F-contractions involving some new rational expressions. *JP J. Fixed Point Theory Appl.* **2016**, *11*, 79–97. [CrossRef]
14. Ahmad, J.; Al-Rawashdeh, A.; Azam, A. New Fixed Point Theorems for Generalized F-Contractions in Complete Metric Spaces. *Fixed Point Theory Appl.* **2015**, *2015*, 80. [CrossRef]

15. Ahmad, J.; Aydi, H.; Mlaiki, N. Fuzzy fixed points of fuzzy mappings via F-contractions and an applications. *J. Intell. Fuzzy Syst.* **2019**, *37*, 5487–5493. [CrossRef]
16. Al-Mazrooei, A.E.; Ahmad, J. Fuzzy fixed point results of generalized almost F-contraction. *J. Math. Comput. Sci.* **2018**, *18*, 206–215 [CrossRef]
17. Shahzad, M.I.; Al-Mazrooei, A.E.; Ahmad, J. Set-valued G-Prešić type F-contractions and fixed point theorems. *J. Math. Anal.* **2019**, *10*, 26–38.
18. Al-Mezel, S.A.; Ahmad, J. Generalized fixed point results for almost (α, F_σ)-contractions with applications to Fredholm integral inclusions. *Symmetry* **2019**, *11*, 1068. [CrossRef]
19. Abdou, A.N.A.; Ahmad, J. Multivalued fixed point theorems for ϑ_ρ-contractions with applications to Volterra integral inclusion. *IEEE Access* **2019**, *7*, 146221–146227. [CrossRef]
20. Altun, G.; Mınak, G.; Dag, H. Multivalued F-contractions on complete metric space. *J. Nonlinear Convex Anal.* **2015**, *16*, 659–666.
21. Cosentino, M.; Vetro, P. Fixed point results for F-contractive mappings of Hardy-Rogers-type. *Filomat* **2014**, *28*, 715–722. [CrossRef]
22. Hussain, N.; Ahmad, J.; Azam, A. On Suzuki-Wardowski type fixed point theorems. *J. Nonlinear Sci. Appl.* **2015**, *8*, 1095–1111. [CrossRef]
23. Budhia, L.B.; Kumam, P.; Martínez-Moreno, J.; Gopal, D. Extensions of almost-F and F-Suzuki contractions with graph and some applications to fractional calculus. *Fixed Point Theory Appl.* **2016**, *2016*, 2. [CrossRef]
24. Gopal, D.; Abbas, M.; Patel, D.K.; Vetro, C. Fixed points of α-type F-contractive mappings with an application to nonlinear fractional differential equation. *Acta Math. Sci.* **2016**, *36*, 957–970. [CrossRef]
25. Mehmood, N.; Ahmad, N. Existence results for fractional order boundary value problem with nonlocal non-separated type multi-point integral boundary conditions. *AIMS Math.* **2019**, *5*, 385–398. [CrossRef]
26. Ahmad, B.; Alsaedi, A.; Alsharif, A. Existence results for fractional-order differential equations with nonlocal multi-point-strip conditions involving Caputo derivative. *Adv. Diff. Eq.* **2015**, *348*. [CrossRef]
27. Liu,W.; Zhuang, H. Existence of solutions for Caputo fractional boundary value problems with integral conditions. *Carpathian J. Math.* 2017, *33*, 207–217.
28. Kilbas, A.A.; Srivastava, H.M.; Trujillo, J.J. *Theory and Applications of Fractional Differential Equations*; Elsevier: Amsterdam, The Netherlands, 2006.

mathematics

MDPI

Article

Optimal Feedback Control Problem for the Fractional Voigt-α Model

Victor Zvyagin [1,*,†], Andrey Zvyagin [2,†] and Anastasiia Ustiuzhaninova [3,†]

1 Department of Algebra and Mathematical Methods of Fluid Dynamics, Voronezh State University, Universitetskaya pl. 1, 394018 Voronezh, Russia
2 Department of Higher Mathematics, Voronezh State Pedagogical University, Lenina st. 86, 394043 Voronezh, Russia; zvyagin.a@mail.ru
3 Research Institute of Mathematics, Voronezh State University, Universitetskaya pl. 1, 394018 Voronezh, Russia; nastyzhka@gmail.com
* Correspondence: zvg_vsu@mail.ru
† These authors contributed equally to this work.

Received: 14 June 2020; Accepted: 20 July 2020; Published: 21 July 2020

Abstract: The study of the existence of an optimal feedback control problem for the initial-boundary value problem that describes the motion of the fractional Voigt-α model of a viscoelastic medium is investigated in this paper. In this model, the Voigt rheological relation is considered with the left-side fractional Riemann-Liouville derivative, which allows to take into account the memory of the medium. Also in this model, the memory is considered along the trajectory of the motion of fluid particles, determined by the velocity field. Due to the insufficient smoothness of the velocity field and, as a consequence, the impossibility of uniquely determining the trajectory for the velocity field for any initial value, a weak solution to the problem under study is introduced using regular Lagrangian flows. Based on the approximation-topological approach to the study of fluid dynamic problems, the existence of an optimal solution that gives a minimum to a given cost functional is proved.

Keywords: optimal feedback control; Voigt model; alpha-model; fractional derivative

MSC: 76D55; 49J20; 35Q35

1. Introduction

The aim of this work is to study the optimal feedback control problem for the alpha-model with the Voigt fractional rheological relation, taking into account the background of a fluid along the trajectory. Note that memory properties in general arise not only in the fluid dynamics field but in many absolute different fields [1]. So the results of this paper can be useful in many fields. A large number of papers have been devoted to the investigation of control problems [2–4]. Although the control problems for linear systems are sufficiently well studied, the situation is not so good for nonlinear systems (even for finite-dimensional cases or local domains). However, due to the complexity of nonlinear systems describing the fluids motion the control of non-Newtonian fluids motion, such as bitumen, polymers, various solutions, emulsions and suspensions, blood, and many others, has not been fully studied. In hydrodynamics the control (optimal control) problems often connected with the fluid control by external forces. Usually in solving such problems, a control is considered from a given (finite) set. In our situation, we consider the external forces control depending on the velocity field. Such types of problems are called feedback control problems [2–5]. In this situation the control is chosen more accurately, since in such a way the control belongs to the image of some multi-valued map. This is more naturally due to the fact that control is not chosen from a finite set of available options.

Also in this paper the alpha model case of fractional Voigt model is considered. Alpha-models are some kind of regularized approximate systems that depend on some positive parameter α, and regularization is carried out by some filtering of the velocity vector, which is contained in the argument of the nonlinear term. The α parameter reflects the width of the spatial filtering scale for the modified speed. The Helmholtz operator $I - \alpha^2\Delta$ is most often used as the filtration kernel. The choice of such an operator is associated with its good mathematical properties. Thus, we ready to proceed to the formulation of the problem. In a bounded domain $\Omega \subset \mathbb{R}^n$ (in 2D and 3D cases, that is, $n = 2,3$) with a sufficiently smooth boundary $\partial\Omega$ on a time interval $[0, T]$, where $T > 0$,, we consider the initial-boundary value problem:

$$\frac{\partial v}{\partial t} + \sum_{i=1}^{n} u_i \frac{\partial v}{\partial x_i} - \mu_0 \Delta v - \frac{\mu_1}{\Gamma(1-\beta)} \mathrm{Div} \int_0^t (t-s)^{-\beta} \mathcal{E}(v)(s, z(s;t,x))\, ds + \nabla p = f, \tag{1}$$

$$u = (I - \alpha^2 \Delta)^{-1} v, \quad t \in [0, T], \quad x \in \Omega, \tag{2}$$

$$z(\tau; t, x) = x + \int_t^\tau v(s, z(s;t,x))\, ds, \quad t, \tau \in [0, T], \quad x \in \Omega, \tag{3}$$

$$\mathrm{div}\, v(t, x) = 0, \quad t \in [0, T], \quad x \in \Omega, \tag{4}$$

$$v|_{t=0} = v_0, \qquad v|_{[0,T]\times\partial\Omega} = 0. \tag{5}$$

Here v is a vector-function of the velocity of a medium particle, u is a vector-function of a modified velocity of a medium particle, defined by equality (2), $z(\tau; t, x)$ is the trajectory of a medium particle, indicating at time τ the location of a medium particle located at time moment t at point x, p is a pressure function, f is a function of the density of external forces, $\alpha > 0$ is scalar parameter, $\mu_0 > 0$, $\mu_1 \geq 0$, $0 < \beta < 1$ are some constants.

$$\mathcal{E} = (\mathcal{E}_{ij}(v)), \quad \mathcal{E}_{ij}(v) = \frac{1}{2}\Big(\frac{\partial v_i}{\partial x_j} + \frac{\partial v_j}{\partial x_i}\Big), \quad i, j = \overline{1, n},$$

is the strain rate tensor. $\Gamma(\beta)$ is the Euler gamma function [6] defined through an absolutely convergent integral

$$\Gamma(\beta) = \int_0^\infty t^{\beta-1} e^{-t}\, dt.$$

This initial-boundary value problem (1)–(5) is an alpha model for the mathematical model of viscoelastic Voigt medium with fractional rheological relation. The idea of using this kind of approximation (the alpha-model) first appeared in paper of J. Leray [7] (in this work, J. Leray used the general form of the filtration kernel) to prove the existence of a weak solution for the Navier-Stokes system of equations. Later, various alpha-models for the Euler equations [8,9], the Navier-Stokes system [10] and others were built on this idea. In general, each alpha model is characterized by its first-order vector differential operator $F(u, v) = (F^1(u, v), \dots, F^n(u, v))$, in which components $F^i(u, v)$ are linear combinations of all kinds of operators of form $u^k \partial_{x_j} v^m$, $v^k \partial_{x_j} u^m$, $u^k \partial_{x_j} u^m$:

$$F^i(u, v) = \sum_{k,j,m=1}^{n} C^i_{kjm} u^k \partial_{x_j} v^m + D^i_{kjm} v^k \partial_{x_j} u^m + E^i_{kjm} u^k \partial_{x_j} u^m, \tag{6}$$

where C^i_{kjm}, D^i_{kjm}, E^i_{kjm} are some real coefficients. Note that in representation (6) monomials of the form $v^k \partial_{x_j} v^m$ are not used, since they do not contain the components of the «smoothed» vector field u.

Interest in the study of alpha-models is primarily associated with their application to the study of turbulence effects for fluid flows. It is also associated with obtaining better numerical results for alpha-models in comparison to the original models. However, most of the works on the solvability of alpha-models are devoted to models of the motion of an ideal or Newtonian fluid [11–14]. Only in the last few years, works began to appear on the study of alpha-models of non-Newtonian fluid [15–18].

This work continues the study of alpha-models for non-Newtonian fluids, namely, for the fractional Voigt model of the viscoelastic medium [19]. This mathematical model describes a viscoelastic fluid flow with a rheological relation $\sigma = \mu_0\mathcal{E}(v) + \mu_1 D^{\beta}_{0t}\mathcal{E}(v) = \mu_0\mathcal{E}(v) + \mu_1 I^{1-\beta}_{0t}\mathcal{E}(v)$, considered along the trajectories of fluid motion. Here D^{β}_{0t} is the left-side fractional Riemann-Liouville derivative and $I^{1-\beta}_{0t}$ is the Riemann-Liouville fractional integral. This model is a fractional analog of the Voigt model, which describes the motion of a linearly elastic-retarded fluid. In order to study a large class of polymers with creep and relaxation effects one must to consider models with fractional derivatives. It turns out that the models with fractional derivatives are most suitable for this [20,21]. Note that the advantage of this model is that, together with the definition of the vector-velocity v of the particle's motion, the trajectory of the particles of this medium motion z is also determined. Also, note that the consideration of fractional derivatives in fluid dynamics has many physical applications [22–24]. One of the possible continuations of this model studies is laid out in References [25] and [26].

2. Preliminary Information and Statement of the Main Results

We introduce the main notation and auxiliary statements.

By $L_p(\Omega)$, $1 \le p < \infty$, we denote the set of measurable vector functions $v : \Omega \to \mathbb{R}^n$, summable with p degree. By $W^m_p(\Omega)$, $m \ge 1$, $p \ge 1$, we denote Sobolev spaces. We consider the space $C^{\infty}_0(\Omega)^n$ of infinitely differentiable vector functions from Ω to $\mathbb{R}^n$ with compact support in Ω. Denote by $\mathcal{V}$ the set $\{v \in C^{\infty}_0(\Omega)^n, \operatorname{div} v = 0\}$. Also by V^0 and V^1 we denote the closure of $\mathcal{V}$ with respect to the norm of $L_2(\Omega)$ and $W^1_2(\Omega)$, respectively, and by V^2 we denote the space $V^2 = W^2_2(\Omega) \cap V^1$.

We introduce from Reference [27] the scale of spaces V^{β}, $\beta \in \mathbb{R}$. For this we consider the Leray projector $P : L_2(\Omega) \to V^0$ and the operator $A = -P\Delta$ defined on $D(A) = V^2$. From this operator we can get a self-adjoint positive operator with compact inverse in V^0. Let $0 < \lambda_1 \le \lambda_2 \le \dots \le \lambda_k \le \dots$ be the eigenvalues of the operator A. We can get an orthonormal basis in V^0 by the eigenfunctions $\{e_j\}$ of the operator A due to the Hilbert theorem on the spectral decomposition of compact operators. Denote by

$$E_{\infty} = \left\{ v = \sum_{j=1}^{N} v_j e_j : v_j \in \mathbb{R},\ N \in \mathbb{N} \right\},$$

the set of finite linear combinations of e_j. Thus, we get the space V^{β}, $\beta \in \mathbb{R}$ as the completion of E_{∞} with respect to the norm

$$\|v\|_{V^{\beta}} = \left(\sum_{k=1}^{\infty} \lambda_k^{\beta} |v_k|^2 \right)^{\frac{1}{2}}. \tag{7}$$

In Reference [27] it is shown that on the space V^{β}, $\beta > -1/2$, norm (7) is equivalent to the ordinary norm $\|\cdot\|_{W^{\beta}_2(\Omega)^n}$ of the space $W^{\beta}_2(\Omega)^n$. In addition, according to Reference [28], the norms in the spaces V^1, V^2 and V^3 can be defined as follows:

$$\|v\|_{V^1} = \left(\int_{\Omega} \nabla v(x) : \nabla v(x)\, dx \right)^{\frac{1}{2}}, \quad \|v\|_{V^2} = \left(\int_{\Omega} \Delta v(x) \Delta v(x)\, dx \right)^{\frac{1}{2}},$$

$$\|v\|_{V^3} = \left(\int_{\Omega} \nabla \Delta v(x) : \nabla \Delta v(x)\, dx \right)^{\frac{1}{2}}.$$

Here the symbol " : " denotes the component-wise matrix product, that is, for $C = (c_{ij})$, $D = (d_{ij})$, $i, j = 1, \dots m$, we put $C : D = \sum_{i,j=1}^{m} c_{ij} d_{ij}$.

Further, through the $V^{-\beta} = (V^{\beta})^*$, $\beta \in \mathbb{N}$, we denote the space dual to V^{β}.

Note that $C([0,T];F)$ is the Banach space of continuous on $[0,T]$ functions, $C_w([0,T];F)$ is the Banach space of weakly continuous on $[0,T]$ functions, $L_p(0,T;F)$ is the Banach spaces of summable on $[0,T]$ with p degree functions with values in a Banach space F, respectively.

The set $C^1D(\overline{\Omega})$ consists of one-to-one mappings $z:\overline{\Omega}\to\overline{\Omega}$ coinciding with the identity mapping on $\partial\Omega$ and having continuous first-order partial derivatives on Ω such that $\det(\partial z/\partial x)=1$ at every point of the domain Ω. For this set the norm of continuous functions $C(\overline{\Omega})$ is used. Further, we will consider the following set $CG=C([0,T]\times[0,T],C^1D(\overline{\Omega}))$. Note that $CG\subset C([0,T]\times[0,T],C^1(\overline{\Omega}))$, therefore, in what follows, CG is considered a metric space with a metric defined by the norm of the space $C([0,T]\times[0,T],C(\overline{\Omega}))$.

We introduce the space in which the solvability of the considered problem will be proved:

$$W_1=\{v\in L_2(0,T;V^1)\cap L_\infty(0,T;V^0),\quad v'\in L_{4/3}(0,T;V^{-1})\}$$

with the norm $\|v\|_{W_1}=\|v\|_{L_2(0,T;V^1)}+\|v\|_{L_\infty(0,T;V^0)}+\|v'\|_{L_{4/3}(0,T;V^{-1})}$.

Denote by $\Delta_\alpha: V^\beta\to V^{\beta-2}$, $\beta\geq 0$ the operator $\Delta_\alpha=(J+\alpha^2A)$, where $J=PI$, and I is the identity operator. By virtue of Reference [28], the operator Δ_α is invertible. If we apply the Leray projection $P:L_2(\Omega)\to V^0$ to the equality $v=(I-\alpha^2\Delta)u$ for $\beta=3$ and express from the last equality u: $u=(J+\alpha^2A)^{-1}v=\Delta_\alpha^{-1}v$. Then, since $v\in V^1$, we get that $u\in V^3$.

Note that for the correct formulation of the considered initial-boundary value problem the trajectories z must be uniquely determined by the velocity field v. In other words, it is necessary that Equation (3) has a unique solution for the velocity field v. However, the existence of solutions to Equation (3) for a fixed v is known in Reference [29] only in case $v\in L_1(0,T;C(\overline{\Omega})^n)$ and this solution is unique for $v\in L_1(0,T;C^1(\overline{\Omega})^n)$ such that $v|_{(0,T)\times\partial\Omega}=0$. Therefore, the trajectories of motion are not uniquely determined even for strong solutions whose partial derivatives that appear in Equation (3) are contained in $L_2(0,T;L_2(\Omega))$. One possible way out of this situation is to regularize the velocity field at each time instant t by averaging over the variable x and determine the trajectories $z(\tau;t,x)$ for the regularized velocity field [30]. However, relatively recently [31,32], the solvability of Cauchy integral problem (3) was investigated in the case when the velocity v belongs to the Sobolev space. Also the existence and uniqueness of regular Lagrangian flows, which are a generalization of the concept of a classical solution, are established.

Definition 1. *Regular Lagrangian flow associated to v is the function $z(\tau;t,x)$, $(\tau;t,x)\in[0,T]\times[0,T]\times\overline{\Omega}$ satisfying conditions:*

1. *the function $\gamma(\tau)=z(\tau;t,x)$ is absolutely continuous and satisfies Equation (3) for almost all $x\in\overline{\Omega}$ and $t\in[0,T]$;*
2. *the equality $m(z(\tau;t,B))=m(B)$ holds for any $t,\tau\in[0,T]$ and an arbitrary Lebesgue measurable set $B\subseteq\overline{\Omega}$ with Lebesgue measure $m(B)$;*
3. *for all $t_i\in[0,T]$, $i=1,2,3$, and almost all $x\in\overline{\Omega}$*

$$z(t_3;t_1,x)=z(t_3;t_2,z(t_2;t_1,x)).$$

We give the necessary results from a regular Lagrangian flow.

Theorem 1. *[31] Let $v\in L_1(0,T;W_p^1(\Omega)^n)$, $1\leq p\leq\infty$ with conditions $\operatorname{div} v(t,x)=0$, $(t,x)\in[0,T]\times\Omega$, and $v|_{[0,T]\times\partial\Omega}=0$. Then there exists a unique regular Lagrangian flow $z\in C(D;L^n)$ associated to v (where $C(D,L)$ is the Banach space of continuous functions on $D=[0,T]\times[0,T]$ with values in the metric space of vector functions L measurable on Ω). Moreover, $z(\tau;t,\overline{\Omega})\subset\overline{\Omega}$ up to a set of measure zero and*

$$\frac{\partial}{\partial\tau}z(\tau;t,x)=v(\tau,z(\tau;t,x)),\quad t,\tau\in[0,T],\quad \text{for almost all}\quad x\in\Omega.$$

Theorem 2. *Let v, $v^m \in L_1(0,T;W_1^p(\Omega)^n)$, $m = 1,2,\ldots$ for some $p > 1$. Let $\operatorname{div} v = 0$, $\operatorname{div} v^m = 0$, $v|_{[0,T]\times\partial\Omega} = 0$, $v^m|_{[0,T]\times\partial\Omega} = 0$. Also, let the inequalities*

$$\|v_x\|_{L_1(0,T;L_p(\Omega)^{n\times n})} + \|v\|_{L_1(0,T;L_p(\Omega)^n)} \leq M,$$
$$\|v_x^m\|_{L_1(0,T;L_p(\Omega)^{n\times n})} + \|v^m\|_{L_1(0,T;L_p(\Omega)^n)} \leq M$$

are valid. Here v_x and v_x^m are the Jacobi matrices of the vector functions v and v^m. Let v^m converges to v in $L_1(Q_T)^N$ as $m \to +\infty$. Let $z^m(\tau;t,x)$ and $z(\tau;t,x)$ are regular Lagrangian flows associated to v^m and v, respectively. Then the sequence z^m converges (up to a subsequence) to z with respect to the Lebesgue measure on the set $[0,T]\times\Omega$ uniformly on $t \in [0,T]$.

This result was proved in Reference [33] in the general case.

Thus, by virtue of Theorem 1 for each $v \in L_2(0,T;V^1)$ and for almost all $x \in \Omega$, the Equation (3) has a unique solution $z(v)$, where $z(v)(\tau;t,x) = z(\tau;t,x)$, in the class of regular Lagrangian flows.

As a control function, we consider the multi-valued map $\Psi : W_1 \multimap L_2(0,T,V^{-1})$. Assume that Ψ satisfies the following conditions:

(Ψ1) Ψ is defined on the space W_1 and has nonempty, compact, and convex values;

(Ψ2) Ψ is compact and upper semicontinuous (that is, for any function $v \in W_1$ and any open set $Y \subset L_2(0,T,V^{-1})$ such that $\Psi(v) \subset Y$, there exists a neighborhood $U(v)$ such that $\Psi(U(v)) \subset Y$);

(Ψ3) Ψ is globally bounded, that is, there exists a constant $R_1 > 0$ such that

$$\|\Psi(v)\|_{L_2(0,T,V^{-1})} := \sup\{\|u\|_{L_2(0,T,V^{-1})} : u \in \Psi(v)\} \leq R_1 \text{ for all } v \in W_1;$$

(Ψ4) Ψ is weakly closed, that is: if $\{v_l\}_{l=1}^{\infty} \subset W_1, v_l \rightharpoonup v_0, u_l \in \Psi(v_l)$ and $u_l \to u_0$ in $L_2(0,T,V^{-1})$ then $u_0 \in \Psi(v_0)$.

In this paper, a weak statement of the feedback control problem for initial-boundary value problem (1)–(5) is considered. By feedback, we mean the condition

$$f \in \Psi(v). \tag{8}$$

We formulate the definition of a weak solution to feedback control problem (1)-(5), (8):

Definition 2. *A pair of functions $(v,f) \in W_1 \times L_2(0,T,V^{-1})$ is called a weak solution of feedback control problem (1)–(5), (8), if it for all $\varphi \in V^1$ and almost all $t \in (0,T)$ satisfies the equality*

$$\langle v',\varphi\rangle - \int_\Omega \sum_{i,j=1}^n (\Delta_\alpha^{-1} v)_i v_j \frac{\partial\varphi_j}{\partial x_i}\,dx + \mu_0 \int_\Omega \nabla v : \nabla\varphi\,dx$$
$$+ \frac{\mu_1}{\Gamma(1-\beta)}\Big(\int_0^t (t-s)^{-\beta}\mathcal{E}(v)(s,z(v)(s;t,x))\,ds, \mathcal{E}(\varphi)\Big) = \langle f,\varphi\rangle, \tag{9}$$

the initial condition $v(0) = v_0$ and feedback condition (8). Here $z(v)$ is a regular Lagrangian flow associated to v.

Remark 1. *It is known that $W \subset C_w(0,T;V^0)$ [34]. Therefore, initial condition (5) has sense.*

The following theorem is the first result of the paper:

Theorem 3. *Let a multi-valued mapping Ψ satisfy conditions (Ψ1) – (Ψ4). Then there is at least one weak solution $(v_*,f_*) \in W_1 \times L_2(0,T,V^{-1})$ of feedback control problem (1)–(5), (8).*

We denote by $\Sigma \subset W_1 \times L_2(0,T;V^{-1})$ the set of all weak solutions of problem (1)–(5), (8). Consider an arbitrary cost functional $\Phi : \Sigma \to \mathbb{R}$, satisfying the following conditions:

($\Phi 1$) For all $(v,f) \in \Sigma$ a number R_2 exists such that $\Phi(v,f) \geq R_2$.
($\Phi 2$) If $v_l \rightharpoonup v_*$ in W_1 and $f_l \to f_*$ in $L_2(0,T;V^{-1})$ then $\Phi(v_*,f_*) \leq \underline{\lim}_{m\to\infty} \Phi(v_l,f_l)$.

As an example of this functional, we can take

$$\Phi(v,f) = \int_0^T \|v(t) - u_*(t)\|_{V^1}^2\,dt + \int_0^T \|f(t)\|_{V^{-1}}^2\,dt.$$

Here u_* is some specified velocity field. This functional characterizes the deviation of velocity from the required, and its minimum yields the minimal deviation of velocity from the one specified by the minimal control. One of the possible applications of the proposed approach is an optimal feedback control problem and the results are in the consideration, analysis and calculation of different such problems with special (necessary in industry) cost functionals Φ.

The following theorem is the second result of this paper.

Theorem 4. *If the mapping Ψ satisfies conditions $(\Psi 1)$–$(\Psi 4)$ and the functional Φ satisfies conditions $(\Phi 1)$, $(\Phi 2)$, then optimal feedback control problem (1)–(5), (8) has at least one weak solution (v_*,f_*) such that*

$$\Phi(v_*,f_*) = \inf_{(v,f)\in\Sigma} \Phi(v,f).$$

The proof of Theorems 3 and 4 is based on the approximation-topological method for investigating fluid dynamics problems [35]. To do this, first, we pass to the operator interpretation of the problem under consideration (operator inclusion) in suitable function spaces. Further, since the operators in the obtained operator inclusion do not have the necessary properties, we consider a problem that approximates the original one (in this case, it is also an operator inclusion, but with a better operator that has the required properties and in better functional spaces). Then, based on a priori estimates of solutions and the theory of the topological degree of multi-valued vector fields, the existence of a solution to the approximation problem is proved. Finally, it is shown that from the sequence of solutions of the approximation problem, one can extract a subsequence that converges in a weak sense to the solution of the original operator inclusion. After proving the solvability of the control problem, it is shown that in the set of solutions there is at least one solution that gives a minimum to a given cost functional (this is why this type of problem is called the optimal feedback control problem for fluid motion).

The work is organized as follows—in Section 3 we consider the family of auxiliary problems and prove the necessary properties of an introduced operators. Also on the basis of the topological degree theory for multivalued maps we prove the solvability of the auxiliary problem and establish necessary estimates for solutions to the auxiliary problem. Section 4 is devoted to the passage, the limit and the proof of Theorem 3. Section 5 is devoted to the proof of Theorem 4. The final Section 6 contains conclusions.

3. The Family of Auxiliary Problems

Throughout this section we will assume that $v_0 \in V^3$.

Consider the following auxiliary family of systems of equations ($0 \le \xi \le 1$) with a small parameter $\varepsilon > 0$:

$$\varepsilon\frac{\partial\Delta^2 v}{\partial t} + \frac{\partial v}{\partial t} + \xi\sum_{i=1}^{n}(\Delta_\alpha^{-1}v)_i\frac{\partial v}{\partial x_i} - \mu_0\Delta v$$

$$-\frac{\mu_1\xi}{\Gamma(1-\beta)}\mathrm{Div}\int_0^t (t-s)^{-\beta}\mathcal{E}(v)\big(s,z(s;t,x)\big)\,ds + \nabla p = \xi f, \tag{10}$$

$$z(\tau;t,x) = x + \int_t^\tau v\big(s,z(s;t,x)\big)\,ds, \quad t,\tau\in[0,T], \quad x\in\Omega, \tag{11}$$

$$\mathrm{div}\,v = 0, \quad t\in[0,T], \quad x\in\Omega, \tag{12}$$

$$v|_{\partial\Omega} = 0, \quad \Delta v|_{\partial\Omega} = 0, \quad t\in[0,T] \tag{13}$$

$$v|_{t=0} = v_0, \quad x\in\Omega. \tag{14}$$

For this family we consider another functional space:

$$W_2 = \{v\in C([0,T];V^3), \quad v'\in L_2(0,T;V^3)\}$$

with the norm $\|v\|_{W_2} = \|v\|_{C(0,T;V^3)} + \|v'\|_{L_2(0,T;V^3)}$.

Equation (10) includes the integral calculated along the trajectories of motion of the fluid particles. As was noted in the previous section, it is necessary that the trajectories are uniquely determined by the velocity field $v(t,x)$. In other words, Equation (11) must have a unique solution for the velocity field $v(t,x)$. Note that for the family of auxiliary problems (10)–(14), the velocity v from the space W_2 has sufficient smoothness (due to the embedding of the space V^3 in $C^1(\overline{\Omega})$ for $n = 2,3$). Thus, it follows from Reference [29] that the Cauchy problem (11) is non-locally uniquely solvable.

Analogously with the definition of a weak solution for feedback control problem (1)–(5), (8), we formulate the definition of a weak solution to auxiliary problem (10)–(14), (8) for fixed $0\le\xi\le 1$.

Definition 3. *A pair of functions $(v,f)\in W_2\times L_2(0,T;V^{-1})$ is called a weak solution to auxiliary problem (10)–(14), (8) if it satisfies for any $\varphi\in V^1$ and almost all $t\in(0,T)$ the equality*

$$\langle v',\varphi\rangle - \xi\int_\Omega\sum_{i,j=1}^{n}(\Delta_\alpha^{-1}v)_i v_j\frac{\partial\varphi_j}{\partial x_i}\,dx + \mu_0\int_\Omega\nabla v:\nabla\varphi\,dx - \varepsilon\int_\Omega\nabla\Delta v':\nabla\varphi\,dx$$

$$+\frac{\mu_1\xi}{\Gamma(1-\beta)}\Big(\int_0^t (t-s)^{-\beta}\mathcal{E}(v)(s,z(s;t,x))\,ds, \mathcal{E}(\varphi)\Big) = \xi\langle f,\varphi\rangle, \tag{15}$$

feedback condition (8) and initial condition (14). Here z is the trajectory associated to the velocity v.

To prove the existence of a weak solution to auxiliary problem (10)–(14),(8) for $\xi = 1$, we rewrite the auxiliary family in operator form. Using the terms in equality (15), we introduce the operators using the following equalities:

$$J : V^3 \to V^{-1}, \quad \langle Jv, \varphi \rangle = \int_\Omega v\varphi\, dx, \quad v \in V^3, \quad \varphi \in V^1;$$

$$A : V^1 \to V^{-1}, \quad \langle Av, \varphi \rangle = \int_\Omega \nabla v : \nabla\varphi\, dx, \quad v \in V^1, \quad \varphi \in V^1;$$

$$A_2 : V^3 \to V^{-1}, \quad \langle A_2 v, \varphi \rangle = -\int_\Omega \nabla\Delta v : \nabla\varphi\, dx, \quad v \in V^3, \quad \varphi \in V^1;$$

$$B : L_4(\Omega) \to V^{-1}, \quad \langle B(v), \varphi \rangle = \int_\Omega \sum_{i,j=1}^{n} (\Delta_\alpha^{-1} v)_i v_j \frac{\partial \varphi_j}{\partial x_i}\, dx, \quad v \in L_4(\Omega), \quad \varphi \in V^1;$$

$$C : V^1 \times CG \to V^{-1}, \quad (C(v,z)(t), \varphi) = \Big(\int_0^t (t-s)^{-\beta} \mathcal{E}(v)(s, z(s;t,x))\, ds, \mathcal{E}(\varphi) \Big),$$
$$v \in V^1, z \in CG, \varphi \in V^1, \text{ for almost all } t \in (0,T).$$

Since the function $\varphi \in V^1$ is arbitrary in (15), for almost all $t \in (0,T)$ this equality is equivalent to the following operator equation in $L_2(0,T;V^{-1})$:

$$Jv' + \varepsilon A_2 v' + \mu_0 Av - \xi B(v) + \frac{\mu_1 \xi}{\Gamma(1-\beta)} C(v,z) = \xi f.$$

Thus, a weak solution to auxiliary problem (10)–(14), (8) for a fixed $0 \le \xi \le 1$ is a solution $v \in W_2$ of the following operator inclusion

$$Jv' + \varepsilon A_2 v' + \mu_0 Av - \xi B(v) + \frac{\mu_1 \xi}{\Gamma(1-\beta)} C(v,z) = \xi f \in \Psi(v), \tag{16}$$

satisfying initial condition (14).

We also define the operators using the following equalities:

$$L : W_2 \to L_2(0,T;V^{-1}) \times V^3, \quad L(v) = ((J + \varepsilon A_2)v' + \mu_0 Av, v|_{t=0});$$

$$K : W_2 \to L_2(0,T;V^{-1}) \times V^3, \quad K(v) = (B(v), 0);$$

$$G : W_2 \to L_2(0,T;V^{-1}) \times V^3, \quad G(v) = (\frac{\mu_1}{\Gamma(1-\beta)} C(v,z), 0);$$

$$\mathcal{Y} : W_2 \to L_2(0,T;V^{-1}) \times V^3, \quad \mathcal{Y}(v) = (\Psi(v), v_0);$$

$$\mathcal{M} : W_2 \to W_2, \quad \mathcal{M}(v) = L^{-1}(\mathcal{Y}(v) + K(v) - G(v)).$$

Thus, from our problem of finding a solution to operator inclusion (16) for a fixed $0 \le \xi \le 1$ satisfying initial condition (14) we get the problem of finding a solution for a fixed $0 \le \xi \le 1$ to the following operator inclusion

$$v \in \xi\mathcal{M}(v) = \xi L^{-1}(\mathcal{Y}(v) + K(v) - G(v)). \tag{17}$$

We need the following properties of the operators from inclusions (16) and (17). In order to not to pile up the notation, we will use the same letter to denote the same operators acting in different function spaces.

Lemma 1. *1. For any function $v \in C([0,T];V^3)$ it holds that the function $Av \in L_2(0,T;V^{-1})$ and the operator $A : C([0,T];V^3) \to L_2(0,T;V^{-1})$ is continuous and the following estimates hold:*

$$\|Av\|_{V^{-1}} \le \|v\|_{V^1}; \quad \|Av\|_{L_2(0,T;V^{-1})} \le \|v\|_{L_2(0,T;V^1)}; \tag{18}$$

$$\|Av\|_{L_2(0,T;V^{-1})} \leq C_1 \|v\|_{C([0,T];V^3)}. \tag{19}$$

2. *The operator $A_2 : V^3 \to V^{-1}$ is linear, continuous, invertible and the following estimate holds:*

$$\|A_2 v\|_{V^{-1}} \leq \|v\|_{V^3}. \tag{20}$$

In addition, the operator $A_2^{-1} : V^{-1} \to V^3$ is also continuous.

3. *For any function $v \in L_p(0,T;V^3)$, $1 \leq p < \infty$ the function $(J + \varepsilon A_2)v$ belongs to $L_p(0,T;V^{-1})$ and the operator $(J + \varepsilon A_2) : L_p(0,T;V^3) \to L_p(0,T;V^{-1})$ is continuous and invertible. In addition, the following estimate holds:*

$$\varepsilon \|v\|_{L_p(0,T;V^3)} \leq \|(J + \varepsilon A_2)v\|_{L_p(0,T;V^{-1})} \leq C_2(1+\varepsilon)\|v\|_{L_p(0,T;V^3)}. \tag{21}$$

Moreover, the inverse to it operator $(J + \varepsilon A_2)^{-1} : L_p(0,T;V^{-1}) \to L_p(0,T;V^3)$is continuous and for any $w \in L_p(0,T;V^{-1})$ we have the estimate

$$\|(J + \varepsilon A_2)^{-1} w\|_{L_p(0,T;V^3)} \leq \frac{1}{\varepsilon}\|w\|_{L_p(0,T;V^{-1})}. \tag{22}$$

4. *The operator $L : W_2 \to L_2(0,T;V^{-1}) \times V^3$ is invertible and the operator $L^{-1} : L_2(0,T;V^{-1}) \times V^3 \to W_2$ is a continuous operator.*

Proof. The proof is carried out in the same way as in Reference [36]. □

Lemma 2. 1. *The map $B : L_4(\Omega) \to V^{-1}$ is continuous and the following estimate holds:*

$$\|B(v)\|_{V^{-1}} \leq C_3 \|v\|^2_{L_4(\Omega)}. \tag{23}$$

2. *For any $v \in L_4(0,T;L_4(\Omega))$ the function $B(v) \in L_2(0,T;V^{-1})$ and the map $B : L_4(0,T;L_4(\Omega)) \to L_2(0,T;V^{-1})$ is continuous.*
3. *For any function $v \in W_2$ the function $B(v) \in L_2(0,T;V^{-1})$ and the map $B : W_2 \to L_2(0,T;V^{-1})$ is compact.*

Proof. 1. For any $v \in L_4(\Omega)$, $\varphi \in V^1$ using Holder's inequality, we obtain

$$|\langle B(v), \varphi\rangle| = \left|\sum_{i,j=1}^{n} \int_\Omega (\Delta_\alpha^{-1} v)_i v_j \frac{\partial \varphi_j}{\partial x_i}\, dx\right| \leq \sum_{i,j=1}^{n} \left(\int_\Omega |(\Delta_\alpha^{-1} v)_i v_j|^2\, dx\right)^{\frac{1}{2}}$$
$$\times \left(\int_\Omega \left|\frac{\partial \varphi_j}{\partial x_i}\right|^2 dx\right)^{\frac{1}{2}} \leq \sum_{i,j=1}^{n} \left(\int_\Omega |(\Delta_\alpha^{-1} v)_i|^4\, dx\right)^{\frac{1}{4}} \left(\int_\Omega |v_j|^4\, dx\right)^{\frac{1}{4}} \|\varphi\|_{V^1}$$
$$\leq C_4 \|\Delta_\alpha^{-1} v\|_{L_4(\Omega)} \|v\|_{L_4(\Omega)} \|\varphi\|_{V^1} \leq C_4 C_5 \|v\|^2_{L_4(\Omega)} \|\varphi\|_{V^1} = C_6 \|v\|^2_{L_4(\Omega)} \|\varphi\|_{V^1}.$$

This implies inequality (23). Note that here we used the following well-known estimate [37,38]:

$$\|\Delta_\alpha^{-1} v\|_{L_p(\Omega)} = \|(I - \alpha^2 \Delta)^{-1} v\|_{L_p(\Omega)} \leq C_5 \|v\|_{L_p(\Omega)}, \quad p > 1. \tag{24}$$

We show the continuity of the map $B : L_4(\Omega) \to V^{-1}$. For arbitrary $v^m, v^0 \in L_4(\Omega)$ we have

$$|\langle B(v^m), \varphi\rangle - \langle B(v^0), \varphi\rangle| = \left| \int_\Omega \sum_{i,j=1}^{n} (\Delta_\alpha^{-1} v^m)_i v_j^m \frac{\partial \varphi_j}{\partial x_i} \, dx - \int_\Omega \sum_{i,j=1}^{n} (\Delta_\alpha^{-1} v^0)_i v_j^0 \frac{\partial \varphi_j}{\partial x_i} \, dx \right|$$
$$\leq \sum_{i,j=1}^{n} \|(\Delta_\alpha^{-1} v^m)_i v_j^m - (\Delta_\alpha^{-1} v^0)_i v_j^0\|_{L_2(\Omega)} \left\| \frac{\partial \varphi_j}{\partial x_i} \right\|_{L_2(\Omega)}$$
$$\leq \|\varphi\|_{V^1} \sum_{i,j=1}^{n} \|(\Delta_\alpha^{-1} v^m)_i v_j^m - (\Delta_\alpha^{-1} v^0)_i v_j^0\|_{L_2(\Omega)}$$
$$= \|\varphi\|_{V^1} \Big(\sum_{i,j=1}^{n} \|(\Delta_\alpha^{-1} v^m)_i v_j^m - (\Delta_\alpha^{-1} v^m)_i v_j^0 + (\Delta_\alpha^{-1} v^m)_i v_j^0 - (\Delta_\alpha^{-1} v^0)_i v_j^0\|_{L_2(\Omega)} \Big)$$
$$\leq \|\varphi\|_{V^1} \Big(\sum_{i,j=1}^{n} \|(\Delta_\alpha^{-1} v^m)_i v_j^m - (\Delta_\alpha^{-1} v^m)_i v_j^0\|_{L_2(\Omega)} + \sum_{i,j=1}^{n} \|(\Delta_\alpha^{-1} v^m)_i v_j^0 - (\Delta_\alpha^{-1} v^0)_i v_j^0\|_{L_2(\Omega)} \Big)$$
$$\leq C_7 \|\varphi\|_{V^1} \Big(\sum_{j=1}^{n} \|\Delta_\alpha^{-1} v^m\|_{L_4(\Omega)} \|v_j^m - v_j^0\|_{L_4(\Omega)} + \sum_{j=1}^{n} \|\Delta_\alpha^{-1}(v^m - v^0)\|_{L_4(\Omega)} \|v_j^0\|_{L_4(\Omega)} \Big)$$
$$\leq C_7 C_5 \|\varphi\|_{V^1} \Big(\sum_{j=1}^{n} \|v^m\|_{L_4(\Omega)} \|v_j^m - v_j^0\|_{L_4(\Omega)} + \sum_{j=1}^{n} \|v^m - v^0\|_{L_4(\Omega)} \|v_j^0\|_{L_4(\Omega)} \Big)$$
$$\leq C_8 (\|v^m\|_{L_4(\Omega)} \|v^m - v^0\|_{L_4(\Omega)} + \|v^m - v^0\|_{L_4(\Omega)} \|v^0\|_{L_4(\Omega)}) \|\varphi\|_{V^1}$$
$$= C_8 (\|v^m\|_{L_4(\Omega)} + \|v^0\|_{L_4(\Omega)}) \|v^m - v^0\|_{L_4(\Omega)} \|\varphi\|_{V^1}.$$

Thereby

$$\|B(v^m) - B(v^0)\|_{V^{-1}} \leq C_8 (\|v^m\|_{L_4(\Omega)} + \|v^0\|_{L_4(\Omega)}) \|v^m - v^0\|_{L_4(\Omega)}.$$

Assuming that $v^m \to v^0$ in $L_4(\Omega)$, we obtain that the map $B : L_4(\Omega) \to V^{-1}$ is continuous.

2. To prove this item, it is necessary to use the last estimate and repeat the proof of Lemma 2.5.4 (item 2) from Reference [28].
3. To prove this item, we use the Aubin-Simon theorem:

Theorem 5. *[28,39,40] Let $X \subset E \subset Y$ are Banach spaces, the embedding $X \subset E$ is compact and the embedding $E \subset Y$ is continuous. Also let $F \subset L_p(0, T; X)$, $1 \leq p \leq \infty$. We assume that for any $f \in F$ its generalized derivative belongs to $L_r(0, T; Y)$, $1 \leq r \leq \infty$. Now let:*

- *F is bounded in $L_p(0, T; X)$;*
- *$\{f' : f \in F\}$ is bounded in $L_r(0, T; Y)$.*

Then for $p < \infty$ the set F is relatively compact in $L_p(0, T; E)$, and for $p = \infty$ and $r > 1$ the set F is relatively compact in $C([0, T]; E)$.

Consider the set $F = \{v \in L_4(0, T; V^3), v' \in L_2(0, T; L_2(\Omega))\}$. Since the embedding $V^3 \subset L_4(\Omega)$ is compact, the embedding $F \subset L_4(0, T; L_4(\Omega))$ is also compact.

From continuity of embeddings

$$C([0, T]; V^3) \subset L_4(0, T; V^3), \quad L_2(0, T; V^3) \subset L_2(0, T; L_2(\Omega))$$

the continuous embedding $W_2 \subset F$ follows. In addition, also we have that the operator $B : L_4(0, T; L_4(\Omega)) \to L_2(0, T; V^{-1})$ is continuous (from the second item of this lemma). Thus, we have the superposition of embeddings:

$$W_2 \subset F \subset L_4(0, T; L_4(\Omega)) \xrightarrow{B} L_2(0, T; V^{-1}),$$

where the first embedding is continuous, the second is compact, and the map B is continuous. Therefore, for any function $v \in W_2$ we obtain that the function $B(v) \in L_2(0,T;V^{-1})$, and the map $B : W_2 \to L_2(0,T;V^{-1})$ is compact. The proof is complete.
□

We proceed to investigate the properties of the map C. We introduce the following norm $\|v\|_{k,L_2(0,T;V^{-1})}$ equal to the norm $\|\overline{v}\|_{L_2(0,T;V^{-1})}$ where $\overline{v}(t) = e^{-kt}v(t)$, $k \geq 0$. Then the following lemma holds.

Lemma 3. *For any $v \in L_2(0,T;V^1)$, $z \in CG$ we have that $C(v,z) \in L_2(0,T;V^{-1})$ and the map $C : L_2(0,T;V^1) \times CG \to L_2(0,T;V^{-1})$ is continuous and bounded. In addition, for any fixed function $z \in CG$ and arbitrary u, $v \in L_2(0,T;V^1)$ the following estimate holds:*

$$\|C(v,z) - C(u,z)\|_{k,L_2(0,T;V^{-1})} \leq C_9 T^{1-\beta}\sqrt{T/k}\|v-u\|_{k,L_2(0,T;V^1)}. \tag{25}$$

Proof. The first part of this lemma is proved similarly to the Lemma 2.2 [30]. We prove necessary estimate (25). Let $\overline{v}(t) = e^{-kt}v(t), \overline{u}(t) = e^{-kt}u(t)$. By definition, for almost all $t \in [0,T]$ we have $\varphi \in L_2(0,T,V^1)$ and obtain

$$\langle e^{-kt}C(v,z)(t) - e^{-kt}C(u,z)(t), \varphi(t)\rangle$$
$$= \int_0^T \int_\Omega \int_0^t e^{-k(t-s)}(t-s)^{-\beta}\mathcal{E}_{ij}(\overline{v}-\overline{u})(s,z(s;t,x))\,ds\,\mathcal{E}_{ij}(\varphi)(t)\,dx\,dt.$$

Then, using the Holder inequality, we obtain

$$\langle e^{-kt}C(v,z)(t) - e^{-kt}C(u,z)(t), \varphi(t)\rangle \leq \int_0^T \int_0^t e^{-k(t-s)}(t-s)^{-\beta}\Big(\int_\Omega \mathcal{E}^2(\overline{v}-\overline{u})(s,z(s;t,x))\,dx\Big)^{1/2}$$
$$\times\Big(\int_\Omega \mathcal{E}^2(\varphi)(t,x)\,dx\Big)^{1/2} ds\,dt = \int_0^T \int_0^t e^{-k(t-s)}(t-s)^{-\beta}\|(\overline{v}-\overline{u})(s,\cdot)\|_{V^1}\|\varphi(t,\cdot)\|_{V^1}\,ds\,dt$$
$$\leq C_9 T^{1-\beta}\|\overline{v}-\overline{u}\|_{L_2(0,T;V^1)}\|\varphi\|_{L_2(0,T;V^1)}\Big(\int_0^T\int_0^t e^{-k(t-s)}\,ds\,dt\Big)^{1/2}.$$

The last inequality holds by virtue of the following estimate [41] (Theorem 2.6)

$$\|\int_0^t (t-s)^{-\beta}\varphi(s)\,ds\|_{L_p(0,T)} \leq C_9 T^{1-\beta}\|\varphi(s)\|_{L_p(0,T)},\ \varphi(s) \in L_p(0,T),\ 1 \leq p < \infty.$$

Estimate the remaining integral:

$$\Big(\int_0^T\int_0^t e^{-k(t-s)}\,ds\,dt\Big)^{1/2} = \frac{1}{k}\int_0^T 1 - e^{-kt}\,dt \leq \frac{1}{k}\int_0^T dt = \frac{T}{k}.$$

Thus, we obtain the estimate

$$\langle e^{-kt}C(v,z)(t) - e^{-kt}C(u,z)(t), \varphi(t)\rangle \leq C_9 T^{1-\beta}\sqrt{T/k}\|\overline{v}-\overline{u}\|_{L_2(0,T;V^1)}\|\varphi\|_{L_2(0,T;V^1)}.$$

From where necessary estimate (25) follows. □

We formulate one more necessary property of the operator C.

But first we define several concepts concerning the measure of noncompactness and L-condensing operators [30,42].

Definition 4. *A non-negative real function ψ defined on a subset of a Banach space F is called a measure of non-compactness if for any subset $\mathcal{M}$ of this space the following properties are satisfied:*

- $\psi(\overline{co}\,\mathcal{M}) = \psi(\mathcal{M})$;
- *for any two sets* $\mathcal{M}_1$ *and* $\mathcal{M}_2$ *from* $\mathcal{M}_1 \subset \mathcal{M}_2$ *follows that* $\psi(\mathcal{M}_1) \leq \psi(\mathcal{M}_2)$.

Here, by $\overline{co}\,\mathcal{M}$ we denote the convex closure of the set $\mathcal{M}$. As an example of a measure of non-compactness, we give the Kuratowski measure of non-compactness: the exact lower bound $d > 0$ for which the set $\mathcal{M}$ can be divided into a finite number of subsets whose diameters are less than d. Kuratowski's non-compactness measure has several important properties:

- $\psi(\mathcal{M}) = 0$, if $\mathcal{M}$ is a relatively compact subset;
- $\psi(\mathcal{M} \cup K) = \psi(\mathcal{M})$ if K is a relatively compact set.

Definition 5. *Let X be bounded subset of a Banach space, and $L : X \to F$ is a map from X into a Banach space F. A map $g : X \to F$ is called L-condensing if $\psi(g(\mathcal{M})) < \psi(L(\mathcal{M}))$ for any set $\mathcal{M} \subseteq X$ such that $\psi(g(\mathcal{M})) \neq 0$.*

Let γ_k be the Kuratowski measure of noncompactness in the space $L_2(0,T;V^{-1})$ with the norm $\|v\|_{k,L_2(0,T;V^{-1})}$. Then the following lemma holds.

Lemma 4. *The map $G : W_2 \to L_2(0,T;V^{-1}) \times V^3$ is L-condensing with respect to the Kuratowski measure of noncompactness γ_k.*

Proof. Let $M \subset W_2 \subset L_2(0,T;V^1)$ be an arbitrary bounded set. By virtue of Theorem 2, the set $z(M)$ is the set of trajectories z that are uniquely determined by the velocities $v \in M$ and this set is relatively compact. Then for any fixed $v \in W_2$ the set $C(v, z(M))$ is relatively compact. In addition, for any $z \in z(M)$, the map $C(\cdot, z)$ satisfies the Lipschitz condition with constant $C_9 T^{1-\beta}\sqrt{T/k}$ in the norms $\|\cdot\|_{k,L_2(0,T,V^1)}$ and $\|\cdot\|_{k,L_2(0,T,V^{-1})}$. Then, by Theorem 1.5.7 [42], the map $C(v,z)$ and, therefore, the map G are $C_9 T^{1-\beta}\sqrt{T/k}$-bounded with respect to the Hausdorff measure χ_k. It is known, see Theorem 1.1.7 [42], that the non-compactness measures of Hausdorff and Kuratowski satisfy the following inequalities $\chi_k(M) \leq \gamma_k(M) \leq 2\chi_k(M)$. Therefore, the estimate

$$\gamma_k(G(M)) \leq C_9 T^{1-\beta}\sqrt{T/k}\,\gamma_k(L(M))$$

hold. Choosing k so that $C_9 T^{1-\beta}\sqrt{T/k} < 1$, we obtain the statement of the lemma. □

Using the above estimates and the properties of the operators, we prove the following a priori estimates for auxiliary family (10)–(14), (8).

Lemma 5. *Let $v_0 \in V^3$. Then for any solution $v \in W_2$ of operator inclusion (16) the following estimates hold:*

$$\|v\|_{L_2(0,T;V^1)} \leq C_{10}(\|v_0\|_{V^0} + \sqrt{\varepsilon}\|v_0\|_{V^2}); \tag{26}$$

$$\|v\|_{C([0,T];V^0)} \leq C_{11}(\|v_0\|_{V^0} + \sqrt{\varepsilon}\|v_0\|_{V^2}); \tag{27}$$

$$\varepsilon\|v\|^2_{C([0,T];V^2)} \leq C_{12}(\|v_0\|^2_{V^0} + \varepsilon\|v_0\|^2_{V^2}), \tag{28}$$

where the constants C_{10}, C_{11}, C_{12} do not depend on ε and ξ.

Proof. Let $v \in W_2$ be a solution of operator inclusion (16). Then for any $\varphi \in V^1$ and almost all $t \in (0,T)$ equality (15) holds. Since it is valid for all $\varphi \in V^1$, we assume that $\varphi = \overline{v}$, where $\overline{v}(t) = e^{-kt}v(t)$. Then

$$\int_\Omega v'(t)\overline{v}(t)\,dx - \xi \int_\Omega \sum_{i,j=1}^n (\Delta_\alpha^{-1}v)_i(t)v_j(t)\frac{\partial \overline{v}_j(t)}{\partial x_i}\,dx + \mu_0 \int_\Omega \nabla v(t) : \nabla \overline{v}(t)\,dx$$
$$+ \frac{\mu_1 \xi}{\Gamma(1-\beta)}\Big(\int_0^t (t-s)^{-\beta}\mathcal{E}(v)(s,z(s;t,x))\,ds, \mathcal{E}(\overline{v}(t))\Big)$$
$$- \varepsilon \int_\Omega \nabla\Delta v'(t) : \nabla\overline{v}(t)\,dx = \xi\langle f(t), \overline{v}(t)\rangle. \quad (29)$$

Let us replace $v(t) = e^{kt}\overline{v}(t)$ and separately transform the terms in the left side of the last equality as follows. Consider the first term:

$$\int_\Omega v'(t)\overline{v}(t)\,dx = \int_\Omega (e^{kt}\overline{v}(t))'\overline{v}(t)\,dx = e^{kt}\int_\Omega \overline{v}'(t)\overline{v}(t)\,dx + ke^{kt}\int_\Omega \overline{v}(t)\overline{v}(t)\,dx$$
$$= \frac{e^{kt}}{2}\int_\Omega \frac{\partial(\overline{v}(t)\overline{v}(t))}{\partial t}\,dx + ke^{kt}\|\overline{v}(t)\|_{V^0}^2 = \frac{e^{kt}}{2}\frac{d}{dt}\|\overline{v}(t)\|_{V^0}^2 + ke^{kt}\|\overline{v}(t)\|_{V^0}^2. \quad (30)$$

Now we turn to the consideration of the following term:

$$\int_\Omega \sum_{i,j=1}^n (\Delta_\alpha^{-1}v)_i(t)v_j(t)\frac{\partial \overline{v}_j(t)}{\partial x_i}\,dx = e^{kt}\int_\Omega \sum_{i,j=1}^n (\Delta_\alpha^{-1}v)_i(t)\overline{v}_j(t)\frac{\partial \overline{v}_j(t)}{\partial x_i}\,dx$$
$$= \frac{e^{kt}}{2}\int_\Omega \sum_{i,j=1}^n (\Delta_\alpha^{-1}v)_i(t)\frac{\partial(\overline{v}_j(t)\overline{v}_j(t))}{\partial x_i}\,dx = -\frac{e^{kt}}{2}\int_\Omega \sum_{i,j=1}^n \frac{\partial(\Delta_\alpha^{-1}v)_i(t)}{\partial x_i}\overline{v}_j^2(t)\,dx$$
$$= -\frac{e^{kt}}{2}\int_\Omega \sum_{j=1}^n \operatorname{div} u(t)\overline{v}_j^2(t)\,dx = 0.$$

Finally, we transform the last term:

$$-\varepsilon\int_\Omega \nabla\Delta v'(t) : \nabla\overline{v}(t)\,dx = -\varepsilon\int_\Omega \nabla\Delta(e^{kt}\overline{v}(t))' : \nabla\overline{v}(t)\,dx - \varepsilon ke^{kt}\int_\Omega \nabla\Delta\overline{v}(t) : \nabla\overline{v}(t)\,dx$$
$$-\varepsilon e^{kt}\int_\Omega \nabla\Delta\overline{v}'(t) : \nabla\overline{v}(t)\,dx = \varepsilon ke^{kt}\int_\Omega \Delta\overline{v}(t)\Delta\overline{v}(t)\,dx + \frac{\varepsilon e^{kt}}{2}\int_\Omega \frac{\partial}{\partial t}\Big(\Delta\overline{v}(t)\Delta\overline{v}(t)\Big)\,dx$$
$$= \varepsilon ke^{kt}\|\overline{v}(t)\|_{V^2}^2 + \frac{\varepsilon e^{kt}}{2}\frac{d}{dt}\|\overline{v}(t)\|_{V^2}^2.$$

Thus, equality (29) can be rewritten as follows:

$$\frac{e^{kt}}{2}\frac{d}{dt}\|\overline{v}(t)\|_{V^0}^2 + ke^{kt}\|\overline{v}(t)\|_{V^0}^2 + \mu_0 e^{kt}\|\overline{v}(t)\|_{V^1}^2 + \varepsilon ke^{kt}\|\overline{v}(t)\|_{V^2}^2 + \frac{\varepsilon e^{kt}}{2}\frac{d}{dt}\|\overline{v}(t)\|_{V^2}^2$$
$$= -\frac{\mu_1\xi}{\Gamma(1-\beta)}\Big(\int_0^t (t-s)^{-\beta}\mathcal{E}(e^{kt}\overline{v})(s,z(s;t,x))\,ds, \mathcal{E}(\overline{v}(t))\Big) + \xi e^{kt}\langle f(t), \overline{v}(t)\rangle. \quad (31)$$

We estimate modulo the right-hand side of the resulting equality. Using the Cauchy inequality

$$bc \le \frac{\delta b^2}{2} + \frac{c^2}{2\delta}$$

for $\delta = 1/\mu_0$, we obtain:

$$\xi e^{kt}\langle f(t), \overline{v}(t)\rangle \le e^{kt}\|f(t)\|_{V^{-1}}\|\overline{v}(t)\|_{V^1} \le \frac{e^{kt}}{2\mu_0}\|f(t)\|_{V^{-1}}^2 + \frac{\mu_0 e^{kt}}{2}\|\overline{v}(t)\|_{V^1}^2.$$

Multiplying both sides of equality (31) on e^{-kt}, for almost all $t \in (0,T)$ we have

$$\frac{1}{2}\frac{d}{dt}\|\overline{v}(t)\|^2_{V^0} + \frac{\varepsilon}{2}\frac{d}{dt}\|\overline{v}(t)\|^2_{V^2} + k\|\overline{v}(t)\|^2_{V^0} + \frac{\mu_0}{2}\|\overline{v}(t)\|^2_{V^1} + \varepsilon k\|\overline{v}(t)\|^2_{V^2}$$

$$\leq \frac{\mu_1}{\Gamma(1-\beta)}\left|\left(e^{-kt}\int_0^t (t-s)^{-\beta}\mathcal{E}(e^{kt}\overline{v})(s,z(s;t,x))\,ds, \mathcal{E}(\overline{v}(t))\right)\right| + \frac{1}{2\mu_0}\|f(t)\|^2_{V^{-1}}.$$

We integrate the last inequality with respect to t from 0 to τ, where $\tau \in [0,T]$. Then

$$\frac{1}{2}\|\overline{v}(t)\|^2_{V^0} + \frac{\varepsilon}{2}\|\overline{v}(t)\|^2_{V^2} + k\int_0^\tau \|\overline{v}(t)\|^2_{V^0}\,dt + \varepsilon k\int_0^\tau \|\overline{v}(t)\|^2_{V^2}\,dt$$
$$+\frac{\mu_0}{2}\int_0^\tau \|\overline{v}(t)\|^2_{V^1}\,dt \leq \frac{1}{2}\|v_0\|^2_{V^0} + \frac{\varepsilon}{2}\|v_0\|^2_{V^2} + \frac{1}{2\mu_0}\int_0^\tau \|f(t)\|^2_{V^{-1}}\,dt$$
$$+\frac{\mu_1}{\Gamma(1-\beta)}\int_0^\tau \left|\left(e^{-kt}\int_0^t (t-s)^{-\beta}\mathcal{E}(e^{kt}\overline{v})(s,z(s;t,x))\,ds, \mathcal{E}(\overline{v}(t))\right)\right| dt.$$

We use estimate (25) for $u = 0$. In this way,

$$\frac{1}{2}\|\overline{v}(t)\|^2_{V^0} + \frac{\varepsilon}{2}\|\overline{v}(t)\|^2_{V^2} + k\int_0^\tau \|\overline{v}(t)\|^2_{V^0}\,dt + \varepsilon k\int_0^\tau \|\overline{v}(t)\|^2_{V^2}\,dt$$
$$+\frac{\mu_0}{2}\int_0^\tau \|\overline{v}(t)\|^2_{V^1}\,dt \leq \frac{1}{2}\|v_0\|^2_{V^0} + \frac{\varepsilon}{2}\|v_0\|^2_{V^2} + \frac{\mu_1 C_9 T^{1-\beta}\sqrt{T/(2k)}}{\Gamma(1-\beta)}\|\overline{v}\|^2_{L_2(0,T;V^1)} + \frac{1}{2\mu_0}\|f\|^2_{L_2(0,T;V^{-1})}.$$

We assume that the number k is sufficiently large such that $\dfrac{\mu_1 C_9 T^{1-\beta}\sqrt{T/(2k)}}{\Gamma(1-\beta)} \leq \mu_0/4$. The nonnegativity of the quantities $\|\overline{v}(t)\|^2_{V^0}$, $\|\overline{v}(t)\|^2_{V^2}$ and $\|\overline{v}(t)\|^2_{V^1}$ yields the following estimates:

$$\frac{\mu_0}{2}\int_0^\tau \|\overline{v}(t)\|^2_{V^1}\,dt \leq \frac{1}{2}\|v_0\|^2_{V^0} + \frac{\varepsilon}{2}\|v_0\|^2_{V^2} + \frac{1}{2\mu_0}\|f\|^2_{L_2(0,T;V^{-1})} + \frac{\mu_0}{4}\|\overline{v}\|^2_{L_2(0,T;V^1)},$$
$$\frac{\varepsilon}{2}\|\overline{v}(t)\|^2_{V^2} \leq \frac{1}{2}\|v_0\|^2_{V^0} + \frac{\varepsilon}{2}\|v_0\|^2_{V^2} + \frac{1}{2\mu_0}\|f\|^2_{L_2(0,T;V^{-1})} + \frac{\mu_0}{4}\|\overline{v}\|^2_{L_2(0,T;V^1)},$$
$$\frac{1}{2}\|\overline{v}(t)\|^2_{V^0} \leq \frac{1}{2}\|v_0\|^2_{V^0} + \frac{\varepsilon}{2}\|v_0\|^2_{V^2} + \frac{1}{2\mu_0}\|f\|^2_{L_2(0,T;V^{-1})} + \frac{\mu_0}{4}\|\overline{v}\|^2_{L_2(0,T;V^1)}.$$

Since the right-hand side in all the above inequalities does not depend on τ, we pass to the maximum in $\tau \in [0,T]$ in the left-hand sides of these inequalities. Then

$$\frac{\mu_0}{2}\|\overline{v}\|^2_{L_2(0,T;V^1)} \leq \frac{1}{2}\|v_0\|^2_{V^0} + \frac{\varepsilon}{2}\|v_0\|^2_{V^2} + \frac{1}{2\mu_0}\|f\|^2_{L_2(0,T;V^{-1})} + \frac{\mu_0}{4}\|\overline{v}\|^2_{L_2(0,T;V^1)},$$
$$\frac{\varepsilon}{2}\|\overline{v}\|^2_{C([0,T];V^2)} \leq \frac{1}{2}\|v_0\|^2_{V^0} + \frac{\varepsilon}{2}\|v_0\|^2_{V^2} + \frac{1}{2\mu_0}\|f\|^2_{L_2(0,T;V^{-1})} + \frac{\mu_0}{4}\|\overline{v}\|^2_{L_2(0,T;V^1)},$$
$$\frac{1}{2}\|\overline{v}\|^2_{C([0,T];V^0)} \leq \frac{1}{2}\|v_0\|^2_{V^0} + \frac{\varepsilon}{2}\|v_0\|^2_{V^2} + \frac{1}{2\mu_0}\|f\|^2_{L_2(0,T;V^{-1})} + \frac{\mu_0}{4}\|\overline{v}\|^2_{L_2(0,T;V^1)}.$$

From this and feedback condition (8) the required estimates (26)-(28) directly follow. The proof is complete. □

Lemma 6. *Let $v_0 \in V^3$. For any solution $v \in W_2$ for operator inclusion (16) we have the following estimates*

$$\varepsilon\|v'\|_{L_2(0,T;V^3)} \leq C_{13}\Big(1+\frac{1}{\varepsilon}\Big)\|v_0\|^2_{V^0} + C_{13}\sqrt{\varepsilon}\|v_0\|_{V^2} + C_{13}\|v_0\|^2_{V^2}; \tag{32}$$

$$\|v\|_{C([0,T];V^3)} \leq \|v_0\|_{V^3} + \frac{C_{13}T^{\frac{1}{2}}}{\varepsilon}\Big(1+\frac{1}{\varepsilon}\Big)\|v_0\|^2_{V^0} + \frac{C_{13}T^{\frac{1}{2}}}{\sqrt{\varepsilon}}\|v_0\|_{V^2} + \frac{C_{13}T^{\frac{1}{2}}}{\varepsilon}\|v_0\|^2_{V^2}; \tag{33}$$

$$\|v'\|_{L_{4/3}(0,T;V^{-1})} \leq C_{14}(\|v_0\|^2_{V^0} + \varepsilon\|v_0\|^2_{V^2} + 1); \tag{34}$$

$$\varepsilon\|v'\|_{L_{4/3}(0,T;V^3)} \leq C_{15}(\|v_0\|^2_{V^0} + \varepsilon\|v_0\|^2_{V^2} + 1); \tag{35}$$

where the constants C_{13}, C_{14}, C_{15} do not depend on ε, v and ξ.

Proof. Let $v \in W_2$ be a solution of (16). Then it satisfies the following operator equality

$$Jv' + \varepsilon A_2 v' + \mu_0 Av - \xi B(v) + \frac{\mu_1\xi}{\Gamma(1-\beta)}C(v,z) = \xi f. \tag{36}$$

Hence,

$$\|(J+\varepsilon A_2)v'\|_{L_2(0,T;V^{-1})} = \|\xi f + \xi B(v) - \mu_0 Av - \frac{\mu_1\xi}{\Gamma(1-\beta)}C(v,z)\|_{L_2(0,T;V^{-1})}.$$

We estimate the right-hand side of the last equality. By estimates (18) and (25) for $u=0$, we get:

$$\|\xi f + \xi B(v) - \mu_0 Av - \frac{\mu_1\xi}{\Gamma(1-\beta)}C(v,z)\|_{L_2(0,T;V^{-1})}$$
$$\leq \|f\|_{L_2(0,T;V^{-1})} + \|B(v)\|_{L_2(0,T;V^{-1})} + \frac{\mu_1 C_9 T^{1-\beta}}{\Gamma(1-\beta)}\|v\|_{L_2(0,T;V^1)} + \mu_0\|v\|_{L_2(0,T;V^1)}. \tag{37}$$

We separately estimate the $\|B(v)\|_{L_2(0,T;V^{-1})}$. Using (23), and the continuity of the embedding $V^2 \subset L_4(\Omega)$, we have:

$$\|B(v)\|_{L_2(0,T;V^{-1})} = \Big(\int_0^T \|B(v)\|^2_{V^{-1}}\,dt\Big)^{\frac{1}{2}} \leq C_3\Big(\int_0^T \|v(t)\|^4_{L_4(\Omega)}\,dt\Big)^{\frac{1}{2}}$$
$$\leq C_{16}\Big(\int_0^T \|v(t)\|^4_{V^2}\,dt\Big)^{\frac{1}{2}} \leq C_{16}T^{\frac{1}{2}}\max_{t\in[0,T]}\|v(t)\|^2_{V^2} = C_{16}T^{\frac{1}{2}}\|v\|^2_{C([0,T];V^2)}.$$

We rewrite inequality (37) as follows

$$\|\xi f + \xi B(v) - \mu_0 Av - \frac{\mu_1\xi}{\Gamma(1-\beta)}C(v,z)\|_{L_2(0,T;V^{-1})}$$
$$\leq C_{17}(\|f\|_{L_2(0,T;V^{-1})} + C_{16}T^{1/2}\|v\|^2_{C([0,T];V^2)} + \|v\|_{L_2(0,T;V^1)}).$$

From the a priori estimates (26) and (28) it immediately follows that

$$\|(J+\varepsilon A_2)v'\|_{L_2(0,T;V^{-1})} \leq C_{13}\Big(1+\frac{1}{\varepsilon}\Big)\|v_0\|^2_{V^0} + C_{13}\sqrt{\varepsilon}\|v_0\|_{V^2} + C_{13}\|v_0\|^2_{V^2}.$$

To prove estimate (32), it remains to use the left (21) for $p=2$:

$$\varepsilon\|v'\|_{L_2(0,T;V^3)} \leq \|(J+\varepsilon A_2)v'\|_{L_2(0,T;V^{-1})}$$
$$\leq C_{13}\Big(1+\frac{1}{\varepsilon}\Big)\Big(\|v_0\|^2_{V^0} + \|f\|^2_{L_2(0,T;V^{-1})}\Big) + C_{13}\sqrt{\varepsilon}\|v_0\|_{V^2} + C_{13}\|v_0\|^2_{V^2}.$$

Hence, inequality (32) is established.

We pass to the proof of estimate (33). Represent the function $v \in W_2$ as follows:

$$v(t) = \int_0^t v'(s)ds + v_0.$$

Then

$$\|v(t)\|_{V^3} \leq \left\| v_0 + \int_0^t v'(s)\,ds \right\|_{V^3} \leq \|v_0\|_{V^3} + \int_0^t \|v'(s)\|_{V^3}\,ds \leq \|v_0\|_{V^3} + T^{\frac{1}{2}} \|v'\|_{L_2(0,T;V^3)}.$$

Since the right-hand side of the resulting inequality does not depend on t, we pass to the maximum in $\tau \in [0,T]$ in the left-hand side. Then, taking into account estimate (32), we obtain

$$\max_{t\in[0,T]} \|v(t)\|_{V^3} \leq \|v_0\|_{V^3} + \frac{C_{13}T^{\frac{1}{2}}}{\varepsilon}\Big(1+\frac{1}{\varepsilon}\Big)\|v_0\|^2_{V^0} + \frac{C_{13}T^{\frac{1}{2}}}{\sqrt{\varepsilon}}\|v_0\|_{V^2} + \frac{C_{13}T^{\frac{1}{2}}}{\varepsilon}\|v_0\|^2_{V^2}.$$

Thus, we received estimate (33).

Now we prove inequality (34). As before, $v \in W_2$ is a solution of operator Equation (36). Then

$$\begin{aligned}\|v'\|_{L_{4/3}(0,T;V^{-1})} &\leq \|\xi f + \xi B(v) - \mu_0 Av - \frac{\mu_1 \xi}{\Gamma(1-\beta)} C(v,z) - \varepsilon A^2 v'\|_{L_{4/3}(0,T;V^{-1})} \\ &\leq \|f\|_{L_{4/3}(0,T;V^{-1})} + \|B(v)\|_{L_{4/3}(0,T;V^{-1})} + \mu_0\|Av\|_{L_{4/3}(0,T;V^{-1})} \\ &\quad + \frac{\mu_1}{\Gamma(1-\beta)}\|C(v,z)\|_{L_{4/3}(0,T;V^{-1})} + \varepsilon\|A^2v'\|_{L_{4/3}(0,T;V^{-1})}. \end{aligned} \quad (38)$$

We separately consider the terms on the right-hand side of the last inequality. First, we estimate $\|B(v)\|_{L_{4/3}(0,T;V^{-1})}$. Given from Reference [34] the well-known inequality for $n = 3$

$$\|u\|_{L_4(\Omega)} \leq 2^{\frac{1}{2}}\|u\|^{\frac{1}{4}}_{L_2(\Omega)}\|\nabla u\|^{\frac{3}{4}}_{L_2(\Omega)}, \quad u \in V^1,$$

and estimate (23), we obtain (for the case $n = 2$ the proof is similar):

$$\begin{aligned}\|B(v)\|_{L_{4/3}(0,T;V^{-1})} &= \Big(\int_0^T \|B(v)\|^{\frac{4}{3}}_{V^{-1}}\,dt\Big)^{\frac{3}{4}} \leq C_3\Big(\int_0^T \|v\|^{\frac{8}{3}}_{L_4(\Omega)}\,dt\Big)^{\frac{3}{4}} \\ &\leq 2C_3\Big(\int_0^T \|v\|^{\frac{2}{3}}_{L_2(\Omega)}\|\nabla v\|^2_{L_2(\Omega)}\,dt\Big)^{\frac{3}{4}} \leq C_{18}\Big(\int_0^T \|v\|^{\frac{2}{3}}_{V^0}\|v\|^2_{V^1}\,dt\Big)^{\frac{3}{4}} \\ &\leq C_{18}\|v\|^{\frac{1}{2}}_{C([0,T];V^0)}\Big(\int_0^T \|v\|^2_{V^1}\,dt\Big)^{\frac{3}{4}} = C_{18}\|v\|^{\frac{1}{2}}_{C([0,T];V^0)}\|v\|^{\frac{3}{2}}_{L_2(0,T;V^1)}. \end{aligned} \quad (39)$$

Consider the following term. We use the Holder inequality and estimate (18). Then

$$\begin{aligned}\|Av\|_{L_{4/3}(0,T;V^{-1})} &= \Big(\int_0^T \|Av\|^{\frac{4}{3}}_{V^{-1}}\,dt\Big)^{\frac{3}{4}} \leq \Big(\int_0^T \|v\|^{\frac{4}{3}}_{V^1}\,dt\Big)^{\frac{3}{4}} \\ &\leq T^{\frac{1}{4}}\Big(\int_0^T \|v\|^2_{V^1}\,dt\Big)^{\frac{1}{2}} = T^{\frac{1}{4}}\|v\|_{L_2(0,T;V^1)}. \end{aligned} \quad (40)$$

Similarly, using the Holder inequality and estimate (25) for $u = 0$, we obtain an estimate for the next term:

$$\begin{aligned}\|C(v,z)\|_{L_{4/3}(0,T;V^{-1})} &= \Big(\int_0^T \|C(v,z)\|^{\frac{4}{3}}_{V^{-1}}\,dt\Big)^{\frac{3}{4}} \leq T^{\frac{1}{4}}\Big(\int_0^T \|C(v,z)\|^2_{V^{-1}}\,dt\Big)^{\frac{1}{2}} \\ &= T^{\frac{1}{4}}\|C(v,z)\|_{L_2(0,T;V^{-1})} \leq T^{\frac{1}{4}}T^{1-\beta}C_9\|v\|_{L_2(0,T;V^1)}. \end{aligned}$$

Finally, we consider the last term. Using inequality (20), we get:

$$\varepsilon\|A^2v'\|_{L_{4/3}(0,T;V^{-1})} = \varepsilon\Big(\int_0^T \|A^2v'\|_{V^{-1}}^{\frac{4}{3}}\,dt\Big)^{\frac{3}{4}} \leq \varepsilon\Big(\int_0^T \|v'\|_{V^3}^{\frac{4}{3}}\,dt\Big)^{\frac{3}{4}} \leq \varepsilon\|v'\|_{L_{4/3}(0,T;V^3)}.$$

Let us estimate the right-hand side of the last inequality. We use the left side of estimate (22) for $p = 4/3$. Thus, to obtain an estimate of $\varepsilon\|v'\|_{L_{4/3}(0,T;V^3)}$, it is necessary to obtain an estimate of $\|(J + \varepsilon A_2)v'\|_{L_{4/3}(0,T;V^{-1})}$. To do this, we again use operator Equation (36). From its appearance, it follows that

$$\varepsilon\|v'\|_{L_{4/3}(0,T;V^3)} \leq \|f\|_{L_{4/3}(0,T;V^{-1})} + \|B(v)\|_{L_{4/3}(0,T;V^{-1})} - \mu_0\|Av\|_{L_{4/3}(0,T;V^{-1})} + \mu_1\|C(v,z)\|_{L_{4/3}(0,T;V^{-1})}.$$

Thus,

$$\begin{aligned}&\varepsilon\|A^2v'\|_{L_{4/3}(0,T;V^{-1})} \leq \varepsilon\|v'\|_{L_{4/3}(0,T;V^3)}\\ &\leq \|f\|_{L_{4/3}(0,T;V^{-1})} + \|B(v)\|_{L_{4/3}(0,T;V^{-1})} + \mu_0\|Av\|_{L_{4/3}(0,T;V^{-1})} + \frac{\mu_1}{\Gamma(1-\beta)}\|C(v,z)\|_{L_{4/3}(0,T;V^{-1})}.\end{aligned} \tag{41}$$

From (38), estimates (39)–(41) and a priori estimates (26) and (27), we get

$$\begin{aligned}&\|v'\|_{L_{4/3}(0,T;V^{-1})} \leq 2(\|f\|_{L_{4/3}(0,T;V^{-1})} + \|B(v)\|_{L_{4/3}(0,T;V^{-1})} + \mu_0\|Av\|_{L_{4/3}(0,T;V^{-1})}\\ &+\frac{\mu_1}{\Gamma(1-\beta)}\|C(v,z)\|_{L_{4/3}(0,T;V^{-1})}) \leq C_{19}(\|f\|_{L_2(0,T;V^{-1})} + \|v\|_{L_2(0,T;V^1)}\\ &+\|v\|^{\frac{1}{2}}_{C([0,T];V^0)}\|v\|^{\frac{3}{2}}_{L_2(0,T;V^1)}) \leq C_{20}(\|f\|_{L_2(0,T;V^{-1})} + \|v_0\|_{V^0} + \sqrt{\varepsilon}\|v_0\|_{V^2}\\ &+(\|v_0\|_{V^0} + \sqrt{\varepsilon}\|v_0\|_{V^2} + \|f\|_{L_2(0,T;V^{-1})})^{\frac{1}{2}}(\|v_0\|_{V^0} + \sqrt{\varepsilon}\|v_0\|_{V^2} + \|f\|_{L_2(0,T;V^{-1})}))^{\frac{3}{2}}\\ &\leq C_{21}(\|v_0\|_{V^0} + \sqrt{\varepsilon}\|v_0\|_{V^2} + \|f\|_{L_2(0,T;V^{-1})} + 1)^2 \leq 4C_{21}(\|v_0\|^2_{V^0} + \varepsilon\|v_0\|^2_{V^2} + 1).\end{aligned}$$

This completes the proof of inequality (34), where $C_{14} = 4C_{21}$.

Finally, applying again estimates (39) and (40), for the right-hand side of (41), as well as a priori estimates (26) and (27), we obtain

$$\begin{aligned}&\varepsilon\|v'\|_{L_{4/3}(0,T;V^3)} \leq 2(\|f\|_{L_{4/3}(0,T;V^{-1})} + \|B(v)\|_{L_{4/3}(0,T;V^{-1})} + \mu_0\|Av\|_{L_{4/3}(0,T;V^{-1})}\\ &+\frac{\mu_1}{\Gamma(1-\beta)}\|C(v,z)\|_{L_{4/3}(0,T;V^{-1})}) \leq C_{22}(\|f\|_{L_2(0,T;V^{-1})} + \|v\|_{L_2(0,T;V^1)}\\ &+\|v\|^{\frac{1}{2}}_{C([0,T];V^0)}\|v\|^{\frac{3}{2}}_{L_2(0,T;V^1)}) \leq C_{23}(\|f\|_{L_2(0,T;V^{-1})} + \|v_0\|_{V^0} + \sqrt{\varepsilon}\|v_0\|_{V^2}\\ &+(\|v_0\|_{V^0} + \sqrt{\varepsilon}\|v_0\|_{V^2} + \|f\|_{L_2(0,T;V^{-1})})^{\frac{1}{2}}(\|v_0\|_{V^0} + \sqrt{\varepsilon}\|v_0\|_{V^2} + \|f\|_{L_2(0,T;V^{-1})})^{\frac{3}{2}})\\ &\leq C_{24}(\|v_0\|_{V^0} + \sqrt{\varepsilon}\|v_0\|_{V^2} + \|f\|_{L_2(0,T;V^{-1})} + 1)^2 \leq 4C_{24}(\|v_0\|^2_{V^0} + \varepsilon\|v_0\|^2_{V^2} + 1).\end{aligned}$$

Thus, inequality (35), where $C_{15} = 4C_{24}$ is established. The proof is complete. □

Lemma 7. *Let $v_0 \in V^3$. Then for any solution $v \in W_2$ of operator Equation (16) we have the following estimate:*

$$\|v\|_{W_2} \leq C_{25}, \tag{42}$$

where $C_{25} > 0$ is a constant that depends on ε.

Theorem 6. *Let $v_0 \in V^3$. Then there is at least one solution $v \in W_2$ of auxiliary problem (10)–(14), (8) for $\xi = 1$.*

Proof. To prove this theorem, we use the topological degree theory for multi-valued vector fields [2,43]. Consider operator inclusion (17). From Corollary 7 it follows that all solutions of inclusion (17) are in the ball $B_R \subset W_2$ of radius $R = C_{25} + 1$ centered at zero. By item 4 of Lemma 1 the operator $L : W_2 \to L_2(0,T;V^{-1}) \times V^3$ is invertible. Then there is no solution of the family of following inclusions

$$v \in \xi \mathcal{M}, \quad \text{where} \quad \xi \in [0,1],$$

on the boundary of the same ball B_R.

By virtue of item 4 of Lemma 1 the operator $L^{-1} : L_2(0,T;V^{-1}) \times V^3 \to W_2$ is continuous. By the Lemmas 2 and 4 the map $(\mathcal{Y}(v) + K(v) - G(v)) : W_2 \to L_2(0,T;V^{-1}) \times V^3$ is L - condensing with respect to the Kuratowski γ_k non-compactness measure. Therefore, the operator $\mathcal{M} : W_2 \to W_2$ is condensing with respect to the Kuratowski γ_k non-compactness measure.

Thus, the vector field $v - \xi\mathcal{M}(v)$ is non-degenerate on the boundary of the ball B_R, which means that the topological degree $\deg(I - \xi\mathcal{M}(v), B_R, 0)$ is defined for this vector field . By the properties of homotopy invariance and normalization of degree we obtain that

$$\deg(I - \mathcal{M}(v), B_R, 0) = \deg(I, B_R, 0) = 1.$$

The non-zero degree of the mapping ensures the existence of at least one solution $v \in W_2$ of inclusion (17) for $\xi = 1$, and therefore of auxiliary problem (8), (10)–(14) for $\xi = 1$. The theorem is proved. □

4. Proof of Theorem 3

We proceed directly to the proof of the solvability of feedback control problem (1)–(5), (8). To do this, we carry out the passage to the limit in auxiliary problem (10)–(14), (8) for $\xi = 1$. Since the space V^3 is dense in V^0, then for each $v_0^* \in V^0$ there exists a sequence $v_0^m \in V^3$ converging to v_0^* in V^0. If $v_0^* \equiv 0$, then we put $v_0^m \equiv 0$, $\varepsilon_m = 1/m$. If $\|v_0^*\|_{V^0} \neq 0$, then starting from some number we have $\|v_0^m\|_{V^2} \neq 0$. Then we put $\varepsilon_m = 1/(m\|v_0^m\|_{V^2}^2)$. Under our choice $\{\varepsilon_m\}$ resulting sequence converges to zero as $m \to \infty$. Moreover, $\varepsilon_m \|v_0^m\|_{V^2}^2 \leq 1$.

By Theorem 6, for each ε_m and v_0^m there exists a solution $v_m \in W_2 \subset W_1$ of auxiliary problem (10)–(14), (8) for $\xi = 1$. Thus, each solution v_m for all $\varphi \in V^1$ for almost all $t \in (0,T)$ satisfies the equality

$$\langle v_m', \varphi \rangle - \int_\Omega \sum_{i,j=1}^{n} (\Delta_\alpha^{-1} v_m)_i (v_m)_j \frac{\partial \varphi_j}{\partial x_i}\, dx + \mu_0 \int_\Omega \nabla v_m : \nabla \varphi \, dx$$

$$-\varepsilon_m \int_\Omega \nabla \Delta v_m' : \nabla \varphi \, dx + \frac{\mu_1}{\Gamma(1-\beta)} \Big(\int_0^t (t-s)^{-\beta} \mathcal{E}(v_m)(s, z_m(s;t,x))\, ds, \mathcal{E}(\varphi) \Big) = \langle f_m, \varphi \rangle, \tag{43}$$

and the initial condition

$$v_m|_{t=0} = v_0^m.$$

Since the sequence $\{v_0^m\}$ converges in V^0, it is bounded by the norm V^0. Hence,

$$\|v_0^m\|_{V^0}^2 + \varepsilon_m \|v_0^m\|_{V^2}^2 \leq C_{26}.$$

Thus, from estimates (26), (27), (34) and (35) we obtain that

$$\|v_m\|^2_{L_2(0,T;V^1)} \leq C_{27}, \tag{44}$$

$$\|v_m\|^2_{C([0,T];V^0)} \leq C_{28}, \tag{45}$$

$$\|v'_m\|_{L_{4/3}(0,T;V^{-1})} \leq C_{29}, \tag{46}$$

$$\varepsilon\|v'_m\|_{L_{4/3}(0,T;V^3)} \leq C_{30}, \tag{47}$$

where the constants $C_{27}, C_{28}, C_{29}, C_{30}$ do not depend on ε. Due to the continuity of the embedding $C([0,T];V^0) \subset L_\infty(0,T;V^0)$ and estimates (44)–(46), without loss of generality (if necessary, passing to a subsequence) we obtain that

$$v_m \to v_* \quad \text{weakly in} \quad L_2(0,T;V^1) \quad \text{as} \quad m \to \infty, \tag{48}$$

$$v_m \to v_* \quad \text{*-weakly in} \quad L_\infty(0,T;V^0) \quad \text{as} \quad m \to \infty, \tag{49}$$

$$v'_m \to v'_* \quad \text{weakly in} \quad L_{4/3}(0,T;V^{-1}) \quad \text{as} \quad m \to \infty, \tag{50}$$

and that the limit function v_* belongs to the space W_1.

Consider Cauchy problem (3) for the limit function v_*. Since $v_* \in W_1$, therefore v_* satisfies the conditions of Theorem 1. Therefore, in $[0,T] \times [0,T] \times \overline{\Omega}$ there exists a Lagrangian regular flow $z_*(\tau;t,x)$ associated to v_*. We denote by $z_m(\tau;t,x)$ the Lagrangian regular flow associated to v_m.

Lemma 8. *The sequence $z_m(\tau;t,x)$ converges to $z(\tau;t,x)$ with respect to the Lebesgue measure on the set $[0,T] \times \Omega$ in (τ,x) for $t \in [0,T]$.*

This lemma follows from the a priori estimate (42) and Theorem 2.

The proofs of the solvability of feedback control problem (8), (1)–(5) are divided into two parts. First, we prove the passage to the limit in auxiliary problem (8), (10)–(14) for $\xi = 1$ and a test function φ from V^1, which is sufficiently smooth, then for the arbitrary function $\varphi \in V^1$.

I part. Let the test function $\varphi \in V^1$ be smooth. We pass to the limit in each term of (43).

For $m \to \infty$, by the definition of weak convergence $v_m \to v^*$ in $L_2(0,T;V^1)$ we get

$$\mu_0 \int_\Omega \nabla v_m : \nabla\varphi\, dx \to \mu_0 \int_\Omega \nabla v_* : \nabla\varphi\, dx$$

for any $\varphi \in V^1$.

Due to weak convergence $v'_m \to v'_*$ in $L_{4/3}(0,T;V^{-1})$ as $m \to \infty$ we obtain that

$$\langle v'_m, \varphi\rangle \to \langle v'_*, \varphi\rangle$$

for any $\varphi \in V^1$.

Further, using estimate (47), without loss of generality and, if necessary, passing to a subsequence, we have that there exists a function $u \in L_{4/3}(0,T;V^3)$ such that

$$\varepsilon_m v'_m \to u \quad \text{weakly in} \quad L_{4/3}(0,T;V^3) \quad \text{as} \quad m \to \infty.$$

Then

$$\varepsilon_m\langle \nabla\Delta v'_m, \nabla\varphi\rangle \to \langle \nabla\Delta u, \nabla\varphi\rangle, \quad \text{as} \quad m \to \infty.$$

However, the sequence $\varepsilon_m v'_m$ converges to zero in the sense of distributions on $[0,T]$ with values in V^{-3}. Indeed, for any smooth scalar function ψ with compact support and for $\varphi \in V^3$, we obtain

$$\lim_{m\to\infty}\left|\varepsilon_m\int_0^T\int_\Omega \nabla\Delta v'_m : \nabla\varphi\, dx\psi(t)\, dt\right| = \lim_{m\to\infty}\varepsilon_m\left|\int_0^T\int_\Omega \Delta v'_m\Delta\varphi\, dx\psi(t)\, dt\right|$$
$$= \lim_{m\to\infty}\varepsilon_m\left|\int_0^T\int_\Omega \nabla v'_m : \nabla\Delta\varphi\, dx\psi(t)\, dt\right| = \lim_{m\to\infty}\varepsilon_m\lim_{m\to\infty}\left|\int_0^T\int_\Omega \nabla v'_m : \nabla\Delta\varphi\, dx\psi(t)\, dt\right|$$
$$= \lim_{m\to\infty}\varepsilon_m\lim_{m\to\infty}\left|\int_\Omega\left(\int_0^T \nabla v'_m\psi(t)\, dt\right) : \nabla\Delta\varphi\, dx\right|$$
$$= \lim_{m\to\infty}\varepsilon_m\lim_{m\to\infty}\left|\int_\Omega\left(\int_0^T \nabla v_m\frac{\partial\psi(t)}{\partial t}\, dt\right) : \nabla\Delta\varphi\, dx\right|$$
$$= \lim_{m\to\infty}\varepsilon_m\lim_{m\to\infty}\left|\int_0^T\int_\Omega \nabla v_m : \nabla\Delta\varphi\, dx\frac{\partial\psi(t)}{\partial t}\, dt\right|.$$

Since v_m weakly converges to v^* in $L_2(0,T;V^1)$ and, therefore, converges to v^* in the sense of distributions, then

$$\lim_{m\to\infty}\varepsilon_m\lim_{m\to\infty}\left|\int_0^T\int_\Omega \nabla v_m : \nabla\Delta\varphi\, dx\psi(t)\, dt\right| = \left|\int_0^T\int_\Omega \nabla v_* : \nabla\Delta\varphi\, dx\frac{\partial\psi(t)}{\partial t}\, dt\right|\lim_{m\to\infty}\varepsilon_m = 0.$$

Thus, due to the uniqueness of the weak limit

$$\varepsilon_m\langle\nabla\Delta v'_m, \nabla\varphi\rangle \to 0 \quad \text{as} \quad m\to\infty.$$

Since the embedding $V^1 \subset L_4(\Omega)$ is completely continuous, and the embedding $L_4(\Omega) \subset V^{-1}$ is continuous, by Theorem 5 it follows, that

$$F = \{v \in L_2(0,T;V^1), v' \in L_{4/3}(0,T;V^{-1})\} \subset L_2(0,T;L_4(\Omega)).$$

Then, taking into account estimates (45) and (46) we conclude that

$$v_m \to v_* \quad \text{strongly in} \quad L_2(0,T;L_4(\Omega)).$$

Since the operator $\Delta_\alpha^{-1} = (I-\alpha^2\Delta)^{-1} : L_2(0,T;V^1) \to L_2(0,T;V^3)$ is continuous, then

$$\int_\Omega\sum_{i,j=1}^n(\Delta_\alpha^{-1}v_m)_i(v_m)_j\frac{\partial\varphi_j}{\partial x_i}\, dx \to \int_\Omega\sum_{i,j=1}^n(\Delta_\alpha^{-1}v_*)_i(v_*)_j\frac{\partial\varphi_j}{\partial x_i}\, dx \quad \text{as} \quad m\to\infty,$$

where the first sequence $(\Delta_\alpha^{-1}v_m)_i$ weakly converges in $L_2(0,T;V^1)$, and the second $(v_m)_j$ strongly in $L_2(0,T;L_4(\Omega))$. Consequently, their product converges weakly to the product of limits.

Now show that

$$\frac{\mu_1}{\Gamma(1-\beta)}\left(\int_0^t(t-s)^{-\beta}\mathcal{E}(v_m)(s, z_m(s;t,x))\, ds, \mathcal{E}(\varphi)\right)$$
$$\to \frac{\mu_1}{\Gamma(1-\beta)}\left(\int_0^t(t-s)^{-\beta}\mathcal{E}(v_*)(s, z_*(s;t,x))\, ds, \mathcal{E}(\varphi)\right). \quad (51)$$

Consider the following difference

$$\frac{\mu_1}{\Gamma(1-\beta)}\Big(\int_0^t (t-s)^{-\beta}\mathcal{E}(v_m)(s, z_m(s;t,x))\,ds, \mathcal{E}(\varphi)\Big)$$
$$-\frac{\mu_1}{\Gamma(1-\beta)}\Big(\int_0^t (t-s)^{-\beta}\mathcal{E}(v_*)(s, z_*(s;t,x))\,ds, \mathcal{E}(\varphi)\Big)$$
$$=\frac{\mu_1}{\Gamma(1-\beta)}\Big(\int_0^t (t-s)^{-\beta}\int_\Omega \Big[\mathcal{E}(v_m)(s, z_m(s;t,x)) - \mathcal{E}(v_*)(s, z_m(s;t,x))\Big] : \mathcal{E}(\varphi)\,dx\,ds\Big)$$
$$+\frac{\mu_1}{\Gamma(1-\beta)}\Big(\int_0^t (t-s)^{-\beta}\int_\Omega \Big[\mathcal{E}(v_*)(s, z_m(s;t,x)) - \mathcal{E}(v_*)(s, z_*(s;t,x))\Big] : \mathcal{E}(\varphi)\,dx\,ds\Big) = Z_1^m + Z_2^m.$$

(1) We show first that $Z_1^m \to 0$ as $m \to \infty$.
Denote the integral over domain Ω in Z_1^m by I:

$$I = \int_\Omega \Big[\mathcal{E}(v_m)(s, z_m(s;t,x)) - \mathcal{E}(v_*)(s, z_m(s;t,x))\Big] : \mathcal{E}(\varphi)\,dx.$$

We make the change of variables $x = z^m(t;s,y)$ in I (where the reverse change is $y = z_m(s;t,x)$):

$$I = \int_\Omega \Big[\mathcal{E}(v_m)(s,y) - \mathcal{E}(v_*)(s,y)\Big] : \mathcal{E}(\varphi)(z_m(t;s,y))\,dy.$$

We rewrite Z_1^m and continue the further expansion:

$$Z_1^m = \frac{\mu_1}{\Gamma(1-\beta)}\Big(\int_0^t (t-s)^{-\beta}\int_\Omega \Big[\mathcal{E}(v_m)(s,y) - \mathcal{E}(v_*)(s,y)\Big] : \mathcal{E}(\varphi)(z_m(t;s,y))\,dy\,ds\Big)$$

$$= \frac{\mu_1}{\Gamma(1-\beta)}\Big(\int_0^t (t-s)^{-\beta}\int_\Omega \Big[\mathcal{E}(v_m)(s,y) - \mathcal{E}(v_*)(s,y)\Big] : \Big[\mathcal{E}(\varphi)(z_m(t;s,y))$$
$$-\mathcal{E}(\varphi)(z_*(t;s,y))\Big]\,dy\,ds\Big) + \frac{\mu_1}{\Gamma(1-\beta)}\Big(\int_0^t (t-s)^{-\beta}\int_\Omega \Big[\mathcal{E}(v_m)(s,y)$$
$$-\mathcal{E}(v_*)(s,y)\Big] : \mathcal{E}(\varphi)(z_*(t;s,y))\,dy\,ds\Big) = Z_{11}^m + Z_{12}^m.$$

(a) Due to the weak convergence v_m to v_* in the space $L_2(0,T;V^1)$, we obtain that $Z_{12}^m \to 0$ as $m \to \infty$.

(b) Applying the Holder and the Cauchy-Bunyakovsky inequalities, we get

$$|Z_{11}^m|^2 \le C_{31}\Big(\int_0^t (t-s)^{-\beta}\|v_m(s,\cdot) - v_*(s,\cdot)\|_{V^1}\|\varphi_x(z_m(t;s,\cdot)) - \varphi_x(z_*(t;s,\cdot))\|_{V^0}\,ds\Big)^2$$
$$\le C_{32}\|v_m(s,\cdot) - v_*(s,\cdot)\|_{L_2(0,T;V^1)} \times \int_0^T \|\varphi_x(z_m(t;s,\cdot)) - \varphi_x(z_*(t;s,\cdot))\|_{V^0}\,ds. \quad (52)$$

We denote the second efficient in the last inequality by $\Phi_m(s)$:

$$\Phi_m(s) = \int_0^T \|\varphi_x(z_m(t;s,\cdot)) - \varphi_x(z_*(t;s,\cdot))\|_{V^0}\,ds.$$

We show the convergence $\Phi_m(s) \to 0$ as $m \to \infty$ for every $s \in [0,T]$. Note, that

$$\Phi_m(s) = \int_0^T \int_\Omega |\varphi_x(z_m(t;s,y)) - \varphi_x(z_*(t;s,y))|^2\,dy\,ds.$$

Let $\varepsilon > 0$ be a sufficiently small number.The continuity of the function φ_x in $\overline{\Omega}$ means that there exists $\delta(\varepsilon)$ such that if $|x'' - x'| \leq \delta(\varepsilon)$, then

$$|\varphi_x(x'') - \varphi_x(x')| \leq \varepsilon. \tag{53}$$

Since the sequence $z_m(t;s,y)$ converges to $z_*(t;s,y)$ in the Lebesgue measure with respect to (t,y), therefore for $\delta(\varepsilon)$ there exists the number $N = N(\delta(\varepsilon))$ which for $m \geq N$ the following inequality holds:

$$m(\{(t,y) :\ |z_m(t;s,y) - z_*(t;s,y)| \geq \delta(\varepsilon)\}) \leq \varepsilon. \tag{54}$$

We denote

$$Q(> \delta(\varepsilon)) = \{(t,y) \in Q_T :\ |z_m(t;s,y) - z_*(t;s,y)| > \delta(\varepsilon)\};$$
$$Q(\leq \delta(\varepsilon)) = \{(t,y) \in Q_T :\ |z_m(t;s,y) - z_*(t;s,y)| \leq \delta(\varepsilon)\}.$$

Then

$$\Phi_m(s) \leq C_{33}\Big(\int_{Q(>\delta(\varepsilon))} |\varphi_x(z_m(t;s,y)) - \varphi_x(z_*(t;s,y))|^2\, dy\, ds$$
$$\cdot \int_{Q(\leq\delta(\varepsilon))} |\varphi_x(z_m(t;s,y)) - \varphi_x(z_*(t;s,y))|^2\, dy\, ds\Big) = C_{33}\Big(\Phi_m^1(s) + \Phi_m^2(s)\Big). \tag{55}$$

By virtue of (53) for $\Phi_m^2(s)$ we have $|z_m(t;s,y) - z_*(t;s,y)| \leq \delta(\varepsilon)$. Hence

$$\Phi_m^2(s) \leq \int_{Q(\leq\delta(\varepsilon))} \varepsilon^2\, dy\, ds = C_{34}\varepsilon^2. \tag{56}$$

By virtue of (54) for $\Phi_m^1(s)$ we have $m(Q(> \delta(\varepsilon))) \leq \varepsilon$. Hence

$$\Phi_m^1(s) \leq C_{35}\|\varphi_x\|_{C(\Omega)} \int_{Q(>\delta(\varepsilon))} dy\, ds = C_{35}\varepsilon\|\varphi_x\|_{C(\Omega)}. \tag{57}$$

Thus, from (55), (56) and (57) it follows that for small $\varepsilon > 0$ and $m \geq N(\delta(\varepsilon))$ the following inequality holds

$$\Phi_m(s) \leq C_{36}\varepsilon.$$

Consequently, convergence $\Phi_m(s) \to 0$ as $m \to \infty$ for all $s \in [0,T]$ is obtained. Consider the right side of inequality (4). Due to the boundedness of the first efficient (since $v_m \in L_2(0,T;V^1)$) and the convergence to 0 of the second efficient as $m \to \infty$, we get that $Z_{11}^m \to 0$ as $m \to \infty$.

Thus, it is proved that $Z_1^m \to 0$ as $m \to \infty$.

(2) Now show that $Z_2^m \to 0$ as $m \to \infty$. Consider the auxiliary function $\tilde{v}(t,x)$ smooth and finite on $[0,T] \times \Omega$ such that $\|v_* - \tilde{v}\|_{L_2(0,T;V^1)} \leq \varepsilon$ for sufficiently small $\varepsilon > 0$. We now estimate Z_2^m through three integrals

$$|Z_2^m| \leq C_{37}\Big(\int_0^t (t-s)^{-\beta} \int_\Omega \|v_*(s,\, z_m(s;t,x)) - \tilde{v}(s, z_m(s;t,x))\|_{V^1}\, ds$$
$$+ \int_0^t (t-s)^{-\beta} \int_\Omega \|\tilde{v}(s,\, z_m(s;t,x)) - \tilde{v}(s, z_*(s;t,x))\|_{V^1}\, ds$$
$$+ \int_0^t (t-s)^{-\beta} \int_\Omega \|\tilde{v}(s,\, z_*(s;t,x)) - v_*(s, z_*(s;t,x))\|_{V^1}\, ds\Big) = C_{37}\big(Z_{21}^m + Z_{22}^m + Z_{23}^m\big).$$

We make a change of variables in the norms under the integrals Z_{21}^m and Z_{23}^m:

$$\|v_*(s, z_m(s;t,x)) - \widetilde{v}(s, z_m(s;t,x))\|_{V^1} = \|v_*(s,y) - \widetilde{v}(s,y)\|_{V^1};$$

$$\|\widetilde{v}(s, z_*(s;t,x)) - v_*(s, z_*(s;t,x))\|_{V^1} = \|\widetilde{v}(s,y) - v_*(s,y)\|_{V^1}.$$

Then we get

$$Z_{21}^m + Z_{23}^m = C_{37}\Big(\int_0^t (t-s)^{-\beta}\|v_*(s,\cdot) - \widetilde{v}(s,\cdot)\|_{V^1}\,ds\Big) \le C_{37}\varepsilon.$$

We estimate also Z_{22}^m

$$Z_{22}^m \le C_{37}\Big(\int_0^t (t-s)^{-\beta}\Big(\int_\Omega |\widetilde{v}_x(s, z_m(s;t,\cdot)) - \widetilde{v}_x(s, z_*(s;t,\cdot))|^2\,dx\Big)^{1/2} ds\Big).$$

By virtue of Lemma 8 $z_m(s;t,x)$ converges to $z(s;t,x)$ and the function $\widetilde{v}_x(t,x)$ is bounded and smooth. Therefore, by the Lebesgue theorem, we obtain that $Z_2^m \to 0$ as $m \to \infty$. Thus, convergence (51) is proved.

Taking into account the a priori estimates (44)–(46) and conditions (Ψ1)-(Ψ4), without loss of generality, we can assume that there exists $f_* \in L_2(0,T;V^{-1})$ such that $f_m \to f_* \in \Psi(v_*)$ as $m \to \infty$.

As a result, it was shown that the functions v_* and f_* with a smooth test function φ from V^1 satisfy the equality:

$$\langle v_*', \varphi\rangle - \int_\Omega \sum_{i,j=1}^n (\Delta_\alpha^{-1} v_*)_i (v_*)_j \frac{\partial \varphi_j}{\partial x_i}\,dx + \mu_0 \int_\Omega \nabla v_* : \nabla\varphi\,dx$$
$$+ \frac{\mu_1}{\Gamma(1-\beta)}\Big(\int_0^t (t-s)^{-\beta}\mathcal{E}(v_*)(s, z_*(s;t,x))\,ds, \mathcal{E}(\varphi)\Big) = \langle f_*, \varphi\rangle. \quad (58)$$

Since the sequence $\{v_m\}$ has a priori estimates (44), (45) and (46), due to the weak convergence properties for v_* we immediately obtain the estimate:

$$\|v_*\|_{L_\infty(0,T;V^0)} + \|v_*\|_{L_2(0,T;V^1)} + \|v_*\|_{L_{4/3}(0,T;V^{-1})} \le C_{38}.$$

Whence it follows that $v_* \in W_1$. Thus, the passage to the limit was proved for a test function $\varphi \in V^1$, which is smooth.

II part. Let us prove this passage to the limit for an arbitrary test function φ from V^1. We rewrite (58) for smooth φ in the form:

$$[G_1, \varphi] - [G_2, \varphi] = 0, \quad (59)$$

where

$$[G_1, \varphi] = \langle v', \varphi\rangle - \int_\Omega \sum_{i,j=1}^n (\Delta_\alpha^{-1} v)_i (v)_j \frac{\partial \varphi_j}{\partial x_i}\,dx + \mu_0 \int_\Omega \nabla v : \nabla\varphi\,dx$$
$$+ \frac{\mu_1}{\Gamma(1-\beta)}\Big(\int_0^t (t-s)^{-\beta}\mathcal{E}(v)(s, z(s;t,x))\,ds, \mathcal{E}(\varphi)\Big); [G_2, \varphi] = \langle f, \varphi\rangle.$$

Lemma 9. *Let the test function φ be smooth. Then*

$$|[G_1, \varphi]| \le C_{39}\|\varphi\|_{V^1}, \quad |[G_2, \varphi]| \le C_{40}\|\varphi\|_{V^1}. \quad (60)$$

The proof of this Lemma is similar to obtaining a priori estimates in section 3.

Since the set of smooth functions is dense in V^1, for $\varphi \in V^1$ there exists a sequence of smooth functions $\varphi^l \in V^1$ such that $|\varphi^l - \varphi|_{V^1} \to 0$ for $l \to \infty$. By virtue of (59) we obtain

$$[G_1, \varphi] - [G_2, \varphi] = [G_1, \varphi - \varphi^l] - [G_2, \varphi - \varphi^l] + [G_1, \varphi^l] - [G_2, \varphi^l]$$
$$= [G_1, \varphi - \varphi^l] - [G_2, \varphi - \varphi^l].$$

From the last equality and estimates (60) we obtain

$$|[G_1, \varphi] - [G_2, \varphi]| \leq C_{41}|\varphi - \varphi^l|.$$

Taking into account the last inequality and passing to the limit as $l \to \infty$ in equality (58) for $\varphi = \varphi^l$ we obtain equality (58) for arbitrary $\varphi \in V^1$, which completes the proof of the existence of weak solutions for feedback control problem (1)–(5), (8).

5. Proof of Theorem 4

From Theorem 3 we obtain that the set of solutions is nonempty. Therefore, there exists a minimizing sequence $(v_l, f_l) \in \Sigma$ such that

$$\lim_{l \to \infty} \Phi(v_l, f_l) = \inf_{(v,f) \in \Sigma} \Phi(v, f).$$

As before, in the proof of Theorem 3 from estimates (44)–(46) it follows:

$$v_l \rightharpoonup v_* \text{ weakly in } L_2(0, T; V^1),$$
$$v_l \rightharpoonup \text{*-weakly in } L_\infty(0, T; V^0),$$
$$v_l' \rightharpoonup v_*' \text{ weakly in } L_{4/3}(0, T; V^{-1}),$$
$$v_l \to v_* \text{ strongly in } L_2(0, T; L_4(\Omega)),$$

$$z_l(\tau; t, x) \to z(\tau; t, x) \text{ in the Lebesgue measure with respect to } (\tau, x) \text{ on } [0, T] \times \Omega,$$
$$f_l \to f_* \in \Psi(v_*) \text{ strongly in } L_2(0, T; V^{-1}).$$

Similarly from inclusion

$$Jv_l' + \mu_0 A v_l - B(v_l) + \frac{\mu_1}{\Gamma(1-\beta)} C(v_l, z_l) = f_l \in \Psi(v_l),$$

passing to the limit, we obtain

$$Jv_*' + \mu_0 A v_* - B(v_*) + \frac{\mu_1}{\Gamma(1-\beta)} C(v_*, z_*) = f_* \in \Psi(v_*).$$

We get that $(v_*, f_*) \in \Sigma$. Since the functional Φ is lower semicontinuous with respect to the relatively weak topology, we have

$$\Phi(v_*, f_*) \leq \inf_{(v,f) \in \Sigma} \Phi(v, f).$$

Thereby (v_*, f_*) is the required solution. The theorem is proved.

6. Conclusions

To summarize all reasonings, calculations and proofs in this paper, the mathematical model describing the motion of viscoelastic mediums was investigated. This model is equipped with the Voigt rheological relation. This relation is considered with the left-side fractional Riemann-Liouville derivative, which allows us to take into account the memory of the medium. This memory is considered along the trajectory of the motion of fluid particles, determined by the velocity field. This allows a more accurate description of the physical process of fluid motion. Also in this paper the model under consideration is called the alpha-model. Interest in the study of alpha-models is primarily associated with their application to the study of turbulence effects for fluid flows.

The main result of this paper is the solutions existence to the feedback control problem for the mathematical model under consideration. Also the existence of an optimal solution to the problem under consideration that gives a minimum to a given bounded quality functional is proved. Results of this paper provide an opportunity for the future investigation of this model. The authors propose the following future research directions for the model under consideration—1) the numerical analysis of the obtained solutions; 2) the consideration of a turbulence case of this problem; 3) the investigation of a II class of alpha-models for this problem and so forth.

Author Contributions: Conceptualization, V.Z.; methodology, V.Z.; supervision, V.Z.; investigation, V.Z., A.Z. and A.U.; writing – original draft, review and editing, V.Z., A.Z. and A.U. All authors have read and agreed to the published version of the manuscript.

Funding: The work of the first author was supported by the Russian Foundation for Basic Research (project no. 20-01-00051, Lemma 4) and by the Ministry of Science and Higher Education of the Russian Federation (project no. FZGU-2020-0035, Theorem 3). The work of the second author was supported by the Russian Foundation for Basic Research (project no. 19-31-60014, Theorem 4). The work of the third author was supported by the Russian Science Foundation (project no. 19-11-00146, Theorem 6).

Conflicts of Interest: The authors declare no conflict of interest.

References

1. Martinez-Garcia, M.; Zhang, Y.; Gordon, T. Memory Pattern Identification for Feedback Tracking Control in Human-Machine Systems. *Hum. Factors* **2019**. doi:10.1177/0018720819881008. [CrossRef]
2. Zvyagin, V.; Obukhovskii, V.; Zvyagin, A. On inclusions with multivalued operators and their applications to some optimization problems. *J. Fixed Point Theory Appl.* **2014**, *16*, 27–82. doi:10.1007/S11784-015-0219-2. [CrossRef]
3. Zvyagin, A.V. Optimal Feedback Control for Leray and Navier–Stokes Alpha Models. *Dokl. Math.* **2019**, *99*, 299–302. doi:10.1134/S1064562419030190. [CrossRef]
4. Zvyagin, V.G.; Zvyagin, A.V.; Turbin, M.V. Optimal Feedback Control Problem for the Bingham Model with Periodical Boundary Conditions on Spatial Variables. *J. Math. Sci.* **2020**, pp. 959–980. doi:10.1007/s10958-020-04667-7. [CrossRef]
5. Zvyagin, A.V. Optimal Feedback Control in the Stationary Mathematical Model of Low Concentrated Aqueous Polymer Solutions. *Appl. Anal.* **2013**, *92*, 1157–1168. doi:10.1080/00036811.2011.653795. [CrossRef]
6. Kilbas, A.A.; Srivastava, H.M.; Trujillo, J.J. *Theory and Applications of Fractional Differential Equations*; Elsevier: Amsterdam, The Netherlands, 2006; Volume 204.
7. Leray, J. Sur le mouvement d'un liquide visqueux emplissant l'espace. *Acta Math.* **1934**, *63*, 193–248. doi:10.1007/BF02547354. [CrossRef]
8. Holm, D.D.; Marsden, J.E.; Ratiu, T.S. The Euler-Poincare models of ideal fluids with nonlinear dispersion. *Phys. Rev. Lett.* **1998**, *80*, 4173–4176. doi:10.1103/PhysRevLett.80.4173. [CrossRef]
9. Holm, D.D.; Marsden, J.E.; Ratiu, T.S. The Euler–Poincaré Equations and Semidirect Products with Applications to Continuum Theories. *Adv. Math.* **1998**, *137*, 1–81. doi:10.1006/aima.1998.1721. [CrossRef]
10. Chen, S.; Foias, C.; Holm, D.D.; Olson, E.; Titi, E.S.; Wynne, S. Camassa-Holm equations as a closure model for turbulent channel and pipe flow. *Phys. Rev. Lett.* **1998**, *81*, 5338–5341. doi:10.1103/PhysRevLett.81.5338. [CrossRef]
11. Lemarie-Rieusset, P.G. *The Navier-Stokes Problem in the 21st Century*; Taylor and Francis Group, CRC Press: Boca Raton, FL, USA, 2016.
12. Cheskidov, A.; Holm, D.D.; Olson, E.; Titi, E.S. On a Leray-α model of turbulence. *Proc. R. Soc. A* **2005**, *461*, 629–649. doi:10.1098/rspa.2004.1373. [CrossRef]
13. Zvyagin, A.V. Solvability of Thermoviscoelastic Problem for Leray Alpha-Model. *Russ. Math.* **2016**, *60*, 629–649. doi:10.3103/S1066369X16100091. [CrossRef]

14. Foias, C.; Holm, D.D.; Titi, E.S. The three dimensional viscous Camassa-Holm equations, and their relation to the Navier-Stokes equations and turbulence theory. *J. Dyn. Differ. Equ.* **2002**, *14*, 1–35. doi:10.1023/A:1012984210582. [CrossRef]
15. Zvyagin, A.V.; Polyakov, D.M. On the solvability of the Jeffreys-Oldroyd-α model. *Differ. Equ.* **2016**, *52*, 761–766. doi:10.1134/S0012266116060069. [CrossRef]
16. Zvyagin, A.V.; Zvyagin, V.G.; Polyakov, D.M. On solvability of a fluid flow alpha-model with memory. *Russ. Math.* **2018**, *6*, 69–74. doi:10.3103/S1066369X18060075. [CrossRef]
17. Zvyagin, A.V.; Zvyagin, V.G.; Polyakov, D.M. Dissipative Solvability of an Alpha Model of Fluid Flow with Memory. *Comput. Math. Math. Phys.* **2019**, *59*, 1185–1198. doi:10.1134/s0965542519070133. [CrossRef]
18. Zvyagin, A.V. Weak solvability and convergence of solutions for the fractional Voigt-model of a viscoelastic medium. *Russ. Math. Surv.* **2019**, *74*, 549–551. doi:10.1070/RM9880. [CrossRef]
19. Zvyagin, V.; Orlov, V. Weak solvability of fractional Voigt model of viscoelasticity. *Discrete Contin. Dyn. Syst. Ser. A* **2018**, *38*, 6327–6350. doi:10.3934/dcds.2018270. [CrossRef]
20. Mainardi, F.; Spada, G. Creep, relaxation and viscosity properties for basic fractional models in rheology. *Eur. Phys. J. Spec. Top.* **2011**, *193*, 133–160. doi:10.1140/epjst/e2011-01387-1. [CrossRef]
21. Caputo, M.; Mainardi, F. A new dissipation model based on memory mechanism. *Pure Appl. Geophys.* **1971**, *91*, 134–147. doi:10.1007/BF00879562. [CrossRef]
22. Gladkov, S.O. Theory of One-Dimensional and Quasi-one-dimensional Heat Conduction. *Tech. Phys.* **1997**, *42*, 724–727. doi:10.1134/1.1258707. [CrossRef]
23. Gladkov, S.O. On the Theory of Hydrodynamic Phenomena in Quasi-one-dimensional Systems. *Tech. Phys.* **2001**, *46*, 1475–1477. doi:10.1134/1.1418518. [CrossRef]
24. Gladkov, S.O.; Bogdanova, S.B. The Heat-transfer Theory for Quasi-n-dimensional System. *Physica B (Amst. Neth.)* **2010**, *405*, 1973–1975. doi:10.1016/j.physb.2010.01.077. [CrossRef]
25. Gladkov, S.O.; Bogdanova, S.B. On Fractional Differentiation. *Vestn. Samara Univ. Nat. Sci. Ser.* **2018**, *24*, 7–13. doi:10.18287/2541-7525-2018-24-3-7-13. [CrossRef]
26. Gladkov, S.O.; Bogdanova, S.B. To the Question of Fractional Differentiation. Part II. *Vestn. Samara Univ. Nat. Sci. Ser.* **2019**, *25*, 7–11. doi:10.18287/2541-7525-2019-25-3-7-11. [CrossRef]
27. Fursikov, A.V. *Optimal Control of Distributed Systems. Theory and Applications*; American Mathematical Society: Providence, RI, USA, 2000; Volume 187.
28. Zvyagin, V.G.; Turbin, M.V. *Mathematical Problems in Viscoelastic Hydrodynamics*; Krasand URSS Russia: Moscow, Russia, 2012. (In Russian)
29. Orlov, V.P.; Sobolevskii, P.E. On mathematical models of a viscoelasticity with a memory. *Differ. Integral Equ.* **1991**, *4*, 103–115.
30. Zvyagin, V.G.; Dmitrienko, V.T. On weak solutions of a regularized model of a viscoelastic fluid. *Differ. Equ.* **2002**, *38*, 1731–1744. doi:10.1023/A:1023860129831. [CrossRef]
31. DiPerna, R.J.; Lions, P.L. Ordinary differential equations, transport theory and Sobolev spaces. *Invent. Math.* **1989**, *98*, 511–547. doi:10.1007/BF01393835. [CrossRef]
32. Crippa, G. The ordinary differential equation with non-Lipschitz vector fields. *Boll. Unione Mat. Ital.* **2008**, *9*, 333–348.
33. Crippa, G.; De Lellis, C. Estimates and regularity results for the DiPerna-Lions flow. *J. Reine Angew. Math.* **2008**, *2008*, 15–46. doi:10.1515/CRELLE.2008.016. [CrossRef]
34. Temam, R. *Navier-Stokes Equations: Theory and Numerical Analysis*; AMS Chelsea: Providence, RI, USA, 2001.
35. Zvyagin, V.G. Topological approximation approach to study of mathematical problems of hydrodynamics. *J. Math. Sci.* **2014**, *201*, 830–858. doi:10.1007/s10958-014-2028-3. [CrossRef]
36. Turbin, M.V.; Ustiuzhaninova, A.S. The existence theorem for a weak solution to initial-boundary value problem for system of equations describing the motion of weak aqueous polymer solutions. *Russ. Math.* **2019**, *63*, 54–69. doi:10.3103/s1066369x19080061. [CrossRef]
37. Agranovich, M.S.; Vishik, M.I. Elliptic problems with a parameter and parabolic problems of general type. *Russ. Math. Surv.* **1964**, *19*, 53–157. doi:10.1070/RM1964v019n03ABEH001149. [CrossRef]
38. Agmon, S. On the eigenfunctions and on the eigenvalues of general elliptic boundary value problems. *Commun. Pure Appl. Math.* **1962**, *15*, 119–147. doi:10.1007/978-3-642-10994-2_1. [CrossRef]
39. Aubin, J.P. Un théorème de compacité. *Comptes Rendus L'AcadéMie Des Sci.* **1963**, *256*, 5042–5044.

40. Simon, J. Compact sets in the spacel $L^p(0,T;B)$. *Ann. di Mat. Pura ed Appl.* **1986**, *146*, 65–96. doi:0.1007/BF01762360. [CrossRef]
41. Samko, S.G.; Kilbas, A.A.; Marichev, O.I. *Integrals and Fractional Derivatives and Some of Their Applications;* Science and Technology: Minsk, Belarus, 1987. (In Russian).
42. Ahmerov, P.P.; Kamenskii, M.I.; Potapov, A.S.; Rodkina, A.E.; Sadovskii, B.N. *Non-Compactness and Compacting Operators;* Nauka: Novosibirsk, Russia, 1986. (In Russian)
43. Borisovich, Y.G.; Gel'man, B.D.; Myshkis, A.D.; Obukhovskii, V.V. *Introduction to the Theory of Multivalued Mappings and Differential Inclusions;* URSS: Moscow, Russia, 2011. (In Russian)

mathematics

MDPI

Article

Existence of Positive Solutions for a System of Singular Fractional Boundary Value Problems with p-Laplacian Operators

Ahmed Alsaedi [1,*], Rodica Luca [2] and Bashir Ahmad [1]

1 Nonlinear Analysis and Applied Mathematics (NAAM)-Research Group, Department of Mathematics, Faculty of Science, King Abdulaziz University, P.O. Box 80203, Jeddah 21589, Saudi Arabia; bashirahmad_qau@yahoo.com

2 Department of Mathematics, Gh. Asachi Technical University, 11 Blvd. Carol I, 700506 Iasi, Romania; rluca@math.tuiasi.ro

* Correspondence: aalsaedi@hotmail.com or alsaedi@kau.edu.sa

Received: 27 September 2020; Accepted: 27 October 2020; Published: 31 October 2020

Abstract: We investigate the existence and multiplicity of positive solutions for a system of Riemann–Liouville fractional differential equations with singular nonnegative nonlinearities and p-Laplacian operators, subject to nonlocal boundary conditions which contain fractional derivatives and Riemann–Stieltjes integrals.

Keywords: Riemann–Liouville fractional differential equations; nonlocal boundary conditions; positive solutions; existence; multiplicity

MSC: 34A08; 34B15; 45G15

1. Introduction

We consider the system of fractional differential equations

$$\begin{cases} D_{0+}^{\alpha_1}(\varphi_{r_1}(D_{0+}^{\beta_1}u(t))) + f(t,u(t),v(t)) = 0, \quad t \in (0,1), \\ D_{0+}^{\alpha_2}(\varphi_{r_2}(D_{0+}^{\beta_2}v(t))) + g(t,u(t),v(t)) = 0, \quad t \in (0,1), \end{cases} \tag{1}$$

with the nonlocal boundary conditions

$$\begin{cases} u^{(j)}(0) = 0,\ j = 0,\ldots,n-2;\ D_{0+}^{\beta_1}u(0) = 0,\ D_{0+}^{\gamma_0}u(1) = \sum_{i=1}^{p}\int_0^1 D_{0+}^{\gamma_i}u(t)\,dH_i(t), \\ v^{(j)}(0) = 0,\ j = 0,\ldots,m-2;\ D_{0+}^{\beta_2}v(0) = 0,\ D_{0+}^{\delta_0}v(1) = \sum_{i=1}^{q}\int_0^1 D_{0+}^{\delta_i}v(t)\,dK_i(t), \end{cases} \tag{2}$$

where $\alpha_1,\ \alpha_2 \in (0,1]$, $\beta_1 \in (n-1,n]$, $\beta_2 \in (m-1,m]$, $n,\ m \in \mathbb{N}$, $n,\ m \geq 3$, $p,q \in \mathbb{N}$, $\gamma_i \in \mathbb{R}$ for all $i = 0,\ldots,p$, $0 \leq \gamma_1 < \gamma_2 < \cdots < \gamma_p \leq \gamma_0 < \beta_1 - 1$, $\gamma_0 \geq 1$, $\delta_i \in \mathbb{R}$ for all $i = 0,\ldots,q$, $0 \leq \delta_1 < \delta_2 < \cdots < \delta_q \leq \delta_0 < \beta_2 - 1$, $\delta_0 \geq 1$, $r_1, r_2 > 1$, $\varphi_{r_i}(\tau) = |\tau|^{r_i-2}\tau$, $\varphi_{r_i}^{-1} = \varphi_{\varrho_i}$, $\varrho_i = \frac{r_i}{r_i-1}$, $i = 1,2$, the functions f and g are nonnegative and they may be singular at $t = 0$ and/or $t = 1$, the integrals from the boundary conditions (2) are Riemann–Stieltjes integrals with H_i, $i = 1,\ldots,p$ and K_j, $j = 1,\cdots,q$ functions of bounded variation, and $D_{0+}^{\theta}u$ denotes the Riemann–Liouville fractional derivative of order θ of function u (for $\theta = \alpha_1,\ \beta_1,\ \alpha_2,\ \beta_2,\ \gamma_i$ for $i = 0,\ldots,p$, δ_j for $j = 0,\ldots,q$). The fractional derivative $D_{0+}^{\theta}u$ is defined by $D_{0+}^{\theta}u(t) = \frac{1}{\Gamma(r-\theta)}\left(\frac{d}{dt}\right)^r \int_0^t (t-s)^{r-\theta-1}u(s)\,ds$, $t > 0$, where $r = \lfloor\theta\rfloor + 1$, $\lfloor\theta\rfloor$ stands for the largest integer not greater than θ, and $\Gamma(\zeta) = \int_0^\infty t^{\zeta-1}e^{-t}\,dt$, $\zeta > 0$, is the gamma function (the Euler function of second type). This work is motivated by the application of p-Laplacian operator in several fields such as nonlinear elasticity, fluid flow through porous media, glaciology, nonlinear electrorheological fluids, etc., for details, see [1] and the references cited therein.

Under some assumptions on the functions f and g, we present existence and multiplicity results for the positive solutions of problem (1) and (2). By a positive solution of problem (1) and (2) we mean a pair of functions $(u,v) \in (C([0,1],\mathbb{R}_+))^2$, satisfying the system (1) and the boundary conditions (2), with $u(t) > 0$ for all $t \in (0,1]$, or $v(t) > 0$ for all $t \in (0,1]$, ($\mathbb{R}_+ = [0,\infty)$). In the proof of our main theorems we use the Guo–Krasnosel'skii fixed point theorem (see [2]). The existence and nonexistence of positive solutions for the system (1) with two positive parameters λ and μ, and nonsingular and nonnegative nonlinearities, supplemented with the multi-point boundary conditions

$$\begin{cases} u^{(j)}(0)=0,\ j=0,\ldots,n-2;\ D_{0+}^{\beta_1}u(0)=0,\ D_{0+}^{p_1}u(1)=\sum_{i=1}^{N}a_iD_{0+}^{q_1}u(\xi_i), \\ v^{(j)}(0)=0,\ j=0,\ldots,m-2;\ D_{0+}^{\beta_2}v(0)=0,\ D_{0+}^{p_2}v(1)=\sum_{i=1}^{M}b_iD_{0+}^{q_2}v(\eta_i), \end{cases}$$

where $p_1, p_2, q_1, q_2 \in \mathbb{R}$, $p_1 \in [1,n-2]$, $p_2 \in [1,m-2]$, $q_1 \in [0,p_1]$, $q_2 \in [0,p_2]$, $\xi_i, a_i \in \mathbb{R}$ for all $i = 1,\ldots,N$ ($N \in \mathbb{N}$), $0 < \xi_1 < \cdots < \xi_N \le 1$, $\eta_i, b_i \in \mathbb{R}$ for all $i = 1,\ldots,M$ ($M \in \mathbb{N}$), $0 < \eta_1 < \cdots < \eta_M \le 1$, was investigated in [3], by applying the Guo–Krasnosel'skii theorem. In the paper [4], the authors studied the system (1) with positive parameters, and nonsingular and nonnegative nonlinearities, subject to the nonlocal coupled boundary conditions

$$\begin{cases} u^{(j)}(0)=0,\ j=0,\ldots,n-2;\ D_{0+}^{\beta_1}u(0)=0,\ D_{0+}^{\gamma_0}u(1)=\displaystyle\sum_{i=1}^{p}\int_0^1 D_{0+}^{\gamma_i}v(t)\,dH_i(t), \\ v^{(j)}(0)=0,\ j=0,\ldots,m-2;\ D_{0+}^{\beta_2}v(0)=0,\ D_{0+}^{\delta_0}v(1)=\displaystyle\sum_{i=1}^{q}\int_0^1 D_{0+}^{\delta_i}u(t)\,dK_i(t), \end{cases}$$

where $p,q \in \mathbb{N}$, $\gamma_i \in \mathbb{R}$ for all $i = 0,1,\ldots,p$, $0 \le \gamma_1 < \gamma_2 < \cdots < \gamma_p \le \delta_0 < \beta_2 - 1$, $\delta_0 \ge 1$, $\delta_i \in \mathbb{R}$ for all $i = 0,1,\ldots,q$, $0 \le \delta_1 < \delta_2 < \cdots < \delta_q \le \gamma_0 < \beta_1 - 1$, $\gamma_0 \ge 1$.

In [5], by applying the fixed point theorem for mixed monotone operators, the authors proved the existence of positive solutions for the multi-point boundary value problem for nonlinear Riemann–Liouville fractional differential equations

$$\begin{cases} D_{0+}^{\beta}\varphi_p(D_{0+}^{\alpha}u(t)) = f(t,u(t)),\ 0<t<1, \\ u(0)=0,\ D_{0+}^{\gamma}u(1)=\sum_{i=1}^{m-2}\xi_iD_{0+}^{\gamma}u(\eta_i),\ D_{0+}^{\alpha}u(0)=0, \\ \varphi_p(D_{0+}^{\alpha}u(1))=\sum_{i=1}^{m-2}\zeta_i\varphi_p(D_{0+}^{\alpha}u(\eta_i)), \end{cases}$$

where $\alpha,\beta \in (1,2]$, $\gamma \in (0,1]$, $\xi_i, \eta_i, \zeta_i \in (0,1)$, $i = 1,\ldots,m-2$, and f is a nonnegative function which may be singular at $x = 0$. In [6], the authors investigated the existence and uniqueness of positive solutions for the fractional boundary value problem

$$\begin{cases} {}^cD_{0+}^{\alpha}\varphi_p\left(D_{0+}^{\beta}u(t)+\varphi_q(I_{0+}^{r}h(t,I_{0+}^{\rho_1}u(t),D_{0+}^{\gamma}u(t)))\right) \\ \quad +f(t,I_{0+}^{\rho_2}u(t),D_{0+}^{\gamma}u(t))=0,\ t\in(0,1), \\ u(0)=D_{0+}^{\delta_1}u(0)=\cdots=D_{0+}^{\delta_{n-2}}u(0)=D_{0+}^{\beta}u(0)=0, \\ D_{0+}^{k_0}u(1)=\lambda_1\int_0^1 l_1(\tau)D_{0+}^{k_1}u(\tau)dA_1(\tau)+\lambda_2\int_0^{\zeta} l_2(\tau)D_{0+}^{k_2}u(\tau)dA_2(\tau) \\ \quad +\lambda_3\sum_{i=1}^{\infty}\mu_iD_{0+}^{k_3}u(\eta_i), \end{cases}$$

where $\alpha \in (0,1]$, $\beta \in (n-1,n]$, $n \ge 3$, ${}^cD_{0+}^{\alpha}u$ denotes the Caputo fractional derivative of order α of function u defined by ${}^cD_{0+}^{\alpha}u(t) = \frac{1}{\Gamma(1-\alpha)}\int_0^t(t-s)^{-\alpha}u'(s)\,ds$, $t > 0$, for $\alpha \in (0,1)$, and ${}^cD_{0+}^{\alpha}u(t) = u'(t)$, $t > 0$, for $\alpha = 1$, and the nonlinear terms f and h may be singular on the time variable and space variables. The authors used in [6] the theory of mixed monotone operators, and they also discussed there the dependence of solutions upon a parameter.

Systems with fractional differential equations without p-Laplacian operators, with parameters or without parameters, subject to various multi-point or Riemann–Stieltjes integral boundary conditions were studied in the last years in [7–27]. For various applications of the fractional differential equations in many scientific and engineering domains we refer the reader to the books [28–34], and their references.

The paper is organized as follows. In Section 2, we study two nonlocal boundary value problems for fractional differential equations with p-Laplacian operators, and we present some properties of the associated Green functions. Section 3 contains the main existence theorems for the positive solutions for our problem (1) and (2), and in Section 4, we give two examples which illustrate our results.

2. Auxiliary Results

We consider firstly the nonlinear fractional differential equation

$$D_{0+}^{\alpha_1}(\varphi_{r_1}(D_{0+}^{\beta_1}u(t))) + h(t) = 0, \ \ t \in (0,1), \tag{3}$$

with the boundary conditions

$$\begin{cases} u^{(j)}(0) = 0, \ j = 0, \dots, n-2; \ \ D_{0+}^{\beta_1}u(0) = 0, \\ D_{0+}^{\gamma_0}u(1) = \sum_{i=1}^{p} \int_0^1 D_{0+}^{\gamma_i}u(t)\,dH_i(t), \end{cases} \tag{4}$$

where $\alpha_1 \in (0,1]$, $\beta_1 \in (n-1,n]$, $n \in \mathbb{N}$, $n \geq 3$, $p \in \mathbb{N}$, $\gamma_i \in \mathbb{R}$ for all $i = 0, \dots, p$, $0 \leq \gamma_1 < \gamma_2 < \cdots < \gamma_p \leq \gamma_0 < \beta_1 - 1$, $\gamma_0 \geq 1$, H_i, $i = 1, \dots, p$ are bounded variation functions, and $h \in C(0,1) \cap L^1(0,1)$. We denote by

$$\Delta_1 = \frac{\Gamma(\beta_1)}{\Gamma(\beta_1 - \gamma_0)} - \sum_{i=1}^{p} \frac{\Gamma(\beta_1)}{\Gamma(\beta_1 - \gamma_i)} \int_0^1 s^{\beta_1 - \gamma_i - 1}\,dH_i(s).$$

Lemma 1. *If $\Delta_1 \neq 0$, then the unique solution $u \in C[0,1]$ of problem (3) and (4) is given by*

$$u(t) = \int_0^1 \mathcal{G}_1(t,s)\varphi_{\varrho_1}(I_{0+}^{\alpha_1}h(s))\,ds, \ \ t \in [0,1], \tag{5}$$

where the Green function $\mathcal{G}_1$ is given by

$$\mathcal{G}_1(t,s) = g_1(t,s) + \frac{t^{\beta_1 - 1}}{\Delta_1} \sum_{i=1}^{p} \left(\int_0^1 g_{2i}(\tau,s)\,dH_i(\tau) \right), \ \ t,s \in [0,1], \tag{6}$$

with

$$\begin{aligned} g_1(t,\zeta) &= \frac{1}{\Gamma(\beta_1)} \begin{cases} t^{\beta_1-1}(1-\zeta)^{\beta_1-\gamma_0-1} - (t-\zeta)^{\beta_1-1}, \ 0 \leq \zeta \leq t \leq 1, \\ t^{\beta_1-1}(1-\zeta)^{\beta_1-\gamma_0-1}, \ \ 0 \leq t \leq \zeta \leq 1, \end{cases} \\ g_{2i}(\tau,\zeta) &= \frac{1}{\Gamma(\beta_1 - \gamma_i)} \begin{cases} \tau^{\beta_1-\gamma_i-1}(1-\zeta)^{\beta_1-\gamma_0-1} - (\tau-\zeta)^{\beta_1-\gamma_i-1}, \\ \qquad 0 \leq \zeta \leq \tau \leq 1, \\ \tau^{\beta_1-\gamma_i-1}(1-\zeta)^{\beta_1-\gamma_0-1}, \ \ 0 \leq \tau \leq \zeta \leq 1, \end{cases} \\ & i = 1, \dots, p. \end{aligned} \tag{7}$$

Proof. We denote by $\varphi_{r_1}(D_{0+}^{\beta_1}u(t)) = x(t)$. Then problem (3) and (4) is equivalent to the following two boundary value problems

$$D_{0+}^{\alpha_1}x(t) + h(t) = 0, \ \ 0 < t < 1; \ \ x(0) = 0, \tag{8}$$

and

$$\begin{cases} D_{0+}^{\beta_1} u(t) = \varphi_{\varrho_1}(x(t)), \; 0 < t < 1; \\ u^{(j)}(0) = 0, \; j = 0, \ldots, n-2; \; D_{0+}^{\gamma_0} u(1) = \sum_{i=1}^{p} \int_0^1 D_{0+}^{\gamma_i} u(t) \, dH_i(t). \end{cases} \tag{9}$$

For the first problem (8), the function

$$x(t) = -I_{0+}^{\alpha_1} h(t) = -\frac{1}{\Gamma(\alpha_1)} \int_0^t (t-s)^{\alpha_1 - 1} h(s) \, ds, \; t \in [0,1], \tag{10}$$

is the unique solution $x \in C[0,1]$ of (8). For the second problem (9), if $\Delta_1 \neq 0$, then by [7] (Lemma 2.2), we deduce that the function

$$u(t) = -\int_0^1 \mathcal{G}_1(t,s) \varphi_{\varrho_1}(x(s)) \, ds, \; t \in [0,1], \tag{11}$$

where $\mathcal{G}_1$ is given by (6), is the unique solution $u \in C[0,1]$ of problem (9). Now, by using relations (10) and (11), we find formula (5) for the unique solution $u \in C[0,1]$ of problem (3) and (4). □

Next we consider the nonlinear fractional differential equation

$$D_{0+}^{\alpha_2}(\varphi_{r_2}(D_{0+}^{\beta_2} v(t))) + k(t) = 0, \; t \in (0,1), \tag{12}$$

with the boundary conditions

$$\begin{cases} v^{(j)}(0) = 0, \; j = 0, \ldots, m-2; \; D_{0+}^{\beta_2} v(0) = 0, \\ D_{0+}^{\delta_0} v(1) = \sum_{i=1}^{q} \int_0^1 D_{0+}^{\delta_i} v(t) \, dK_i(t), \end{cases} \tag{13}$$

where $\alpha_2 \in (0,1]$, $\beta_2 \in (m-1, m]$, $m \in \mathbb{N}$, $m \geq 3$, $q \in \mathbb{N}$, $\delta_i \in \mathbb{R}$ for all $i = 0, \ldots, q$, $0 \leq \delta_1 < \delta_2 < \cdots < \delta_q \leq \delta_0 < \beta_2 - 1$, $\delta_0 \geq 1$, K_i, $i = 1, \ldots, q$ are bounded variation functions, and $k \in C(0,1) \cap L^1(0,1)$. We denote by

$$\Delta_2 = \frac{\Gamma(\beta_2)}{\Gamma(\beta_2 - \delta_0)} - \sum_{i=1}^{q} \frac{\Gamma(\beta_2)}{\Gamma(\beta_2 - \delta_i)} \int_0^1 s^{\beta_2 - \delta_i - 1} \, dK_i(s).$$

In a similar manner as above we obtain the following result.

Lemma 2. *If $\Delta_2 \neq 0$, then the unique solution $v \in C[0,1]$ of problem (12) and (13) is given by*

$$v(t) = \int_0^1 \mathcal{G}_2(t,s) \varphi_{\varrho_2}(I_{0+}^{\alpha_2} k(s)) \, ds, \; t \in [0,1], \tag{14}$$

where the Green function $\mathcal{G}_2$ is given by

$$\mathcal{G}_2(t,s) = g_3(t,s) + \frac{t^{\beta_2 - 1}}{\Delta_2} \sum_{i=1}^{q} \left(\int_0^1 g_{4i}(\tau, s) \, dK_i(\tau) \right), \; t, s \in [0,1], \tag{15}$$

with

$$\begin{aligned} g_3(t,\zeta) &= \frac{1}{\Gamma(\beta_2)} \begin{cases} t^{\beta_2-1}(1-\zeta)^{\beta_2-\delta_0-1} - (t-\zeta)^{\beta_2-1}, \; 0 \leq \zeta \leq t \leq 1, \\ t^{\beta_2-1}(1-\zeta)^{\beta_2-\delta_0-1}, \; 0 \leq t \leq \zeta \leq 1, \end{cases} \\ g_{4i}(\tau,\zeta) &= \frac{1}{\Gamma(\beta_2-\delta_i)} \begin{cases} \tau^{\beta_2-\delta_i-1}(1-\zeta)^{\beta_2-\delta_0-1} - (\tau-\zeta)^{\beta_2-\delta_i-1}, \\ \qquad 0 \leq \zeta \leq \tau \leq 1, \\ \tau^{\beta_2-\delta_i-1}(1-\zeta)^{\beta_2-\delta_0-1}, \; 0 \leq \tau \leq \zeta \leq 1, \end{cases} \\ i &= 1, \ldots, q. \end{aligned} \tag{16}$$

By using the properties of the functions $g_1, g_{2i}, i = 1, \ldots, p, g_3, g_{4i}, i = 1, \ldots, q$ given by (7) and (16) (see [7,17]), we obtain the following properties of the Green functions $\mathcal{G}_1$ and $\mathcal{G}_2$ that we will use in the next section.

Lemma 3. *Assume that $H_i : [0,1] \to \mathbb{R}, i = 1, \ldots, p$, and $K_j : [0,1] \to \mathbb{R}, j = 1, \ldots, q$ are nondecreasing functions and $\Delta_1 > 0,\ \Delta_2 > 0$. Then the Green functions $\mathcal{G}_1$ and $\mathcal{G}_2$ given by (6) and (15) have the properties:*

(a) $\mathcal{G}_1, \mathcal{G}_2 : [0,1] \times [0,1] \to [0, \infty)$ *are continuous functions;*
(b) $\mathcal{G}_1(t,s) \le \mathcal{J}_1(s)$ *for all* $t,\ s \in [0,1]$, *where*
$\mathcal{J}_1(s) = h_1(s) + \frac{1}{\Delta_1} \sum_{i=1}^{p} \int_0^1 g_{2i}(\tau, s)\, dH_i(\tau)$, *with*
$h_1(s) = \frac{1}{\Gamma(\beta_1)}[(1-s)^{\beta_1 - \gamma_0 - 1} - (1-s)^{\beta_1 - 1}], s \in [0,1]$;
(c) $\mathcal{G}_1(t,s) \ge t^{\beta_1 - 1}\mathcal{J}_1(s)$ *for all* $t,\ s \in [0,1]$;
(d) $\mathcal{G}_2(t,s) \le \mathcal{J}_2(s)$ *for all* $t,\ s \in [0,1]$, *where*
$\mathcal{J}_2(s) = h_2(s) + \frac{1}{\Delta_2} \sum_{i=1}^{q} \int_0^1 g_{4i}(\tau, s)\, dK_i(\tau)$, *with*
$h_2(s) = \frac{1}{\Gamma(\beta_2)}[(1-s)^{\beta_2 - \delta_0 - 1} - (1-s)^{\beta_2 - 1}], s \in [0,1]$;
(e) $\mathcal{G}_2(t,s) \ge t^{\beta_2 - 1}\mathcal{J}_2(s)$ *for all* $t,\ s \in [0,1]$.

By similar arguments used in the proof of [17] (Lemma 2.5), we deduce the next lemma.

Lemma 4. *Assume that $H_i : [0,1] \to \mathbb{R}, i = 1, \ldots, p$ and $K_j : [0,1] \to \mathbb{R}, j = 1, \ldots, q$ are nondecreasing functions, $\Delta_1 > 0$, $\Delta_2 > 0$, $h \in C(0,1) \cap L^1(0,1)$, $k \in C(0,1) \cap L^1(0,1)$, $h(t) \ge 0$ for all $t \in (0,1)$, $k(t) \ge 0$ for all $t \in (0,1)$. Then the solutions u and v of problems (3), (4), (12) and (13), respectively, satisfy the inequalities $u(t) \ge 0$, $v(t) \ge 0$ for all $t \in [0,1]$. In addition, we have the inequalities $u(t) \ge t^{\beta_1 - 1} u(\tau)$, $v(t) \ge t^{\beta_2 - 1} v(\tau)$ for all $t, \tau \in [0,1]$.*

3. Existence of Positive Solutions

In this section, we investigate the existence of positive solutions for problem (1) and (2) under various assumptions on the functions f and g which may be singular at $t = 0$ and/or $t = 1$. We present the basic assumptions that we will use in the main theorems.

(I1) $\alpha_1, \alpha_2 \in (0,1]$, $\beta_1 \in (n-1, n]$, $\beta_2 \in (m-1, m]$, $n, m \in \mathbb{N}$, $n, m \ge 3$, $p, q \in \mathbb{N}$, $\gamma_i \in \mathbb{R}$ for all $i = 0, \ldots, p$, $0 \le \gamma_1 < \gamma_2 < \cdots < \gamma_p \le \gamma_0 < \beta_1 - 1$, $\gamma_0 \ge 1$, $\delta_i \in \mathbb{R}$ for all $i = 0, \ldots, q$, $0 \le \delta_1 < \delta_2 < \cdots < \delta_q \le \delta_0 < \beta_2 - 1$, $\delta_0 \ge 1$, H_i, $i = 1, \ldots, p$, K_j, $j = 1, \ldots, q$ are nondecreasing functions, $\Delta_1 > 0$, $\Delta_2 > 0$, $r_i > 1$, $\varphi_{r_i}(s) = |s|^{r_i - 2}s$, $\varphi_{r_i}^{-1} = \varphi_{\varrho_i}$, $\varrho_i = \frac{r_i}{r_i - 1}$, $i = 1, 2$.

(I2) The functions $f, g \in C((0,1) \times \mathbb{R}_+ \times \mathbb{R}_+, \mathbb{R}_+)$ and there exist the functions $\zeta_i \in C((0,1), \mathbb{R}_+)$ and $\chi_i \in C([0,1] \times \mathbb{R}_+ \times \mathbb{R}_+, \mathbb{R}_+)$, $i = 1, 2$, with $\Lambda_1, \Lambda_2 \in (0, \infty)$ such that

$$f(t,x,y) \le \zeta_1(t)\chi_1(t,x,y), \quad g(t,x,y) \le \zeta_2(t)\chi_2(t,x,y), \quad \forall t \in (0,1),\ x, y \in \mathbb{R}_+, \tag{17}$$

where $\Lambda_1 = \int_0^1 (1-s)^{\beta_1 - \gamma_0 - 1} \varphi_{\varrho_1}(I_{0+}^{\alpha_1} \zeta_1(s))\, ds$, $\Lambda_2 = \int_0^1 (1-s)^{\beta_2 - \delta_0 - 1} \varphi_{\varrho_2}(I_{0+}^{\alpha_2} \zeta_2(s))\, ds$.

Remark 1. *We present below two cases in which $\Lambda_1, \Lambda_2 \in (0, \infty)$; for other cases see the examples from Section 4.*
a) If $f, g \in C([0,1] \times \mathbb{R}_+ \times \mathbb{R}_+, \mathbb{R}_+)$, that is $\zeta_i(s) = 1$ for all $s \in [0,1]$, $i = 1, 2$, $\chi_1 = f$, $\chi_2 = g$, then the inequalities (17) are satisfied with equality. In addition, the conditions $\Lambda_1, \Lambda_2 \in (0, \infty)$ are also satisfied, because in this nonsingular case, we obtain

$$\begin{aligned}\Lambda_1 &= \int_0^1 (1-s)^{\beta_1-\gamma_0-1}\varphi_{\varrho_1}(I_{0+}^{\alpha_1}\zeta_1(s))\,ds = \int_0^1 (1-s)^{\beta_1-\gamma_0-1}(I_{0+}^{\alpha_1}\zeta_1(s))^{\varrho_1-1}\,ds \\ &= \frac{1}{(\Gamma(\alpha_1))^{\varrho_1-1}}\int_0^1 (1-s)^{\beta_1-\gamma_0-1}\left(\int_0^s (s-\tau)^{\alpha_1-1}\,d\tau\right)^{\varrho_1-1} ds \\ &= \frac{1}{(\Gamma(\alpha_1+1))^{\varrho_1-1}}\int_0^1 (1-s)^{\beta_1-\gamma_0-1}s^{\alpha_1(\varrho_1-1)}\,ds \\ &= \frac{1}{(\Gamma(\alpha_1+1))^{\varrho_1-1}}B(\alpha_1(\varrho_1-1)+1,\beta_1-\gamma_0)\in(0,\infty),\end{aligned}$$

where $B(\theta_1,\theta_2)=\int_0^1 t^{\theta_1-1}(1-t)^{\theta_2-1}\,dt$ is the beta function (the Euler function of first type), with $\theta_1,\theta_2>0$. In a similar manner we have $\Lambda_2=\frac{1}{(\Gamma(\alpha_2+1))^{\varrho_2-1}}B(\alpha_2(\varrho_2-1)+1,\beta_2-\delta_0)\in(0,\infty)$.
b) If $\zeta_1,\zeta_2\in L^2(0,1)$, $\zeta_1\not\equiv 0$, $\zeta_2\not\equiv 0$, and $\alpha_1,\alpha_2\in(1/2,1]$, then by using the Cauchy inequality we find

$$\begin{aligned}0<\Lambda_1 &\le \frac{1}{(\Gamma(\alpha_1))^{\varrho_1-1}}\int_0^1 (1-s)^{\beta_1-\gamma_0-1}\left(\int_0^s (s-\tau)^{2(\alpha_1-1)}\,d\tau\right)^{\frac{\varrho_1-1}{2}}\left(\int_0^s \zeta_1^2(\tau)\,d\tau\right)^{\frac{\varrho_1-1}{2}} ds \\ &\le \frac{\|\zeta_1\|_2^{\varrho_1-1}}{(\Gamma(\alpha_1))^{\varrho_1-1}(2\alpha_1-1)^{\frac{\varrho_1-1}{2}}}\int_0^1 s^{\frac{(2\alpha_1-1)(\varrho_1-1)}{2}}(1-s)^{\beta_1-\gamma_0-1}\,ds \\ &= \frac{\|\zeta_1\|_2^{\varrho_1-1}}{(\Gamma(\alpha_1))^{\varrho_1-1}(2\alpha_1-1)^{\frac{\varrho_1-1}{2}}}B\left(\frac{(2\alpha_1-1)(\varrho_1-1)}{2}+1,\beta_1-\gamma_0\right)<\infty,\end{aligned}$$

where $\|\zeta_1\|_2$ is the norm of ζ_1 in the space $L^2(0,1)$. In a similar manner we obtain $\Lambda_2\in(0,\infty)$.

By using Lemmas 1 and 2 (the relations (5) and (14)), (u,v) is a solution of problem (1) and (2) if and only if (u,v) is a solution of the nonlinear system of integral equations

$$\begin{cases} u(t)=\int_0^1 \mathcal{G}_1(t,s)\varphi_{\varrho_1}(I_{0+}^{\alpha_1}f(s,u(s),v(s)))\,ds, & t\in[0,1], \\ v(t)=\int_0^1 \mathcal{G}_2(t,s)\varphi_{\varrho_2}(I_{0+}^{\alpha_2}g(s,u(s),v(s)))\,ds, & t\in[0,1]. \end{cases}$$

We consider the Banach space $\mathcal{X}=C[0,1]$ with supremum norm $\|u\|=\sup_{t\in[0,1]}|u(t)|$, and the Banach space $\mathcal{Y}=\mathcal{X}\times\mathcal{X}$ with the norm $\|(u,v)\|_{\mathcal{Y}}=\|u\|+\|v\|$. We define the cone $\mathcal{Q}\subset\mathcal{Y}$ by

$$\mathcal{Q}=\{(u,v)\in\mathcal{Y},\ u(t)\ge 0,\ v(t)\ge 0,\ \forall t\in[0,1]\}.$$

We also define the operators $\mathcal{A}_1,\mathcal{A}_2:\mathcal{Y}\to\mathcal{X}$ and $\mathcal{A}:\mathcal{Y}\to\mathcal{Y}$ by

$$\begin{cases} \mathcal{A}_1(u,v)(t)=\int_0^1 \mathcal{G}_1(t,s)\varphi_{\varrho_1}(I_{0+}^{\alpha_1}f(s,u(s),v(s)))\,ds, & t\in[0,1], \\ \mathcal{A}_2(u,v)(t)=\int_0^1 \mathcal{G}_2(t,s)\varphi_{\varrho_2}(I_{0+}^{\alpha_2}g(s,u(s),v(s)))\,ds, & t\in[0,1], \end{cases}$$

and $\mathcal{A}(u,v)=(\mathcal{A}_1(u,v),\mathcal{A}_2(u,v))$, $(u,v)\in\mathcal{Y}$. Then (u,v) is a solution of problem (1) and (2) if and only if (u,v) is a fixed point of operator $\mathcal{A}$.

Lemma 5. *Assume that (I1) and (I2) hold. Then $\mathcal{A}:\mathcal{Q}\to\mathcal{Q}$ is a completely continuous operator (continuous, and it maps bounded sets into relatively compact sets).*

Proof. We denote by $M_i = \int_0^1 \mathcal{J}_i(s)\varphi_{\varrho_i}(I_{0+}^{\alpha_i}\zeta_i(s))\,ds$, $i=1,2$. Using $(I2)$ and Lemma 3, we deduce that $M_i > 0$, $i=1,2$. In addition, we find

$$\begin{aligned}
M_1 &= \int_0^1 \left[h_1(s) + \frac{1}{\Delta_1}\sum_{i=1}^{p}\int_0^1 g_{2i}(\tau,s)\,dH_i(\tau)\right]\varphi_{\varrho_1}(I_{0+}^{\alpha_1}\zeta_1(s))\,ds\\
&= \int_0^1 \frac{1}{\Gamma(\beta_1)}(1-s)^{\beta_1-\gamma_0-1}(1-(1-s)^{\gamma_0})\varphi_{\varrho_1}(I_{0+}^{\alpha_1}\zeta_1(s))\,ds\\
&\quad + \frac{1}{\Delta_1}\int_0^1\left(\sum_{i=1}^{p}\int_0^1 g_{2i}(\tau,s)\,dH_i(\tau)\right)\varphi_{\varrho_1}(I_{0+}^{\alpha_1}\zeta_1(s))\,ds\\
&\le \frac{1}{\Gamma(\beta_1)}\int_0^1 (1-s)^{\beta_1-\gamma_0-1}\varphi_{\varrho_1}(I_{0+}^{\alpha_1}\zeta_1(s))\,ds\\
&\quad + \frac{1}{\Delta_1}\int_0^1\left(\sum_{i=1}^{p}\int_0^1 \frac{1}{\Gamma(\beta_1-\gamma_i)}\tau^{\beta_1-\gamma_i-1}(1-s)^{\beta_1-\gamma_0-1}\,dH_i(\tau)\right)\varphi_{\varrho_1}(I_{0+}^{\alpha_1}\zeta_1(s))\,ds\\
&= \Lambda_1\left(\frac{1}{\Gamma(\beta_1)} + \frac{1}{\Delta_1}\sum_{i=1}^{p}\frac{1}{\Gamma(\beta_1-\gamma_i)}\int_0^1 \tau^{\beta_1-\gamma_i-1}\,dH_i(\tau)\right) < \infty,\\
M_2 &= \int_0^1 \left[h_2(s) + \frac{1}{\Delta_2}\sum_{i=1}^{q}\int_0^1 g_{4i}(\tau,s)\,dK_i(\tau)\right]\varphi_{\varrho_2}(I_{0+}^{\alpha_2}\zeta_2(s))\,ds\\
&= \int_0^1 \frac{1}{\Gamma(\beta_2)}(1-s)^{\beta_2-\delta_0-1}(1-(1-s)^{\delta_0})\varphi_{\varrho_2}(I_{0+}^{\alpha_2}\zeta_2(s))\,ds\\
&\quad + \frac{1}{\Delta_2}\int_0^1\left(\sum_{i=1}^{q}\int_0^1 g_{4i}(\tau,s)\,dK_i(\tau)\right)\varphi_{\varrho_2}(I_{0+}^{\alpha_2}\zeta_2(s))\,ds\\
&\le \frac{1}{\Gamma(\beta_2)}\int_0^1 (1-s)^{\beta_2-\delta_0-1}\varphi_{\varrho_2}(I_{0+}^{\alpha_2}\zeta_2(s))\,ds\\
&\quad + \frac{1}{\Delta_2}\int_0^1\left(\sum_{i=1}^{q}\int_0^1 \frac{1}{\Gamma(\beta_2-\delta_i)}\tau^{\beta_2-\delta_i-1}(1-s)^{\beta_2-\delta_0-1}\,dK_i(\tau)\right)\varphi_{\varrho_2}(I_{0+}^{\alpha_2}\zeta_2(s))\,ds\\
&= \Lambda_2\left(\frac{1}{\Gamma(\beta_2)} + \frac{1}{\Delta_2}\sum_{i=1}^{q}\frac{1}{\Gamma(\beta_2-\delta_i)}\int_0^1 \tau^{\beta_2-\delta_i-1}\,dK_i(\tau)\right) < \infty.
\end{aligned}$$

By Lemma 3 we conclude that $\mathcal{A}$ maps $\mathcal{Q}$ into $\mathcal{Q}$.

We will show that $\mathcal{A}$ maps bounded sets into relatively compact sets. Suppose $\mathcal{S} \subset \mathcal{Q}$ is an arbitrary bounded set. Then there exists $L_1 > 0$ such that $\|(u,v)\|_{\mathcal{Y}} \le L_1$ for all $(u,v) \in \mathcal{S}$. By the continuity of χ_1 and χ_2 we deduce that there exists $L_2 > 0$ such that $L_2 = \max\{\sup_{t\in[0,1],\,u,v\in[0,L_1]}\chi_1(t,u,v),\ \sup_{t\in[0,1],\,u,v\in[0,L_1]}\chi_2(t,u,v)\}$. By using Lemma 3, for any $(u,v)\in\mathcal{S}$ and $t\in[0,1]$, we obtain

$$\begin{aligned}
\mathcal{A}_1(u,v)(t) &\le \int_0^1 \mathcal{J}_1(s)\varphi_{\varrho_1}(I_{0+}^{\alpha_1}f(s,u(s),v(s)))\,ds\\
&\le \int_0^1 \mathcal{J}_1(s)\varphi_{\varrho_1}\left(\frac{1}{\Gamma(\alpha_1)}\int_0^s (s-\tau)^{\alpha_1-1}\zeta_1(\tau)\chi_1(\tau,u(\tau),v(\tau))\,d\tau\right)ds\\
&\le L_2^{\varrho_1-1}\int_0^1 \mathcal{J}_1(s)\varphi_{\varrho_1}\left(\frac{1}{\Gamma(\alpha_1)}\int_0^s (s-\tau)^{\alpha_1-1}\zeta_1(\tau)\,d\tau\right)ds\\
&= L_2^{\varrho_1-1}\int_0^1 \mathcal{J}_1(s)\varphi_{\varrho_1}(I_{0+}^{\alpha_1}\zeta_1(s))\,ds = M_1L_2^{\varrho_1-1},\\
\mathcal{A}_2(u,v)(t) &\le \int_0^1 \mathcal{J}_2(s)\varphi_{\varrho_2}(I_{0+}^{\alpha_2}g(s,u(s),v(s)))\,ds\\
&\le \int_0^1 \mathcal{J}_2(s)\varphi_{\varrho_2}\left(\frac{1}{\Gamma(\alpha_2)}\int_0^s (s-\tau)^{\alpha_2-1}\zeta_2(\tau)\chi_2(\tau,u(\tau),v(\tau))\,d\tau\right)ds\\
&\le L_2^{\varrho_2-1}\int_0^1 \mathcal{J}_2(s)\varphi_{\varrho_2}\left(\frac{1}{\Gamma(\alpha_2)}\int_0^s (s-\tau)^{\alpha_2-1}\zeta_2(\tau)\,d\tau\right)ds\\
&= L_2^{\varrho_2-1}\int_0^1 \mathcal{J}_2(s)\varphi_{\varrho_2}(I_{0+}^{\alpha_2}\zeta_2(s))\,ds = M_2L_2^{\varrho_2-1}.
\end{aligned}$$

Then $\|\mathcal{A}_1(u,v)\| \le M_1 L_2^{\varrho_1-1}$, $\|\mathcal{A}_2(u,v)\| \le M_2 L_2^{\varrho_2-1}$ for all $(u,v) \in \mathcal{S}$, and so $\mathcal{A}_1(\mathcal{S})$, $\mathcal{A}_2(\mathcal{S})$ and $\mathcal{A}(\mathcal{S})$ are bounded.

We will prove next that $\mathcal{A}(\mathcal{S})$ is equicontinuous. By using Lemma 1, for $(u,v) \in \mathcal{S}$ and $t \in [0,1]$ we deduce

$$\begin{aligned}
\mathcal{A}_1(u,v)(t) &= \int_0^1 \left(g_1(t,s) + \frac{t^{\beta_1-1}}{\Delta_1} \sum_{i=1}^p \int_0^1 g_{2i}(\tau,s)\,dH_i(\tau) \right) \varphi_{\varrho_1}(I_{0+}^{\alpha_1} f(s,u(s),v(s)))\,ds \\
&= \int_0^t \frac{1}{\Gamma(\beta_1)} \left[t^{\beta_1-1}(1-s)^{\beta_1-\gamma_0-1} - (t-s)^{\beta_1-1} \right] \varphi_{\varrho_1}(I_{0+}^{\alpha_1} f(s,u(s),v(s)))\,ds \\
&+ \int_t^1 \frac{1}{\Gamma(\beta_1)} t^{\beta_1-1}(1-s)^{\beta_1-\gamma_0-1} \varphi_{\varrho_1}(I_{0+}^{\alpha_1} f(s,u(s),v(s)))\,ds \\
&+ \frac{t^{\beta_1-1}}{\Delta_1} \int_0^1 \sum_{i=1}^p \left(\int_0^1 g_{2i}(\tau,s)\,dH_i(\tau) \right) \varphi_{\varrho_1}(I_{0+}^{\alpha_1} f(s,u(s),v(s)))\,ds.
\end{aligned}$$

Hence for any $t \in (0,1)$ we find

$$\begin{aligned}
(\mathcal{A}_1(u,v))'(t) &= \int_0^t \frac{1}{\Gamma(\beta_1)} \left[(\beta_1-1)t^{\beta_1-2}(1-s)^{\beta_1-\gamma_0-1} - (\beta_1-1)(t-s)^{\beta_1-2} \right] \\
&\times \varphi_{\varrho_1}(I_{0+}^{\alpha_1} f(s,u(s),v(s)))\,ds \\
&+ \int_t^1 \frac{1}{\Gamma(\beta_1)} (\beta_1-1)t^{\beta_1-2}(1-s)^{\beta_1-\gamma_0-1} \varphi_{\varrho_1}(I_{0+}^{\alpha_1} f(s,u(s),v(s)))\,ds \\
&+ \frac{(\beta_1-1)t^{\beta_1-2}}{\Delta_1} \int_0^1 \sum_{i=1}^p \left(\int_0^1 g_{2i}(\tau,s)\,dH_i(\tau) \right) \varphi_{\varrho_1}(I_{0+}^{\alpha_1} f(s,u(s),v(s)))\,ds.
\end{aligned}$$

Then for any $t \in (0,1)$ we obtain

$$\begin{aligned}
|(\mathcal{A}_1(u,v))'(t)| &\le \frac{1}{\Gamma(\beta_1-1)} \int_0^t [t^{\beta_1-2}(1-s)^{\beta_1-\gamma_0-1} + (t-s)^{\beta_1-2}] \\
&\times \varphi_{\varrho_1}(I_{0+}^{\alpha_1}(\zeta_1(s)\chi_1(s,u(s),v(s))))\,ds \\
&+ \frac{1}{\Gamma(\beta_1-1)} \int_t^1 t^{\beta_1-2}(1-s)^{\beta_1-\gamma_0-1} \varphi_{\varrho_1}(I_{0+}^{\alpha_1}(\zeta_1(s)\chi_1(s,u(s),v(s))))\,ds \\
&+ \frac{(\beta_1-1)t^{\beta_1-2}}{\Delta_1} \int_0^1 \sum_{i=1}^p \left(\int_0^1 g_{2i}(\tau,s)\,dH_i(\tau) \right) \varphi_{\varrho_1}(I_{0+}^{\alpha_1}(\zeta_1(s)\chi_1(s,u(s),v(s))))\,ds.
\end{aligned}$$

Therefore for any $t \in (0,1)$ we deduce

$$\begin{aligned}
|(\mathcal{A}_1(u,v))'(t)| &\le L_2^{\varrho_1-1} \Bigg[\frac{1}{\Gamma(\beta_1-1)} \int_0^t [t^{\beta_1-2}(1-s)^{\beta_1-\gamma_0-1} + (t-s)^{\beta_1-2}] \varphi_{\varrho_1}(I_{0+}^{\alpha_1}\zeta_1(s))\,ds \\
&+ \frac{1}{\Gamma(\beta_1-1)} \int_t^1 t^{\beta_1-2}(1-s)^{\beta_1-\gamma_0-1} \varphi_{\varrho_1}(I_{0+}^{\alpha_1}\zeta_1(s))\,ds \\
&+ \frac{(\beta_1-1)t^{\beta_1-2}}{\Delta_1} \int_0^1 \sum_{i=1}^p \left(\int_0^1 g_{2i}(\tau,s)\,dH_i(\tau) \right) \varphi_{\varrho_1}(I_{0+}^{\alpha_1}\zeta_1(s))\,ds \Bigg].
\end{aligned} \tag{18}$$

We denote by

$$\begin{aligned}
\theta_1(t) &= \frac{1}{\Gamma(\beta_1-1)} \int_0^t [t^{\beta_1-2}(1-s)^{\beta_1-\gamma_0-1} + (t-s)^{\beta_1-2}] \varphi_{\varrho_1}(I_{0+}^{\alpha_1}\zeta_1(s))\,ds \\
&+ \frac{1}{\Gamma(\beta_1-1)} \int_t^1 t^{\beta_1-2}(1-s)^{\beta_1-\gamma_0-1} \varphi_{\varrho_1}(I_{0+}^{\alpha_1}\zeta_1(s))\,ds, \\
\theta_2(t) &= \theta_1(t) + \frac{(\beta_1-1)t^{\beta_1-2}}{\Delta_1} \int_0^1 \sum_{i=1}^p \left(\int_0^1 g_{2i}(\tau,s)\,dH_i(\tau) \right) \varphi_{\varrho_1}(I_{0+}^{\alpha_1}\zeta_1(s))\,ds.
\end{aligned}$$

We compute the integral of function θ_1, by exchanging the order of integration, and we have

$$\int_0^1 \theta_1(t)\,dt = \frac{1}{\Gamma(\beta_1)}\int_0^1 (1-s)^{\beta_1-\gamma_0-1}(1+(1-s)^{\gamma_0})\varphi_{\varrho_1}(I_{0+}^{\alpha_1}\zeta_1(s))\,ds$$
$$\leq \frac{2}{\Gamma(\beta_1)}\int_0^1 (1-s)^{\beta_1-\gamma_0-1}\varphi_{\varrho_1}(I_{0+}^{\alpha_1}\zeta_1(s))\,ds = \frac{2\Lambda_1}{\Gamma(\beta_1)} < \infty.$$

For the integral of the function θ_2, we obtain

$$\begin{aligned}\int_0^1 \theta_2(t)\,dt &= \int_0^1 \theta_1(t)\,dt + \left(\int_0^1 \frac{(\beta_1-1)t^{\beta_1-2}}{\Delta_1}dt\right)\\
&\quad\times\left(\int_0^1 \sum_{i=1}^{p}\left(\int_0^1 g_{2i}(\tau,s)\,dH_i(\tau)\right)\varphi_{\varrho_1}(I_{0+}^{\alpha_1}\zeta_1(s))\,ds\right)\\
&\leq \frac{2}{\Gamma(\beta_1)}\int_0^1 (1-s)^{\beta_1-\gamma_0-1}\varphi_{\varrho_1}(I_{0+}^{\alpha_1}\zeta_1(s))\,ds\\
&\quad+\frac{1}{\Delta_1}\left(\int_0^1 \sum_{i=1}^{p}\left(\int_0^1 \frac{1}{\Gamma(\beta_1-\gamma_i)}\tau^{\beta_1-\gamma_i-1}(1-s)^{\beta_1-\gamma_0-1}\,dH_i(\tau)\right)\varphi_{\varrho_1}(I_{0+}^{\alpha_1}\zeta_1(s))\,ds\right)\\
&= \frac{2}{\Gamma(\beta_1)}\int_0^1 (1-s)^{\beta_1-\gamma_0-1}\varphi_{\varrho_1}(I_{0+}^{\alpha_1}\zeta_1(s))\,ds\\
&\quad+\frac{1}{\Delta_1}\left(\int_0^1 (1-s)^{\beta_1-\gamma_0-1}\varphi_{\varrho_1}(I_{0+}^{\alpha_1}\zeta_1(s))\,ds\right)\left(\sum_{i=1}^{p}\frac{1}{\Gamma(\beta_1-\gamma_i)}\int_0^1 \tau^{\beta_1-\gamma_i-1}dH_i(\tau)\right).\end{aligned}$$

Then we deduce

$$\int_0^1 \theta_2(t)\,dt \leq \Lambda_1\left(\frac{2}{\Gamma(\beta_1)}+\frac{1}{\Delta_1}\sum_{i=1}^{p}\frac{1}{\Gamma(\beta_1-\gamma_i)}\int_0^1 \tau^{\beta_1-\gamma_i-1}\,dH_i(\tau)\right) < \infty. \tag{19}$$

We conclude that $\theta_2 \in L^1(0,1)$. Hence for any $t_1, t_2 \in [0,1]$ with $t_1 \leq t_2$ and $(u,v) \in \mathcal{S}$, by (18) and (19), we find

$$|\mathcal{A}_1(u,v)(t_1) - \mathcal{A}_1(u,v)(t_2)| = \left|\int_{t_1}^{t_2} (\mathcal{A}_1(u,v))'(t)\,dt\right| \leq L_2^{\varrho_1-1}\int_{t_1}^{t_2}\theta_2(t)\,dt. \tag{20}$$

By (19), (20) and the absolute continuity of the integral function, we deduce that $\mathcal{A}_1(\mathcal{S})$ is equicontinuous. By a similar approach, we obtain that $\mathcal{A}_2(\mathcal{S})$ is also equicontinuous, and so $\mathcal{A}(\mathcal{S})$ is equicontinuous. Using the Ascoli–Arzela theorem, we conclude that $\mathcal{A}_1(\mathcal{S})$ and $\mathcal{A}_2(\mathcal{S})$ are relatively compact sets, and so $\mathcal{A}(\mathcal{S})$ is also relatively compact. Besides, we can prove that $\mathcal{A}_1$, $\mathcal{A}_2$ and $\mathcal{A}$ are continuous on $\mathcal{Q}$ (see [16] (Lemma 1.4.1)). Then $\mathcal{A}$ is a completely continuous operator on $\mathcal{Q}$. □

We define now the cone

$$\mathcal{Q}_0 = \{(u,v) \in \mathcal{Q},\ \min_{t\in[0,1]} u(t) \geq t^{\beta_1-1}\|u\|,\ \min_{t\in[0,1]} v(t) \geq t^{\beta_2-1}\|v\|\}.$$

Under the assumptions $(I1)$ and $(I2)$, by using Lemma 4, we obtain $\mathcal{A}(\mathcal{Q}) \subset \mathcal{Q}_0$, and so $\mathcal{A}|_{\mathcal{Q}_0} : \mathcal{Q}_0 \to \mathcal{Q}_0$ (denoted again by $\mathcal{A}$) is also a completely continuous operator. For $r > 0$ we denote by B_r the open ball centered at zero of radius r, and by $\overline{B}_r$ and ∂B_r its closure and its boundary, respectively.

Theorem 1. *Assume that $(I1)$ and $(I2)$ hold. In addition, the functions χ_1, χ_2, f and g satisfy the conditions*

$(I3)$ *There exist $\mu_1 \geq 1$ and $\mu_2 \geq 1$ such that*

$$\chi_{10} = \lim_{\substack{x+y\to 0\\ x,y\geq 0}}\sup_{t\in[0,1]}\frac{\chi_1(t,x,y)}{\varphi_{r_1}((x+y)^{\mu_1})} = 0 \ \ and\ \ \chi_{20} = \lim_{\substack{x+y\to 0\\ x,y\geq 0}}\sup_{t\in[0,1]}\frac{\chi_2(t,x,y)}{\varphi_{r_2}((x+y)^{\mu_2})} = 0;$$

(I4) *There exists* $[a_1, a_2] \subset [0,1], 0 < a_1 < a_2 < 1$ *such that*

$$f^i_\infty = \lim_{\substack{x+y\to\infty \\ x,y\geq 0}} \inf_{t\in[a_1,a_2]} \frac{f(t,x,y)}{\varphi_{r_1}(x+y)} = \infty \ \text{or} \ g^i_\infty = \lim_{\substack{x+y\to\infty \\ x,y\geq 0}} \inf_{t\in[a_1,a_2]} \frac{g(t,x,y)}{\varphi_{r_2}(x+y)} = \infty.$$

Then problem (1) and (2) has at least one positive solution $(u(t), v(t)),\ t \in [0,1]$.

Proof. We consider the above cone $\mathcal{Q}_0$. By (I3) we deduce that for $\epsilon_1 = \frac{1}{(2M_1)^{r_1-1}}$ and $\epsilon_2 = \frac{1}{(2M_2)^{r_2-1}}$, there exists $R_1 \in (0,1)$ such that

$$\chi_i(t,x,y) \leq \epsilon_i (x+y)^{\mu_i(r_i-1)}, \ \forall t \in [0,1], \ x+y \leq R_1, \ i = 1,2, \tag{21}$$

where $M_i,\ i = 1,2$ are defined in the proof of Lemma 5. Then by (21) and Lemma 3, for any $(u,v) \in \partial B_{R_1} \cap \mathcal{Q}_0$ and $t \in [0,1]$, we obtain

$$\begin{aligned}
\mathcal{A}_i(u,v)(t) &\leq \int_0^1 \mathcal{J}_i(s)\varphi_{\varrho_i}(I^{\alpha_i}_{0+}(\zeta_i(s)\chi_i(s,u(s),v(s))))\,ds \\
&\leq \int_0^1 \mathcal{J}_i(s)\varphi_{\varrho_i}(I^{\alpha_i}_{0+}(\zeta_i(s)\epsilon_i(u(s)+v(s))^{\mu_i(r_i-1)}))\,ds \\
&\leq \epsilon_i^{\varrho_i-1}\int_0^1 \mathcal{J}_i(s)\varphi_{\varrho_i}(I^{\alpha_i}_{0+}\zeta_i(s)(\|u\|+\|v\|)^{\mu_i(r_i-1)})\,ds \\
&= \epsilon_i^{\varrho_i-1}\|(u,v)\|_{\mathcal{Y}}^{\mu_i}\int_0^1 \mathcal{J}_i(s)\varphi_{\varrho_i}(I^{\alpha_i}_{0+}\zeta_i(s))\,ds \\
&= M_i\epsilon_i^{\varrho_i-1}\|(u,v)\|_{\mathcal{Y}}^{\mu_i} \leq M_i\epsilon_i^{\varrho_i-1}\|(u,v)\|_{\mathcal{Y}} = \tfrac{1}{2}\|(u,v)\|_{\mathcal{Y}}, \ i=1,2.
\end{aligned}$$

So we deduce that

$$\|\mathcal{A}(u,v)\|_{\mathcal{Y}} = \|\mathcal{A}_1(u,v)\| + \|\mathcal{A}_2(u,v)\| \leq \|(u,v)\|_{\mathcal{Y}}, \ \forall (u,v) \in \partial B_{R_1} \cap \mathcal{Q}_0. \tag{22}$$

By (I4), we suppose that $f^i_\infty = \infty$ (in a similar manner we can study the case $g^i_\infty = \infty$). Then for $\epsilon_3 = 2(A\min\{a_1^{\beta_1-1}, a_1^{\beta_2-1}\})^{1-r_1}$, where $A = \frac{a_1^{\beta_1-1}}{(\Gamma(\alpha_1+1))^{\varrho_1-1}}\int_{a_1}^{a_2}\mathcal{J}_1(s)(s-a_1)^{\alpha_1(\varrho_1-1)}\,ds$, there exists $C_1 > 0$ such that

$$f(t,x,y) \geq \epsilon_3(x+y)^{r_1-1} - C_1, \ \forall t \in [a_1,a_2], \ x,y \geq 0. \tag{23}$$

Then by (23), for any $(u,v) \in \mathcal{Q}_0$ and $t \in [a_1,a_2]$, we find

$$\begin{aligned}
\mathcal{A}_1(u,v)(t) &\geq \int_{a_1}^{a_2}\mathcal{G}_1(t,s)\left(\frac{1}{\Gamma(\alpha_1)}\int_{a_1}^{s}(s-\tau)^{\alpha_1-1}f(\tau,u(\tau),v(\tau))\,d\tau\right)^{\varrho_1-1}ds \\
&\geq a_1^{\beta_1-1}\int_{a_1}^{a_2}\frac{\mathcal{J}_1(s)}{(\Gamma(\alpha_1))^{\varrho_1-1}}\left(\int_{a_1}^{s}(s-\tau)^{\alpha_1-1}(\epsilon_3(u(\tau)+v(\tau))^{r_1-1}-C_1)\,d\tau\right)^{\varrho_1-1}ds \\
&= a_1^{\beta_1-1}\int_{a_1}^{a_2}\frac{\mathcal{J}_1(s)}{(\Gamma(\alpha_1))^{\varrho_1-1}}\left(\int_{a_1}^{s}(s-\tau)^{\alpha_1-1}(\epsilon_3(a_1^{\beta_1-1}\|u\|+a_1^{\beta_2-1}\|v\|)^{r_1-1}-C_1)d\tau\right)^{\varrho_1-1}ds \\
&= \frac{a_1^{\beta_1-1}}{(\Gamma(\alpha_1))^{\varrho_1-1}}\int_{a_1}^{a_2}\mathcal{J}_1(s)\left[\epsilon_3(a_1^{\beta_1-1}\|u\|+a_1^{\beta_2-1}\|v\|)^{r_1-1}-C_1\right]^{\varrho_1-1}\frac{(s-a_1)^{\alpha_1(\varrho_1-1)}}{\alpha_1^{\varrho_1-1}}\,ds \\
&\geq \frac{a_1^{\beta_1-1}}{(\Gamma(\alpha_1))^{\varrho_1-1}}\left[\epsilon_3\left(\min\{a_1^{\beta_1-1},a_1^{\beta_2-1}\}\right)^{r_1-1}\|(u,v)\|_{\mathcal{Y}}^{r_1-1}-C_1\right]^{\varrho_1-1} \\
&\quad\times\int_{a_1}^{a_2}\mathcal{J}_1(s)\frac{(s-a_1)^{\alpha_1(\varrho_1-1)}}{\alpha_1^{\varrho_1-1}}\,ds \\
&= \left[A^{\frac{1}{\varrho_1-1}}\epsilon_3\left(\min\{a_1^{\beta_1-1},a_1^{\beta_2-1}\}\right)^{r_1-1}\|(u,v)\|_{\mathcal{Y}}^{r_1-1}-A^{\frac{1}{\varrho_1-1}}C_1\right]^{\varrho_1-1} \\
&= \left(2\|(u,v)\|_{\mathcal{Y}}^{r_1-1}-C_2\right)^{\varrho_1-1}, \ C_2 = A^{r_1-1}C_1.
\end{aligned}$$

Hence we deduce

$$\|\mathcal{A}_1(u,v)\| \geq (2\|(u,v)\|_{\mathcal{Y}}^{r_1-1} - C_2)^{\varrho_1-1}, \ \forall\,(u,v) \in \mathcal{Q}_0.$$

We can choose $R_2 \geq \max\left\{1, C_2^{\varrho_1-1}\right\}$ and then we conclude

$$\|\mathcal{A}(u,v)\|_{\mathcal{Y}} \geq \|\mathcal{A}_1(u,v)\| \geq \|(u,v)\|\|_{\mathcal{Y}}, \ \forall\,(u,v) \in \partial B_{R_2} \cap \mathcal{Q}_0. \tag{24}$$

By using Lemma 5, the relations (22), (24), and the Guo–Krasnosel'skii fixed point theorem, we deduce that $\mathcal{A}$ has a fixed point $(u,v) \in (\overline{B}_{R_2} \setminus B_{R_1}) \cap \mathcal{Q}_0$, that is $R_1 \leq \|(u,v)\|_{\mathcal{Y}} \leq R_2$, and $u(t) \geq t^{\beta_1-1}\|u\|$ and $v(t) \geq t^{\beta_2-1}\|v\|$ for all $t \in [0,1]$. Then $\|u\| > 0$ or $\|v\| > 0$, that is $u(t) > 0$ for all $t \in (0,1]$ or $v(t) > 0$ for all $t \in (0,1]$. Hence $(u(t),v(t)),\ t \in [0,1]$ is a positive solution of problem (1) and (2). □

Remark 2. *Theorem 1 remains valid if the functions χ_1, χ_2 and f satisfy the inequalities (21) and (23), instead of (I3) and (I4).*

Theorem 2. *Assume that (I1) and (I2) hold. In addition the functions χ_1, χ_2, f and g satisfy the conditions*

(I5)

$$\chi_{1\infty} = \lim_{\substack{x+y\to\infty \\ x,y\geq 0}} \sup_{t\in[0,1]} \frac{\chi_1(t,x,y)}{\varphi_{r_1}(x+y)} = 0 \ \text{and} \ \chi_{2\infty} = \lim_{\substack{x+y\to\infty \\ x,y\geq 0}} \sup_{t\in[0,1]} \frac{\chi_2(t,x,y)}{\varphi_{r_2}(x+y)} = 0;$$

(I6) There exist $[a_1,a_2] \subset [0,1]$, $0 < a_1 < a_2 < 1$, $\nu_1 \in (0,1]$ and $\nu_2 \in (0,1]$ such that

$$f_0^i = \lim_{\substack{x+y\to 0 \\ x,y\geq 0}} \inf_{t\in[a_1,a_2]} \frac{f(t,x,y)}{\varphi_{r_1}((x+y)^{\nu_1})} = \infty \ \text{or} \ g_0^i = \lim_{\substack{x+y\to 0 \\ x,y\geq 0}} \inf_{t\in[a_1,a_2]} \frac{g(t,x,y)}{\varphi_{r_2}((x+y)^{\nu_2})} = \infty.$$

Then problem (1) and (2) has at least one positive solution $(u(t),v(t)),\ t \in [0,1]$.

Proof. We consider again the cone $\mathcal{Q}_0$. By (I5) we deduce that for $0 < \epsilon_4 < \min\left\{\frac{1}{(2M_1)^{r_1-1}}, \frac{1}{2M_1^{r_1-1}}\right\}$, $0 < \epsilon_5 < \min\left\{\frac{1}{(2M_2)^{r_2-1}}, \frac{1}{2M_2^{r_2-1}}\right\}$, there exist $C_3 > 0$, $C_4 > 0$ such that

$$\chi_1(t,x,y) \leq \epsilon_4(x+y)^{r_1-1} + C_3, \ \chi_2(t,x,y) \leq \epsilon_5(x+y)^{r_2-1} + C_4, \ \forall t \in [0,1],\ x,y \geq 0. \tag{25}$$

By using (I2) and (25), for any $(u,v) \in \mathcal{Q}_0$, we obtain

$$\begin{aligned}
\mathcal{A}_1(u,v)(t) &\leq \int_0^1 \mathcal{J}_1(s)\varphi_{\varrho_1}(I_{0+}^{\alpha_1}(\zeta_1(s)\chi_1(s,u(s),v(s))))\,ds \\
&\leq \int_0^1 \mathcal{J}_1(s)\varphi_{\varrho_1}(I_{0+}^{\alpha_1}(\zeta_1(s)(\epsilon_4(u(s)+v(s))^{r_1-1} + C_3)))\,ds \\
&\leq \int_0^1 \mathcal{J}_1(s)\frac{1}{(\Gamma(\alpha_1))^{\varrho_1-1}}\left(\epsilon_4\|(u,v)\|_{\mathcal{Y}}^{r_1-1} + C_3\right)^{\varrho_1-1}\left(\int_0^s (s-\tau)^{\alpha_1-1}\zeta_1(\tau)\,d\tau\right)^{\varrho_1-1} ds \\
&= M_1\left(\epsilon_4\|(u,v)\|_{\mathcal{Y}}^{r_1-1} + C_3\right)^{\varrho_1-1}, \ \forall t \in [0,1], \\
\mathcal{A}_2(u,v)(t) &\leq \int_0^1 \mathcal{J}_2(s)\varphi_{\varrho_2}(I_{0+}^{\alpha_2}(\zeta_2(s)\chi_2(s,u(s),v(s))))\,ds \\
&\leq \int_0^1 \mathcal{J}_2(s)\varphi_{\varrho_2}(I_{0+}^{\alpha_2}(\zeta_2(s)(\epsilon_5(u(s)+v(s))^{r_2-1} + C_4)))\,ds \\
&\leq \int_0^1 \mathcal{J}_2(s)\frac{1}{(\Gamma(\alpha_2))^{\varrho_2-1}}\left(\epsilon_5\|(u,v)\|_{\mathcal{Y}}^{r_2-1} + C_4\right)^{\varrho_2-1}\left(\int_0^s (s-\tau)^{\alpha_2-1}\zeta_2(\tau)\,d\tau\right)^{\varrho_2-1} ds \\
&= M_2\left(\epsilon_5\|(u,v)\|_{\mathcal{Y}}^{r_2-1} + C_4\right)^{\varrho_2-1}, \ \forall t \in [0,1].
\end{aligned}$$

Then we find

$$\|\mathcal{A}_1(u,v)\| \le M_1 \left(\epsilon_4 \|(u,v)\|_{\mathcal{Y}}^{r_1-1} + C_3\right)^{\varrho_1-1},$$
$$\|\mathcal{A}_2(u,v)\| \le M_2 \left(\epsilon_5 \|(u,v)\|_{\mathcal{Y}}^{r_2-1} + C_4\right)^{\varrho_2-1},$$

and so

$$\|\mathcal{A}(u,v)\|_{\mathcal{Y}} \le M_1 \left(\epsilon_4 \|(u,v)\|_{\mathcal{Y}}^{r_1-1} + C_3\right)^{\varrho_1-1} + M_2 \left(\epsilon_5 \|(u,v)\|_{\mathcal{Y}}^{r_2-1} + C_4\right)^{\varrho_2-1},$$

for all $(u,v) \in \mathcal{Q}_0$. We choose

$$R_3 > \max\left\{1, \frac{M_1 C_3^{\varrho_1-1} + M_2 C_4^{\varrho_2-1}}{1-\left(M_1\epsilon_4^{\varrho_1-1} + M_2\epsilon_5^{\varrho_2-1}\right)}, \frac{M_1 2^{\varrho_1-2} C_3^{\varrho_1-1} + M_2 2^{\varrho_2-2} C_4^{\varrho_2-1}}{1-\left(M_1 2^{\varrho_1-2}\epsilon_4^{\varrho_1-1} + M_2 2^{\varrho_2-2}\epsilon_5^{\varrho_2-1}\right)}, \frac{M_1 C_3^{\varrho_1-1} + M_2 2^{\varrho_2-2} C_4^{\varrho_2-1}}{1-\left(M_1\epsilon_4^{\varrho_1-1} + M_2 2^{\varrho_2-2}\epsilon_5^{\varrho_2-1}\right)}, \frac{M_1 2^{\varrho_1-2} C_3^{\varrho_1-1} + M_2 C_4^{\varrho_2-1}}{1-\left(M_1 2^{\varrho_1-2}\epsilon_4^{\varrho_1-1} + M_2\epsilon_5^{\varrho_2-1}\right)}\right\}, \tag{26}$$

and then we deduce

$$\|\mathcal{A}(u,v)\|_{\mathcal{Y}} \le \|(u,v)\|_{\mathcal{Y}}, \ \forall\,(u,v) \in \partial B_{R_3} \cap \mathcal{Q}_0. \tag{27}$$

The choosing of R_3 above is based on the inequalities $(a+b)^p \le 2^{p-1}(a^p+b^p)$ for $p \ge 1$ and $a,b \ge 0$, and $(a+b)^p \le a^p + b^p$ for $p \in (0,1]$ and $a,b \ge 0$. Here $p = \varrho_1 - 1$ or $\varrho_2 - 1$. We explain the above inequality (27) in one case, namely $\varrho_1 \ge 2$ and $\varrho_2 \ge 2$. In this situation, by using (26), and the relations $M_1 2^{\varrho_1-2}\epsilon_4^{\varrho_1-1} < \frac{1}{2}$, $M_2 2^{\varrho_2-2}\epsilon_5^{\varrho_2-1} < \frac{1}{2}$ (from the definition of ϵ_4 and ϵ_5), we have the inequalities

$$\begin{aligned}
&M_1(\epsilon_4 R_3^{r_1-1} + C_3)^{\varrho_1-1} + M_2(\epsilon_5 R_3^{r_2-1} + C_4)^{\varrho_2-1} \\
&\le M_1 2^{\varrho_1-2}(\epsilon_4^{\varrho_1-1} R_3 + C_3^{\varrho_1-1}) + M_2 2^{\varrho_2-2}(\epsilon_5^{\varrho_2-1} R_3 + C_4^{\varrho_2-1}) \\
&= (M_1 2^{\varrho_1-2}\epsilon_4^{\varrho_1-1} + M_2 2^{\varrho_2-2}\epsilon_5^{\varrho_2-1}) R_3 + M_1 2^{\varrho_1-2} C_3^{\varrho_1-1} + M_2 2^{\varrho_2-1} C_4^{\varrho_2-1} < R_3.
\end{aligned}$$

In a similar manner we treat the cases: $\varrho_1 \in (1,2]$ and $\varrho_2 \in (1,2]$; $\varrho_1 \ge 2$ and $\varrho_2 \in (1,2]$; $\varrho_1 \in (1,2]$ and $\varrho_2 \ge 2$.

By $(I6)$, we suppose that $g_0^i = \infty$ (in a similar manner we can study the case $f_0^i = \infty$). We deduce that for $\epsilon_6 = (\min\{a_1^{\beta_1-1}, a_1^{\beta_2-1}\})^{\nu_2(1-r_2)} \widetilde{A}^{1-r_2}$, where $\widetilde{A} = \frac{a_1^{\beta_2-1}}{(\Gamma(\alpha_2+1))^{\varrho_2-1}} \int_{a_1}^{a_2} (s-a_1)^{\alpha_2(\varrho_2-1)} \mathcal{J}_2(s)\,ds$, there exists $R_4 \in (0,1]$ such that

$$g(t,x,y) \ge \epsilon_6 (x+y)^{\nu_2(r_2-1)}, \ \forall\, t \in [a_1,a_2],\ x,y \ge 0,\ x+y \le R_4. \tag{28}$$

Then by using (28), for any $(u,v) \in \partial B_{R_4} \cap \mathcal{Q}_0$ and $t \in [a_1,a_2]$, we find

$$\begin{aligned}
&\mathcal{A}_2(u,v)(t) \ge \int_{a_1}^{a_2} \mathcal{G}_2(t,s) \left(\frac{1}{\Gamma(\alpha_2)} \int_{a_1}^{s} (s-\tau)^{\alpha_2-1} g(\tau,u(\tau),v(\tau))\,d\tau\right)^{\varrho_2-1} ds \\
&\ge a_1^{\beta_2-1} \int_{a_1}^{a_2} \mathcal{J}_2(s) \frac{1}{(\Gamma(\alpha_2))^{\varrho_2-1}} \left(\int_{a_1}^{s} (s-\tau)^{\alpha_2-1} \epsilon_6 (u(\tau)+v(\tau))^{\nu_2(r_2-1)} d\tau\right)^{\varrho_2-1} ds \\
&\ge a_1^{\beta_2-1} \int_{a_1}^{a_2} \mathcal{J}_2(s) \frac{1}{(\Gamma(\alpha_2))^{\varrho_2-1}} \left(\int_{a_1}^{s} (s-\tau)^{\alpha_2-1} \epsilon_6 (a_1^{\beta_1-1}\|u\| + a_1^{\beta_2-1}\|v\|)^{\nu_2(r_2-1)} d\tau\right)^{\varrho_2-1} ds \\
&\ge a_1^{\beta_2-1} \epsilon_6^{\varrho_2-1} \left(\min\{a_1^{\beta_1-1}, a_1^{\beta_2-1}\}\right)^{\nu_2} \frac{\|(u,v)\|_{\mathcal{Y}}^{\nu_2}}{(\Gamma(\alpha_2+1))^{\varrho_2-1}} \int_{a_1}^{a_2} (s-a_1)^{\alpha_2(\varrho_2-1)} \mathcal{J}_2(s)\,ds \\
&= \|(u,v)\|_{\mathcal{Y}}^{\nu_2} \ge \|(u,v)\|_{\mathcal{Y}}.
\end{aligned}$$

Therefore $\|\mathcal{A}_2(u,v)\| \ge \|(u,v)\|_{\mathcal{Y}}$ for all $(u,v) \in \partial B_{R_4} \cap \mathcal{Q}_0$, and then

$$\|\mathcal{A}(u,v)\|_{\mathcal{Y}} \ge \|\mathcal{A}_2(u,v)\| \ge \|(u,v)\|_{\mathcal{Y}}, \ \forall\,(u,v) \in \partial B_{R_4} \cap \mathcal{Q}_0. \tag{29}$$

By Lemma 5, the relations (27), (29), and the Guo–Krasnosl'skii fixed point theorem, we conclude that $\mathcal{A}$ has at least one fixed point $(u,v) \in (\overline{B}_{R_3} \setminus B_{R_4}) \cap \mathcal{Q}_0$, that is $R_4 \le \|(u,v)\|_{\mathcal{Y}} \le R_3$, which is a positive solution of problem (1) and (2). □

Remark 3. *Theorem 2 remains valid if the functions χ_1, χ_2 and g satisfy the inequalities (25) and (28), instead of $(I5)$ and $(I6)$.*

Theorem 3. *Assume that $(I1)$, $(I2)$, $(I4)$ and $(I6)$ hold. In addition, the functions χ_1 and χ_2 satisfy the condition*

$(I7)$ $D_0^{\varrho_1-1}M_1 < \frac{1}{2}$, $D_0^{\varrho_2-1}M_2 < \frac{1}{2}$, *where*

$$D_0 = \max\{\max_{t,x,y\in[0,1]} \chi_1(t,x,y), \max_{t,x,y\in[0,1]} \chi_2(t,x,y)\}.$$

Then problem (1) and (2) has at least two positive solutions $(u_1(t),v_1(t))$, $(u_2(t),v_2(t))$, $t \in [0,1]$.

Proof. We consider the operators $\mathcal{A}_1$, $\mathcal{A}_2$, $\mathcal{A}$, and the cone $\mathcal{Q}_0$ defined in this section. If $(I1)$, $(I2)$ and $(I4)$ hold, then by the proof of Theorem 1, we deduce that there exists $R_2 > 1$ (we can consider $R_2 > 1$) such that

$$\|\mathcal{A}(u,v)\|_{\mathcal{Y}} \ge \|(u,v)\|_{\mathcal{Y}}, \ \forall (u,v) \in \partial B_{R_2} \cap \mathcal{Q}_0. \tag{30}$$

If $(I1)$, $(I2)$ and $(I6)$ hold, then by the proof of Theorem 2 we find that there exists $R_4 < 1$ (we can consider $R_4 < 1$) such that

$$\|\mathcal{A}(u,v)\|_{\mathcal{Y}} \ge \|(u,v)\|_{\mathcal{Y}}, \ \forall (u,v) \in \partial B_{R_4} \cap \mathcal{Q}_0. \tag{31}$$

We consider now the set $B_1 = \{(u,v) \in \mathcal{Y}, \ \|(u,v)\|_{\mathcal{Y}} < 1\}$. By $(I7)$, for any $(u,v) \in \partial B_1 \cap \mathcal{Q}_0$ and $t \in [0,1]$, we obtain

$$\begin{aligned} \mathcal{A}_i(u,v)(t) &\le \int_0^1 \mathcal{J}_i(s) \left(\frac{1}{\Gamma(\alpha_i)} \int_0^s (s-\tau)^{\alpha_i-1} \zeta_i(\tau) \chi_i(\tau, u(\tau), v(\tau))\, d\tau \right)^{\varrho_i-1} ds \\ &\le D_0^{\varrho_i-1} \int_0^1 \mathcal{J}_i(s) \left(\frac{1}{\Gamma(\alpha_i)} \int_0^s (s-\tau)^{\alpha_i-1} \zeta_i(\tau)\, d\tau \right)^{\varrho_i-1} ds \\ &= D_0^{\varrho_i-1} \int_0^1 \mathcal{J}_i(s) \varphi_{\varrho_i}(I_{0+}^{\alpha_i} \zeta_i(s))\, ds = D_0^{\varrho_i-1} M_i < \frac{1}{2}, \ i=1,2. \end{aligned}$$

So $\|\mathcal{A}_i(u,v)\| < \frac{1}{2}$, for all $(u,v) \in \partial B_1 \cap \mathcal{Q}_0$, $i=1,2$. Then

$$\|\mathcal{A}(u,v)\|_{\mathcal{Y}} = \|\mathcal{A}_1(u,v)\| + \|\mathcal{A}_2(u,v)\| < 1 = \|(u,v)\|_{\mathcal{Y}}, \ \forall (u,v) \in \partial B_1 \cap \mathcal{Q}_0. \tag{32}$$

Therefore, by (30), (32) and the Guo–Krasnosel'skii fixed point theorem, we conclude that problem (1), (2) has one positive solution $(u_1,v_1) \in \mathcal{Q}_0$ with $1 < \|(u_1,v_1)\|_{\mathcal{Y}} \le R_2$. By (31), (32) and the Guo–Krasnosel'skii fixed point theorem, we deduce that problem (1), (2) has another positive solution $(u_2,v_2) \in \mathcal{Q}_0$ with $R_4 \le \|(u_2,v_2)\|_{\mathcal{Y}} < 1$. Then problem (1) and (2) has at least two positive solutions $(u_1(t),v_1(t))$, $(u_2(t),v_2(t))$, $t \in [0,1]$. □

Remark 4. *Theorem 3 remains valid if the functions f and g satisfy the inequalities (23) and (28), instead of $(I4)$ and $(I6)$.*

4. Examples

Let $\alpha_1 = 1/3$, $\alpha_2 = 1/2$, $\beta_1 = 5/2$, $(n=3)$, $\beta_2 = 13/4$, $(m=4)$, $r_1 = 4$, $\varrho_1 = 4/3$, $r_2 = 5$, $\varrho_2 = 5/4$, $p = 2$, $q = 1$, $\gamma_0 = 4/3$, $\gamma_1 = 1/4$, $\gamma_2 = 6/5$, $\delta_0 = 11/5$, $\delta_1 = 7/6$, $H_1(t) = t/3$ for all $t \in [0,1]$, $H_2(t) = \{1/6, \ t \in [0,2/3); \ 2/3, \ t \in [2/3,1]\}$, $K_1(t) = \{1/4, \ t \in [0,1/3); \ 9/4, \ t \in [1/3,1]\}$.

We consider the system of fractional differential equations

$$\begin{cases} D_{0+}^{1/3}(\varphi_4(D_{0+}^{5/2}u(t))) + f(t,u(t),v(t)) = 0, \ t \in (0,1), \\ D_{0+}^{1/2}(\varphi_5(D_{0+}^{13/4}v(t))) + g(t,u(t),v(t)) = 0, \ t \in (0,1), \end{cases} \tag{33}$$

with the nonlocal boundary conditions

$$\begin{cases} u(0) = u'(0) = 0, \ D_{0+}^{5/2}u(0) = 0, \ D_{0+}^{4/3}u(1) = \dfrac{1}{3}\displaystyle\int_0^1 D_{0+}^{1/4}u(t)\,dt + \dfrac{1}{2}D_{0+}^{6/5}u\left(\dfrac{2}{3}\right), \\ v(0) = v'(0) = v''(0) = 0, \ D_{0+}^{13/4}v(0) = 0, \ D_{0+}^{11/5}v(1) = 2D_{0+}^{7/6}v\left(\dfrac{1}{3}\right). \end{cases} \tag{34}$$

We obtain here $\Delta_1 \approx 0.60331103 > 0$ and $\Delta_2 \approx 1.12479609 > 0$. We also find

$$\begin{aligned}
g_1(t,s) &= \frac{1}{\Gamma(5/2)}\begin{cases} t^{3/2}(1-s)^{1/6} - (t-s)^{3/2}, \ 0 \le s \le t \le 1, \\ t^{3/2}(1-s)^{1/6}, \ 0 \le t \le s \le 1, \end{cases} \\
g_{21}(t,s) &= \frac{1}{\Gamma(9/4)}\begin{cases} t^{5/4}(1-s)^{1/6} - (t-s)^{5/4}, \ 0 \le s \le t \le 1, \\ t^{5/4}(1-s)^{1/6}, \ 0 \le t \le s \le 1, \end{cases} \\
g_{22}(t,s) &= \frac{1}{\Gamma(13/10)}\begin{cases} t^{3/10}(1-s)^{1/6} - (t-s)^{3/10}, \ 0 \le s \le t \le 1, \\ t^{3/10}(1-s)^{1/6}, \ 0 \le t \le s \le 1, \end{cases} \\
g_3(t,s) &= \frac{1}{\Gamma(13/4)}\begin{cases} t^{9/4}(1-s)^{1/20} - (t-s)^{9/4}, \ 0 \le s \le t \le 1, \\ t^{9/4}(1-s)^{1/20}, \ 0 \le t \le s \le 1, \end{cases} \\
g_{41}(t,s) &= \frac{1}{\Gamma(25/12)}\begin{cases} t^{13/12}(1-s)^{1/20} - (t-s)^{13/12}, \ 0 \le s \le t \le 1, \\ t^{13/12}(1-s)^{1/20}, \ 0 \le t \le s \le 1, \end{cases} \\
\mathcal{G}_1(t,s) &= g_1(t,s) + \frac{t^{3/2}}{\Delta_1}\left[\frac{1}{3}\int_0^1 g_{21}(\tau,s)\,d\tau + \frac{1}{2}g_{22}\left(\frac{2}{3},s\right)\right], \ t,s \in [0,1], \\
\mathcal{G}_2(t,s) &= g_3(t,s) + \frac{2t^{9/4}}{\Delta_2}g_{41}\left(\frac{1}{3},s\right), \ t,s \in [0,1], \\
h_1(s) &= \frac{1}{\Gamma(5/2)}[(1-s)^{1/6} - (1-s)^{3/2}], \ s \in [0,1], \\
h_2(s) &= \frac{1}{\Gamma(13/4)}[(1-s)^{1/20} - (1-s)^{9/4}], \ s \in [0,1].
\end{aligned}$$

In addition we deduce

$$\mathcal{J}_1(s) = \begin{cases} h_1(s) + \frac{1}{\Delta_1}\left\{\frac{4}{27\Gamma(9/4)}(1-s)^{1/6} - \frac{4}{27\Gamma(9/4)}(1-s)^{9/4} + \frac{1}{2\Gamma(13/10)}\right. \\ \quad \left.\times\left[\left(\frac{2}{3}\right)^{3/10}(1-s)^{1/6} - \left(\frac{2}{3}-s\right)^{3/10}\right]\right\}, \ 0 \le s < \frac{2}{3}, \\ h_1(s) + \frac{1}{\Delta_1}\left[\frac{4}{27\Gamma(9/4)}(1-s)^{1/6} - \frac{4}{27\Gamma(9/4)}(1-s)^{9/4} + \frac{1}{2\Gamma(13/10)}\right. \\ \quad \left.\times\left(\frac{2}{3}\right)^{3/10}(1-s)^{1/6}\right], \ 2/3 \le s \le 1, \end{cases}$$

$$\mathcal{J}_2(s) = \begin{cases} h_2(s) + \frac{2}{\Delta_2\Gamma(25/12)}\left[\left(\frac{1}{3}\right)^{13/12}(1-s)^{1/20} - \left(\frac{1}{3}-s\right)^{13/12}\right], \ 0 \le s < \frac{1}{3}, \\ h_2(s) + \frac{2}{\Delta_2\Gamma(25/12)}\left(\frac{1}{3}\right)^{13/12}(1-s)^{1/20}, \ \frac{1}{3} \le s \le 1. \end{cases}$$

Example 1. *We consider the functions*

$$f(t,x,y) = \frac{(x+y)^{3a}}{t^{\eta_1}(1-t)^{\eta_2}}, \ g(t,x,y) = \frac{(x+y)^{4b}}{t^{\eta_3}(1-t)^{\eta_4}}, \ t \in (0,1), \ x,y \ge 0, \tag{35}$$

where $a > 1$, $b > 1$, $\eta_1, \eta_2 \in (0,1/4)$, $\eta_3, \eta_4 \in (0,1/3)$. Here $f(t,x,y) = \zeta_1(t)\chi_1(t,x,y)$, $g(t,x,y) = \zeta_2(t)\chi_2(t,x,y)$, $\zeta_1(t) = \frac{1}{t^{\eta_1}(1-t)^{\eta_2}}$, $\zeta_2(t) = \frac{1}{t^{\eta_3}(1-t)^{\eta_4}}$ for all $t \in (0,1)$, $\chi_1(t,x,y) = (x+y)^{3a}$, $\chi_2(t,x,y) = (x+y)^{4b}$ for all $t \in [0,1]$, $x,y \ge 0$. By using the Hölder inequality, we obtain

$$
\begin{aligned}
0 < \Lambda_1 &= \int_0^1 (1-s)^{\beta_1-\gamma_0-1}\varphi_{\varrho_1}(I_{0+}^{\alpha_1}\zeta_1(s))\,ds = \int_0^1 (1-s)^{1/6}\left(I_{0+}^{1/3}\zeta_1(s)\right)^{1/3} ds \\
&= \frac{1}{(\Gamma(1/3))^{1/3}}\int_0^1 (1-s)^{1/6}\left(\int_0^s (s-\tau)^{-2/3}\frac{1}{\tau^{\eta_1}(1-\tau)^{\eta_2}}\,d\tau\right)^{1/3} ds \\
&\le \frac{1}{(\Gamma(1/3))^{1/3}}\int_0^1 (1-s)^{1/6}\left[\left(\int_0^s (s-\tau)^{-8/9}d\tau\right)^{3/4}\left(\int_0^s\left(\frac{1}{\tau^{\eta_1}(1-\tau)^{\eta_2}}\right)^4 d\tau\right)^{1/4}\right]^{1/3} ds \\
&\le \frac{1}{(\Gamma(1/3))^{1/3}}\int_0^1 (1-s)^{1/6}\left[(9s^{1/9})^{3/4}(B(1-4\eta_1,1-4\eta_2))^{1/4}\right]^{1/3} ds \\
&= \frac{3^{1/2}}{(\Gamma(1/3))^{1/3}}(B(1-4\eta_1,1-4\eta_2))^{1/12}B\left(\frac{37}{36},\frac{7}{6}\right) < \infty, \\
0 < \Lambda_2 &= \int_0^1 (1-s)^{\beta_2-\delta_0-1}\varphi_{\varrho_2}(I_{0+}^{\alpha_2}\zeta_2(s))\,ds = \int_0^1 (1-s)^{1/20}\left(I_{0+}^{1/2}\zeta_2(s)\right)^{1/4} ds \\
&= \frac{1}{(\Gamma(1/2))^{1/4}}\int_0^1 (1-s)^{1/20}\left(\int_0^s (s-\tau)^{-1/2}\frac{1}{\tau^{\eta_3}(1-\tau)^{\eta_4}}d\tau\right)^{1/4} ds \\
&\le \frac{1}{(\Gamma(1/2))^{1/4}}\int_0^1 (1-s)^{1/20}\left[\left(\int_0^s (s-\tau)^{-3/4}d\tau\right)^{2/3}\left(\int_0^s \tau^{-3\eta_3}(1-\tau)^{-3\eta_4}d\tau\right)^{1/3}\right]^{1/4} ds \\
&\le \frac{1}{(\Gamma(1/2))^{1/4}}\int_0^1 (1-s)^{1/20}[(4s^{1/4})^{2/3}(B(1-3\eta_3,1-3\eta_4))^{1/3}]^{1/4} ds \\
&= \frac{2^{1/3}}{(\Gamma(1/2))^{1/4}}(B(1-3\eta_3,1-3\eta_4))^{1/12}B\left(\frac{25}{24},\frac{21}{20}\right) < \infty.
\end{aligned}
$$

Hence assumptions $(I1)$ and $(I2)$ are satisfied.

In addition, in $(I3)$, for $\mu_1 = \mu_2 = 1$, we obtain $\chi_{10} = \chi_{20} = 0$, and in $(I4)$ for $[a_1, a_2] \subset (0,1)$ we have $f_\infty^i = \infty$ (and $g_\infty^i = \infty$). Then by Theorem 1, we conclude that problem (33) and (34) with the nonlinearities (35) has at least one positive solution $(u(t), v(t))$, $t \in [0,1]$.

Example 2. *We consider the functions*

$$
\begin{aligned}
f(t,x,y) &= \frac{c_0(t+1)}{(t^2+4)\sqrt[5]{t}}[(x+y)^{\sigma_1} + (x+y)^{\sigma_2}], \quad t \in (0,1],\ x,y \ge 0, \\
g(t,x,y) &= \frac{d_0(2+\sin t)}{(t+3)^4\sqrt[4]{1-t}}(x^{\sigma_3} + y^{\sigma_4}), \quad t \in [0,1),\ x,y \ge 0,
\end{aligned}
\tag{36}
$$

where $c_0 > 0$, $d_0 > 0$, $\sigma_1 > 3$, $\sigma_2 \in (0,3)$, $\sigma_3 > 0$, $\sigma_4 > 0$. *Here we have* $\zeta_1(t) = \frac{1}{\sqrt[5]{t}}$, $t \in (0,1]$, $\chi_1(t,x,y) = \frac{c_0(t+1)}{t^2+4}[(x+y)^{\sigma_1} + (x+y)^{\sigma_2}]$, $t \in [0,1]$, $x, y \ge 0$, $\zeta_2(t) = \frac{1}{\sqrt[4]{1-t}}$, $t \in [0,1)$, $\chi_2(t,x,y) = \frac{d_0(2+\sin t)}{(t+3)^4}(x^{\sigma_3} + y^{\sigma_4})$, $t \in [0,1]$, $x, y \ge 0$. *By using a computer program, we obtain*

$$
\begin{aligned}
\Lambda_1 &= \int_0^1 (1-s)^{\beta_1-\gamma_0-1}\varphi_{\varrho_1}(I_{0+}^{\alpha_1}\zeta_1(s))\,ds = \int_0^1 (1-s)^{1/6}(I_{0+}^{1/3}\zeta_1(s))^{1/3} ds \\
&= \frac{1}{(\Gamma(1/3))^{1/3}}\int_0^1 (1-s)^{1/6}\left(\int_0^s \tau^{-1/5}(s-\tau)^{-2/3}d\tau\right)^{1/3} ds \\
&\overset{\tau=sx}{=} \frac{1}{(\Gamma(1/3))^{1/3}}\int_0^1 (1-s)^{1/6}\left(\int_0^1 (sx)^{-1/5}(s-sx)^{-2/3}s\,dx\right)^{1/3} ds \\
&= \frac{1}{(\Gamma(1/3))^{1/3}}\left(\int_0^1 s^{2/45}(1-s)^{1/6}ds\right)\left(\int_0^1 x^{-1/5}(1-x)^{-2/3}dx\right)^{1/3} \\
&= \frac{1}{(\Gamma(1/3))^{1/3}}B\left(\frac{47}{45},\frac{7}{6}\right)\left(B\left(\frac{4}{5},\frac{1}{3}\right)\right)^{1/3} \approx 0.877777,
\end{aligned}
$$

$$\begin{aligned}\Lambda_2 &= \int_0^1 (1-s)^{\beta_2-\delta_0-1}\varphi_{\varrho_2}(I_{0+}^{\alpha_2}\zeta_2(s))\,ds = \int_0^1 (1-s)^{1/20}(I_{0+}^{1/2}\zeta_2(s))^{1/4}\,ds \\ &= \frac{1}{(\Gamma(1/2))^{1/4}}\int_0^1 (1-s)^{1/20}\left(\int_0^s (s-\tau)^{-1/2}(1-\tau)^{-1/4}d\tau\right)^{1/4} ds \\ &\overset{\tau=sx}{=} \frac{1}{(\Gamma(1/2))^{1/4}}\int_0^1 (1-s)^{1/20}\left(\int_0^1 (s-sx)^{-1/2}(1-sx)^{-1/4}s\,dx\right)^{1/4} ds \\ &= \frac{1}{(\Gamma(1/2))^{1/4}}\int_0^1 (1-s)^{1/20}\left(s^{1/2}\int_0^1 (1-x)^{-1/2}(1-sx)^{-1/4}\,dx\right)^{1/4} ds \\ &= \frac{1}{(\Gamma(1/2))^{1/4}}\int_0^1 (1-s)^{1/20}\left(s^{1/2}\sqrt{\pi}\,{}_2\tilde{F}_1\left[\frac{1}{4},1,\frac{3}{2},s\right]\right)^{1/4} ds \approx 0.901313,\end{aligned}$$

where ${}_2\tilde{F}_1[a,b,c,z] = \frac{1}{\Gamma(b)\Gamma(c-b)}\int_0^1 s^{b-1}(1-s)^{c-b-1}(1-sz)^{-a}\,ds$ *is the regularized hypergeometric function. So* $\Lambda_i \in (0,\infty)$, $i = 1,2$, *and then assumptions* $(I1)$ *and* $(I2)$ *are satisfied.*

For $[a_1, a_2] \subset (0,1)$, we find $f_\infty^i = \infty$, and if we consider $0 < \nu_1 \le 1$, $3\nu_1 > \sigma_2$ we obtain $f_0^i = \infty$, and then assumptions $(I4)$ and $(I6)$ are satisfied. After some computations we deduce

$$\begin{aligned}M_1 &= \int_0^1 \mathcal{J}_1(s)\varphi_{\varrho_1}(I_{0+}^{\alpha_1}\zeta_1(s))\,ds = \int_0^1 \mathcal{J}_1(s)\varphi_{4/3}\left(I_{0+}^{1/3}\frac{1}{\sqrt[5]{s}}\right) ds \\ &= \frac{1}{(\Gamma(1/3))^{1/3}}\int_0^1 \mathcal{J}_1(s)\left(\int_0^s (s-\tau)^{-2/3}\tau^{-1/5}d\tau\right)^{1/3} ds \\ &= \frac{1}{(\Gamma(1/3))^{1/3}}\int_0^1 \mathcal{J}_1(s)\left(s^{2/15}B\left(\frac{4}{5},\frac{1}{3}\right)\right)^{1/3} ds \approx 0.78160052, \\ M_2 &= \int_0^1 \mathcal{J}_2(s)\varphi_{\varrho_2}(I_{0+}^{\alpha_2}\zeta_2(s))\,ds = \int_0^1 \mathcal{J}_2(s)\varphi_{5/4}(I_{0+}^{1/2}\zeta_2(s))\,ds \\ &= \int_0^1 \mathcal{J}_2(s)\left(\frac{1}{\Gamma(1/2)}\int_0^s (s-\tau)^{-1/2}(1-\tau)^{-1/4}d\tau\right)^{1/4} ds \\ &= \frac{1}{(\Gamma(1/2))^{1/4}}\int_0^1 \mathcal{J}_2(s)\left(s^{1/2}\sqrt{\pi}\,{}_2\tilde{F}_1\left[\frac{1}{4},1,\frac{3}{2},s\right]\right)^{1/4} ds \approx 0.65997289.\end{aligned}$$

In addition, we find $D_0 = \max\left\{\frac{2c_0}{5}(2^{\sigma_1}+2^{\sigma_2}), \frac{4d_0}{81}\right\}$. If $c_0 < \min\left\{\frac{5}{16M_1^3(2^{\sigma_1}+2^{\sigma_2})}, \frac{5}{32M_2^4(2^{\sigma_1}+2^{\sigma_2})}\right\}$ and $d_0 < \min\left\{\frac{81}{32M_1^3}, \frac{81}{64M_2^4}\right\}$, then the inequalities $D_0^{1/3}M_1 < \frac{1}{2}$ and $D_0^{1/4}M_2 < \frac{1}{2}$ are satisfied (that is, assumption $(I7)$ is satisfied). For example, if $\sigma_1 = 4$ and $\sigma_2 = 2$, and $c_0 \le 0.032$ and $d_0 \le 5.301$, then the above inequalities are satisfied. By Theorem 3, we conclude that problem (33) and (34) with the nonlinearities (36) has at least two positive solutions $(u_1(t), v_1(t))$, $(u_2(t), v_2(t))$, $t \in [0,1]$.

5. Conclusions

In this paper, we have discussed the existence and multiplicity of positive solutions for a system of Riemann–Liouville fractional differential equations with singular nonnegative nonlinearities and p-Laplacian operators, complemented with nonlocal boundary conditions involving fractional derivatives and Riemann–Stieltjes integrals. Some properties of the associated Green functions are also presented. Two examples are constructed for the illustration of the obtained results.

Author Contributions: Conceptualization, R.L.; Formal analysis, A.A., R.L. and B.A.; Funding acquisition, A.A.; Methodology, A.A., R.L. and B.A. All authors have read and agreed to the published version of the manuscript.

Funding: The Deanship of Scientific Research (DSR) at King Abdulaziz University, Jeddah, Saudi Arabia funded this project, under grant no. FP-18-42.

Acknowledgments: The Deanship of Scientific Research (DSR) at King Abdulaziz University, Jeddah, Saudi Arabia funded this project, under grant no. FP-18-42. The authors thank the reviewers for their constructive remarks on our work.

Conflicts of Interest: The authors declare no conflict of interest.

References

1. Wang, G.; Ren, X.; Zhang, L.; Ahmad, B. Explicit iteration and unique positive solution for a Caputo-Hadamard fractional turbulent flow model. *IEEE Access* **2019**, *7*, 109833–109839. [CrossRef]
2. Guo, D.; Lakshmikantham, V. *Nonlinear Problems in Abstract Cones*; Academic Press: New York, NY, USA, 1988.
3. Luca, R. Positive solutions for a system of fractional differential equations with p-Laplacian operator and multi-point boundary conditions. *Nonlinear Anal. Model. Control* **2018**, *23*, 771–801. [CrossRef]
4. Tudorache, A.; Luca, R. Positive solutions for a system of Riemann–Liouville fractional boundary value problems with p-Laplacian operators. *Adv. Differ. Equ.* **2020**, *2020*, 1–30. [CrossRef]
5. Jong, K.S.; Choi, H.C.; Ri, Y.H. Existence of positive solutions of a class of multi-point boundary value problems for p-Laplacian fractional differential equations with singular source terms. *Commun. Nonlinear Sci. Numer. Simulat.* **2019**, *72*, 272–281. [CrossRef]
6. Wang, F.; Liu, L.; Wu, Y. A numerical algorithm for a class of fractional BVPs with p-Laplacian operator and singularity-the convergence and dependence analysis. *Appl. Math. Comput.* **2020**, *382*, 1–13. [CrossRef]
7. Agarwal, R.P.; Luca, R. Positive Solutions for a Semipositone Singular Riemann–Liouville Fractional Differential Problem. *Int. J. Nonlinear Sci. Numer. Simul.* **2019**, *20*, 823–831. [CrossRef]
8. Ahmad, B.; Alsaedi, A.; Aljoudi, S.; Ntouyas, S.K. A six-point nonlocal boundary value problem of nonlinear coupled sequential fractional integro-differential equations and coupled integral boundary conditions. *J. Appl. Math. Comput.* **2018**, *56*, 367–389. [CrossRef]
9. Ahmad, B.; Alsaedi, A.; Ntouyas, S.K.; Tariboon, J. *Hadamard-Type Fractional Differential Equations, Inclusions and Inequalities*; Springer: Cham, Switzerland, 2017.
10. Ahmad, B.; Luca, R. Existence of solutions for a sequential fractional integro-differential system with coupled integral boundary conditions. *Chaos Solitons Fractals* **2017**, *104*, 378–388. [CrossRef]
11. Bashir, A.; Luca, R. Existence of solutions for a system of fractional differential equations with coupled nonlocal boundary conditions. *Fract. Calc. Appl. Anal.* **2018**, *21*, 423–441.
12. Bashir, A.; Luca, R. Existence of solutions for sequential fractional integro-differential equations and inclusions with nonlocal boundary conditions. *Appl. Math. Comput.* **2018**, *339*, 516–534.
13. Ahmad, B.; Ntouyas, S.; Alsaedi, A. On a coupled system of fractional differential equations with coupled nonlocal and integral boundary conditions. *Chaos Solitons Fractals* **2016**, *83*, 234–241. [CrossRef]
14. Ahmad, B.; Ntouyas, S.K.; Alsaedi, A. Sequential fractional differential equations and inclusions with semi-periodic and nonlocal integro-multipoint boundary conditions. *J. King Saud Univ. Sci.* **2019**, *31*, 184–193. [CrossRef]
15. Aljoudi, S.; Ahmad, B.; Nieto, J.J.; Alsaedi, A. A coupled system of Hadamard type sequential fractional differential equations with coupled strip conditions. *Chaos Solitons Fractals* **2016**, *91*, 39–46. [CrossRef]
16. Henderson, J.; Luca, R. *Boundary Value Problems for Systems of Differential, Difference and Fractional Equations. Positive Solutions*; Elsevier: Amsterdam, The Netherlands, 2016.
17. Henderson, J.; Luca, R. Existence of positive solutions for a singular fractional boundary value problem. *Nonlinear Anal. Model. Control* **2017**, *22*, 99–114. [CrossRef]
18. Henderson, J.; Luca, R.; Tudorache, A. On a system of fractional differential equations with coupled integral boundary conditions. *Fract. Calc. Appl. Anal.* **2015**, *18*, 361–386. [CrossRef]
19. Henderson, J.; Luca, R.; Tudorache, A. Existence and nonexistence of positive solutions for coupled Riemann–Liouville fractional boundary value problems. *Discrete Dyn. Nature Soc.* **2016**, *2016*, 1–12. [CrossRef]
20. Henderson, J.; Luca, R.; Tudorache, A. Existence of positive solutions for a system of fractional boundary value problems. In *Differential and Difference Equations with Applications: ICDDEA, Amadora, Portugal, May 2015, Selected Contributions*; Sandra, P., Zuzana, D., Ondrej, D., Peter, K., Eds.; Springer: Berlin/Heidelberg, Germany, 2016; pp. 349–357.
21. Jiang, J.; Liu, L.; Wu, Y. Symmetric positive solutions to singular system with multi-point coupled boundary conditions. *Appl. Math. Comp.* **2013**, *220*, 536–548. [CrossRef]

22. Jiang, J.; Liu, L.; Wu, Y. Positive solutions to singular fractional differential system with coupled boundary conditions. *Commun. Nonlinear Sci. Numer. Simul.* **2013**, *18*, 3061–3074. [CrossRef]
23. Shen, C.; Zhou, H.; Yang, L. Positive solution of a system of integral equations with applications to boundary value problems of differential equations. *Adv. Differ. Equ.* **2016**, *2016*, 1–26. [CrossRef]
24. Tudorache, A.; Luca, R. Positive solutions for a singular fractional boundary value problem. *Math. Meth. Appl. Sci.* **2020**. [CrossRef]
25. Wang, Y.; Liu, L.; Wu, Y. Positive solutions for a class of higher-order singular semipositone fractional differential systems with coupled integral boundary conditions and parameters. *Adv. Differ. Equ.* **2014**, *2014*, 1–24.
26. Yuan, C. Two positive solutions for $(n-1,1)$-type semipositone integral boundary value problems for coupled systems of nonlinear fractional differential equations. *Commun. Nonlinear Sci. Numer. Simulat.* **2012**, *17*, 930–942.
27. Yuan, C.; Jiang, D.; O'Regan, D.; Agarwal, R.P. Multiple positive solutions to systems of nonlinear semipositone fractional differential equations with coupled boundary conditions. *Electron. J. Qual. Theory Differ. Equ.* **2012**, *2012*, 1–17.
28. Baleanu, D.; Diethelm, K.; Scalas, E.; Trujillo, J.J. *Fractional Calculus Models and Numerical Methods*; Series on Complexity, Nonlinearity and Chaos; World Scientific: Boston, MA, USA, 2012.
29. Das, S. *Functional Fractional Calculus for System Identification and Controls*; Springer: New York, NY, USA, 2008.
30. Kilbas, A.A.; Srivastava, H.M.; Trujillo, J.J. *Theory and Applications of Fractional Differential Equations*; North-Holland Mathematics Studies; Elsevier Science B.V.: Amsterdam, The Netherlands, 2006; Volume 204.
31. Klafter, J.; Lim, S.C.; Metzler, R. (Eds.) *Fractional Dynamics in Physics*; World Scientific: Singapore, 2011.
32. Podlubny, I. *Fractional Differential Equations*; Academic Press: San Diego, CA, USA, 1999.
33. Sabatier, J.; Agrawal, O.P.; Machado, J.A.T. (Eds.) *Advances in Fractional Calculus: Theoretical Developments and Applications in Physics and Engineering*; Springer: Dordrecht, The Netherlands, 2007.
34. Samko, S.G.; Kilbas, A.A.; Marichev, O.I. *Fractional Integrals and Derivatives*; Theory and Applications; Gordon and Breach: Yverdon, Switzerland, 1993.

mathematics

MDPI

Article

On Caputo–Riemann–Liouville Type Fractional Integro-Differential Equations with Multi-Point Sub-Strip Boundary Conditions

Ahmed Alsaedi [1], Amjad F. Albideewi [1], Sotiris K. Ntouyas [1,2] and Bashir Ahmad [1,*]

1 Nonlinear Analysis and Applied Mathematics (NAAM)-Research Group, Department of Mathematics, Faculty of Science, King Abdulaziz University, P.O. Box 80203, Jeddah 21589, Saudi Arabia; aalsaedi@hotmail.com (A.A.); amjad.f.b@hotmail.com (A.F.A.); sntouyas@uoi.gr (S.K.N.)

2 Department of Mathematics, University of Ioannina, 45110 Ioannina, Greece

* Correspondence: bashirahmad_qau@yahoo.com or bahmad@kau.edu.sa

Received: 16 September 2020; Accepted: 26 October 2020; Published: 31 October 2020

Abstract: In this paper, we derive existence and uniqueness results for a nonlinear Caputo–Riemann–Liouville type fractional integro-differential boundary value problem with multi-point sub-strip boundary conditions, via Banach and Krasnosel'skiĭ's fixed point theorems. Examples are included for the illustration of the obtained results.

Keywords: Caputo derivative; Riemann–Liouville integral; multipoint and sub-strip boundary conditions; existence; fixed point theorem

MSC: 34A08; 34B10; 34B15

1. Introduction

The subject of fractional order boundary value problems has been addressed by many researchers in recent years. The interest in the subject owes to its extensive applications in natural and social sciences. Examples include bio-engineering [1], ecology [2], financial economics [3], chaos and fractional dynamics [4], etc. One can find many interesting results using boundary value problems dealing with Caputo, Riemann–Liouville and Hadamard type fractional derivatives and equipped with a variety of boundary conditions in [5–17].

Integro-differential equations constitute an important area of investigation due to their occurrence in several applied fields, such as heat transfer phenomena [18,19], fractional power law [20], etc. Fractional integro-differential equations complemented with different kinds of boundary conditions have also been studied by many researchers, for example, [21–28]. In a recent paper [29], a nonlocal boundary value problem containing left Caputo and right Riemann–Liouville fractional derivatives, and both left and right Riemann–Liouville fractional integral operators was discussed.

Motivated by aforementioned work on integro-differential equations, we introduce and investigate a nonlinear Caputo–Riemann–Liouville type fractional integro-differential boundary value problem involving multi-point sub-strip boundary conditions given by

$$\begin{cases} {}^cD^q x(t) + \sum_{i=1}^{k} I^{p_i} g_i(t, x(t)) = f(t, x(t)), \ 0 < t < 1, \\ x(0) = a, \ \ x'(0) = 0, \ \ x''(0) = 0, \dots, x^{(m-2)}(0) = 0, \\ \alpha x(1) + \beta x'(1) = \gamma_1 \int_0^{\zeta} x(s)ds + \sum_{j=1}^{p} \alpha_j \, x(\eta_j) + \gamma_2 \int_{\xi}^{1} x(s)ds, \end{cases} \tag{1}$$

where ${}^cD^q$ represents the Caputo fractional derivative operator of order $q \in (m-1, m], m \in \mathbb{N}, m \geq 2$, $p_i > 0$, $0 < \zeta, \eta_1, \eta_2, \ldots, \eta_p, \xi < 1$, $f, g_i : [0,1] \times \mathbb{R} \to \mathbb{R}$, $(i = 1, \ldots, k)$ are continuous functions $a, \alpha, \beta, \gamma_1, \gamma_2 \in \mathbb{R}$ and $\alpha_j \in \mathbb{R}$, $j = 1, 2, \ldots, p$. Notice that the fixed/nonlocal points involved in the problem (1) are non-singular.

We emphasize that the problem considered in this paper is novel in the sense that the fractional integro-differential equation involves many finitely Riemann–Liouville fractional integral type nonlinearities together with a non-integral nonlinearity. In the literature, one can find results on linear integro-differential equations [30], fractional integro-differential equations with nonlinearity depending on the linear integral terms [31,32], and initial value problems involving two nonlinear integral terms [33]. In contrast to the aforementioned work, our problem contains many finitely nonlinear integral terms of fractional order, which reduce to the nonlinear integral terms by fixing $p_i = 1, \forall i = 1, \ldots, k$. For specific applications of integral-differential equations in the mathematical modeling of physical problems such as the spreading of disease by the dispersal of infectious individuals, and reaction-–diffusion models in ecology, see [1,2]. In particular, one can find more details on the topic in [34] and the references cited therein. For some recent work on fractional integro-differential equations, see [35,36]. In a more recent work [37], the authors studied the existence of solutions for a fractional integro-differential equation supplemented with dual anti-periodic boundary conditions. Concerning the boundary condition at the terminal position $t = 1$, the linear combination of the unknown function and its derivative is associated with the contribution due to two sub-strips $(0, \zeta)$ and $(\xi, 1)$ and finitely many nonlocal positions between them within the domain $[0, 1]$. This boundary condition covers many interesting situations, for example, it corresponds to the two-strip aperture condition for all $\alpha_j = 0, j = 1, \ldots, p$. By taking $\gamma_1 = 0 = \gamma_2$, this condition takes the form of a multi-point nonlocal boundary condition. It is interesting to note that the role of integral boundary conditions in studying practical problems such as blood flow problems [38] and bacterial self-regularization [39], etc., is crucial. For the application of strip conditions in engineering and real world problems, see [40,41]. On the other hand, the concept of nonlocal boundary conditions plays a significant role when physical, chemical or other processes depend on the interior positions (non-fixed points or segments) of the domain, for instance, see [42–45] and the references therein.

The rest of the paper is arranged as follows. Section 2 contains some related concepts of fractional calculus and an auxiliary result concerning a linear version of the problem (1). We prove the existence and uniqueness of solutions for the problem (1) by applying Banach and Krasnosel'skiǐ's fixed point theorems in Section 3. Finally, examples illustrating the main results are demonstrated in Section 4.

2. Preliminaries

Let us first outline some preliminary concepts of fractional calculus [5].

Definition 1. *The Caputo derivative for a function $h \in AC^n[a, b]$ of fractional order $q \in (n-1, n], n \in \mathbb{N}$, existing almost everywhere on $[a, b]$, is defined by*

$$ {}^cD^q h(t) = \frac{1}{\Gamma(n-q)} \int_a^t (t-u)^{n-q-1} h^{(n)}(u) du, \quad t \in [a, b]. $$

Definition 2. *The Riemann–Liouville fractional integral for a function $h \in L_1[a, b]$ of order $r > 0$, which exists almost everywhere on $[a, b]$, is defined by*

$$ I^r h(t) = \frac{1}{\Gamma(r)} \int_a^t \frac{h(u)}{(t-u)^{1-r}} du, \ t \in [a, b]. $$

Lemma 1. *For $m - 1 < r \leq m$, the general solution of the fractional differential equation ${}^cD^r x(t) = 0$ can be written as*

$$ x(t) = b_0 + b_1 t + b_2 t^2 + \ldots + b_{m-1} t^{m-1}, \tag{2} $$

where $b_i \in \mathbb{R}, i = 0,1,2,\ldots,m-1$.

It follows by Lemma 1 that

$$I^r\, {}^cD^r x(t) = x(t) + b_0 + b_1 t + b_2 t^2 + \ldots + b_{m-1} t^{m-1}, \tag{3}$$

for some $b_i \in \mathbb{R}, i = 0,1,2,\ldots,m-1$, are arbitrary constants.

The following lemma deals with the linear version of the problem (1).

Lemma 2. *For a given function $h \in C([0,1],\mathbb{R})$ the unique solution of the following boundary value problem*

$$\begin{cases} {}^cD^q x(t) = h(t),\ 0 < t < 1, \\ x(0) = a,\ \ x'(0) = 0,\ \ x''(0) = 0,\ldots,x^{(m-2)}(0) = 0, \\ \alpha x(1) + \beta x'(1) = \gamma_1 \displaystyle\int_0^\zeta x(s)ds + \sum_{j=1}^{p} \alpha_j\, x(\eta_j) + \gamma_2 \int_\xi^1 x(s)ds, \end{cases} \tag{4}$$

is given by

$$\begin{aligned} x(t) = {}& \int_0^t \frac{(t-s)^{q-1}}{\Gamma(q)} h(s)ds + a - \frac{t^{m-1}}{\Lambda_1}\Bigg[\alpha \int_0^1 \frac{(1-s)^{q-1}}{\Gamma(q)} h(s)ds + \beta \int_0^1 \frac{(1-s)^{q-2}}{\Gamma(q-1)} h(s)ds \\ & - \gamma_1 \int_0^\zeta \int_0^s \frac{(s-u)^{q-1}}{\Gamma(q)} h(u)duds - \sum_{j=1}^{p} \alpha_j \int_0^{\eta_j} \frac{(\eta_j - s)^{q-1}}{\Gamma(q)} h(s)ds \\ & - \gamma_2 \int_\xi^1 \int_0^s \frac{(s-u)^{q-1}}{\Gamma(q)} h(u)duds + \Lambda_2\Bigg], \end{aligned} \tag{5}$$

where it is assumed that

$$\Lambda_1 = \left(\alpha + \beta(m-1) - \gamma_1\left(\frac{\zeta^m}{m}\right) - \sum_{j=1}^{p} \alpha_j \eta_j^{m-1} - \gamma_2\left(\frac{1-\xi^m}{m}\right)\right) \neq 0, \tag{6}$$

$$\Lambda_2 = a\Big[\alpha - \gamma_1\zeta - \sum_{j=1}^{p} \alpha_j - \gamma_2(1-\xi)\Big]. \tag{7}$$

Proof. Using (3), we can write the general solution of the fractional differential equation in (4) as

$$x(t) = \int_0^t \frac{(t-s)^{q-1}}{\Gamma(q)} h(s)ds - c_0 - c_1 t - c_2 t^2 - \ldots - c_{m-1} t^{m-1}, \tag{8}$$

where $c_0, c_1, c_2, \ldots, c_{m-1} \in \mathbb{R}$ are arbitrary constants. From (8), we have

$$\begin{aligned} x'(t) &= \int_0^t \frac{(t-s)^{q-2}}{\Gamma(q-1)} h(s)ds - c_1 - 2c_2 t - \ldots - (m-1)c_{m-1} t^{m-2}, \\ x''(t) &= \int_0^t \frac{(t-s)^{q-3}}{\Gamma(q-2)} h(s)ds - 2c_2 - \ldots - (m-1)(m-2)c_{m-1} t^{m-3}, \\ &\ \vdots \end{aligned} \tag{9}$$

Applying the conditions $x(0) = a, x'(0) = 0, \ldots, x^{(m-2)}(0) = 0$ in (8), it is found that $c_0 = -a$, $c_1 = 0, \ldots, c_{m-2} = 0$. Then (8) becomes

$$x(t) = \int_0^t \frac{(t-s)^{q-1}}{\Gamma(q)} h(s)ds + a - c_{m-1}t^{m-1}, \tag{10}$$

and

$$x'(t) = \int_0^t \frac{(t-s)^{q-2}}{\Gamma(q-1)} h(s)ds - (m-1)c_{m-1}t^{m-2}. \tag{11}$$

Combining (10) and (11) with the condition $\alpha x(1) + \beta x'(1) = \gamma_1 \int_0^\zeta x(s)ds + \sum_{j=1}^{p} \alpha_j\, x(\eta_j)$ $+ \gamma_2 \int_\xi^1 x(s)ds$, we get

$$\begin{aligned} & \alpha\left[\int_0^1 \frac{(1-s)^{q-1}}{\Gamma(q)} h(s)ds + a - c_{m-1}\right] + \beta\left[\int_0^1 \frac{(1-s)^{q-2}}{\Gamma(q-1)} h(s)ds - (m-1)c_{m-1}\right] \\ = \; & \gamma_1 \int_0^\zeta \left[\int_0^s \frac{(s-u)^{q-1}}{\Gamma(q)} h(u)du + a - c_{m-1}s^{m-1}\right] ds + \sum_{j=1}^{p} \alpha_j \left[\int_0^{\eta_j} \frac{(\eta_j - s)^{q-1}}{\Gamma(q)} h(s)ds \right. \\ & \left. + a - c_{m-1}\eta_j^{m-1}\right] + \gamma_2 \int_\xi^1 \left[\int_0^s \frac{(s-u)^{q-1}}{\Gamma(q)} h(u)du + a - c_{m-1}s^{m-1}\right] ds, \end{aligned}$$

which, together with (6) and (7), yields

$$\begin{aligned} c_{m-1} \; = \; & \frac{1}{\Lambda_1}\left[\alpha \int_0^1 \frac{(1-s)^{q-1}}{\Gamma(q)} h(s)ds + \beta \int_0^1 \frac{(1-s)^{q-2}}{\Gamma(q-1)} h(s)ds \right. \\ & - \gamma_1 \int_0^\zeta \int_0^s \frac{(s-u)^{q-1}}{\Gamma(q)} h(u)duds - \sum_{j=1}^{p} \alpha_j \int_0^{\eta_j} \frac{(\eta_j - s)^{q-1}}{\Gamma(q)} h(s)ds \\ & \left. - \gamma_2 \int_\xi^1 \int_0^s \frac{(s-u)^{q-1}}{\Gamma(q)} h(u)duds + a\alpha - a\gamma_1\zeta - a\sum_{j=1}^{p} \alpha_j - a\gamma_2(1-\xi)\right]. \end{aligned}$$

Substituting the value of c_{m-1} in (10), we obtain

$$\begin{aligned} & x(t) \\ = \; & \int_0^t \frac{(t-s)^{q-1}}{\Gamma(q)} h(s)ds + a - \frac{t^{m-1}}{\Lambda_1}\left[\alpha \int_0^1 \frac{(1-s)^{q-1}}{\Gamma(q)} h(s)ds + \beta \int_0^1 \frac{(1-s)^{q-2}}{\Gamma(q-1)} h(s)ds \right. \\ & - \gamma_1 \int_0^\zeta \int_0^s \frac{(s-u)^{q-1}}{\Gamma(q)} h(u)duds - \sum_{j=1}^{p} \alpha_j \int_0^{\eta_j} \frac{(\eta_i - s)^{q-1}}{\Gamma(q)} h(s)ds \\ & \left. - \gamma_2 \int_\xi^1 \int_0^s \frac{(s-u)^{q-1}}{\Gamma(q)} h(u)duds + \Lambda_2\right]. \end{aligned}$$

We can obtain the converse of this lemma by direct computation. This finishes the proof. □

Using Lemma 2, we can transform problem (1) into a fixed point problem as $x = \mathcal{F}x$, where the operator $\mathcal{F} : C \to C$ is defined by

$$\mathcal{F}x(t) \; = \; \int_0^t \frac{(t-s)^{q-1}}{\Gamma(q)} f(s, x(s))ds - \sum_{i=1}^{k} \int_0^t \frac{(t-s)^{q+p_i-1}}{\Gamma(q+p_i)} g_i(s, x(s))ds + a$$

$$-\frac{t^{m-1}}{\Lambda_1}\Bigg[\alpha\int_0^1\left(\frac{(1-s)^{q-1}}{\Gamma(q)}f(s,x(s))-\sum_{i=1}^{k}\frac{(1-s)^{q+p_i-1}}{\Gamma(q+p_i)}g_i(s,x(s))\right)ds$$
$$+\beta\int_0^1\left(\frac{(1-s)^{q-2}}{\Gamma(q-1)}f(s,x(s))-\sum_{i=1}^{k}\frac{(1-s)^{q+p_i-2}}{\Gamma(q+p_i-1)}g_i(s,x(s))\right)ds$$
$$-\gamma_1\int_0^{\zeta}\int_0^{s}\frac{(s-u)^{q-1}}{\Gamma(q)}f(u,x(u))\,du\,ds \tag{12}$$
$$+\gamma_1\sum_{i=1}^{k}\int_0^{\zeta}\int_0^{s}\frac{(s-u)^{q-1}}{\Gamma(q)}\int_0^{u}\frac{(u-w)^{p_i-1}}{\Gamma(p_i)}g_i(w,x(w))dw\,du\,ds$$
$$-\sum_{j=1}^{p}\alpha_j\left(\int_0^{\eta_j}\frac{(\eta_j-s)^{q-1}}{\Gamma(q)}f(s,x(s))ds-\sum_{i=1}^{k}\int_0^{\eta_j}\frac{(\eta_j-s)^{q+p_i-1}}{\Gamma(q+p_i)}g_i(s,x(s))ds\right)$$
$$-\gamma_2\int_{\xi}^{1}\int_0^{s}\frac{(s-u)^{q-1}}{\Gamma(q)}f(u,x(u))\,du\,ds$$
$$+\gamma_2\sum_{i=1}^{k}\int_{\xi}^{1}\int_0^{s}\frac{(s-u)^{q-1}}{\Gamma(q)}\int_0^{u}\frac{(u-w)^{p_i-1}}{\Gamma(p_i)}g_i(w,x(w))dw\,du\,ds+\Lambda_2\Bigg].$$

Here $\mathcal{C}$, represents the Banach space of all continuous functions $x:[0,1]\to\mathbb{R}$ equipped with the norm $\|x\|=\sup_{t\in[0,1]}|x(t)|$.

By a solution of (1), we mean a function $x\in\mathcal{C}$ of class $C^m[0,1]$ satisfying the nonlocal integro-multipoint boundary value problem (1).

For computational convenience, we set

$$\Omega = \left[\frac{1}{\Gamma(q+1)}+\frac{1}{|\Lambda_1|}\left(\frac{|\alpha|}{\Gamma(q+1)}+\frac{|\beta|}{\Gamma(q)}+\frac{|\gamma_1|\zeta^{q+1}}{\Gamma(q+2)}+\frac{\sum_{j=1}^{p}|\alpha_j|\eta_j^{q}}{\Gamma(q+1)}+\frac{|\gamma_2||1-\xi^{q+1}|}{\Gamma(q+2)}\right)\right], \tag{13}$$

$$\Omega_i = \left[\frac{1}{\Gamma(q+p_i+1)}+\frac{1}{|\Lambda_1|}\left(\frac{|\alpha|}{\Gamma(q+p_i+1)}+\frac{|\beta|}{\Gamma(q+p_i)}+\frac{|\gamma_1|\zeta^{q+p_i+1}}{\Gamma(q+p_i+2)}+\frac{\sum_{j=1}^{p}|\alpha_j|\eta_j^{q+p_i}}{\Gamma(q+p_i+1)}+\frac{|\gamma_2||1-\xi^{q+p_i+1}|}{\Gamma(q+p_i+2)}\right)\right],\ i=1,2,\ldots,k. \tag{14}$$

3. Existence and Uniqueness Results

In the following theorem, we make use of Banach's fixed point theorem.

Theorem 1. *Let $f,g_i:[0,1]\times\mathbb{R}\to\mathbb{R}$ be continuous functions and let there exist constants $L,L_i>0$, $(i=1,\ldots,k)$ such that:*

(A_1) $|f(t,x)-f(t,y)|\le L|x-y|$, and $|g_i(t,x)-g_i(t,y)|\le L_i|x-y|$ for all $t\in[0,1]$, $x,y\in\mathbb{R}$.

Then, the boundary value problem (1) has a unique solution on $[0,1]$ if

$$L\Omega+\sum_{i=1}^{k}L_i\Omega_i<1, \tag{15}$$

where $\Omega,\Omega_i, i=1,2,\ldots,k$ are given by (13) and (14), respectively.

Proof. The proof will be given in two steps.

Step 1. We show that $\mathcal{F}B_r \subset B_r$, where $B_r = \{x \in \mathcal{C} : \|x\| \le r\}$ with $r \ge \Big(M\Omega + \sum_{i=1}^{k} M_i\Omega_i + |a| + (\Lambda_2/\Lambda_1)\Big/\Big(1 - (L\Omega + \sum_{i=1}^{k} L_i\Omega_i)\Big)$ and M, M_i are positive numbers such that $M = \sup_{t\in[0,1]} |f(t,0)|$ and $M_i = \sup_{t\in[0,1]} |g_i(t,0)|, i = 1,2,\ldots,k$.

For $x \in B_r$ and $t \in [0,1]$, it follows by (A_1) that

$$|f(t,x(t))| \le |f(t,x(t)) - f(t,0)| + |f(t,0)| \le L\|x\| + M \le Lr + M. \tag{16}$$

In a similar manner, we have $|g_i(t,x(t))| \le L_i r + M_i,\ i = 1,2,\ldots,k$. Then

$$\begin{aligned}
&\|\mathcal{F}x\| \\
\le\ & \sup_{t\in[0,1]} \Bigg\{ \int_0^t \frac{(t-s)^{q-1}}{\Gamma(q)} |f(s,x(s))| ds + \sum_{i=1}^{k} \int_0^t \frac{(t-s)^{q+p_i-1}}{\Gamma(q+p_i)} |g_i(s,x(s))| ds + |a| \\
& + \frac{t^{m-1}}{|\Lambda_1|} \Bigg[|\alpha| \int_0^1 \left(\frac{(1-s)^{q-1}}{\Gamma(q)} |f(s,x(s))| + \sum_{i=1}^{k} \frac{(1-s)^{q+p_i-1}}{\Gamma(q+p_i)} |g_i(s,x(s))| \right) ds \\
& + |\beta| \int_0^1 \left(\frac{(1-s)^{q-2}}{\Gamma(q-1)} |f(s,x(s))| + \sum_{i=1}^{k} \frac{(1-s)^{q+p_i-2}}{\Gamma(q+p_i-1)} |g_i(s,x(s))| \right) ds \\
& + |\gamma_1| \int_0^\zeta \int_0^s \frac{(s-u)^{q-1}}{\Gamma(q)} |f(u,x(u))| \, du\, ds \\
& + |\gamma_1| \sum_{i=1}^{k} \int_0^\zeta \int_0^s \frac{(s-u)^{q-1}}{\Gamma(q)} \int_0^u \frac{(u-w)^{p_i-1}}{\Gamma(p_i)} |g_i(w,x(w))| \, dw\, du\, ds \\
& + \sum_{j=1}^{p} |\alpha_j| \left(\int_0^{\eta_j} \frac{(\eta_j-s)^{q-1}}{\Gamma(q)} |f(s,x(s))| ds + \sum_{i=1}^{k} \int_0^{\eta_j} \frac{(\eta_j-s)^{q+p_i-1}}{\Gamma(q+p_i)} |g_i(s,x(s))| ds \right) \\
& + |\gamma_2| \int_\xi^1 \int_0^s \frac{(s-u)^{q-1}}{\Gamma(q)} |f(u,x(u))| \, du\, ds \\
& + |\gamma_2| \sum_{i=1}^{k} \int_\xi^1 \int_0^s \frac{(s-u)^{q-1}}{\Gamma(q)} \int_0^u \frac{(u-w)^{p_i-1}}{\Gamma(p_i)} |g_i(w,x(w))| \, dw\, du\, ds + |\Lambda_2| \Bigg] \Bigg\} \\
\le\ & (Lr+M) \left[\sup_{t\in[0,1]} \left\{ \frac{t^q}{\Gamma(q+1)} + \frac{t^{m-1}}{|\Lambda_1|} \left(\frac{|\alpha|}{\Gamma(q+1)} + \frac{|\beta|}{\Gamma(q)} + \frac{|\gamma_1|\zeta^{q+1}}{\Gamma(q+2)} + \frac{\sum_{j=1}^{p} |\alpha_j| \eta_j^q}{\Gamma(q+1)} \right.\right.\right. \\
& \left.\left.\left. + \frac{|\gamma_2|(1-\xi^{q+1})}{\Gamma(q+2)} \right) \right] + \sum_{i=1}^{k} (L_i r + M_i) \left[\frac{t^{q+p_i}}{\Gamma(q+p_i+1)} + \frac{t^{m-1}}{|\Lambda_1|} \left(\frac{|\alpha|}{\Gamma(q+p_i+1)} \right.\right. \\
& \left.\left. + \frac{|\beta|}{\Gamma(q+p_i)} + \frac{|\gamma_1|\zeta^{q+p_i+1}}{\Gamma(q+p_i+2)} + \frac{\sum_{j=1}^{p} |\alpha_i| \eta_j^{q+p_i}}{\Gamma(q+p_i+1)} + \frac{|\gamma_2|(1-\xi^{q+p_i+1})}{\Gamma(q+p_i+2)} \right) \right] \\
& \left. + |a| + \frac{t^{m-1}|\Lambda_2|}{|\Lambda_1|} \right\} \\
=\ & (Lr+M)\Omega + \sum_{i=1}^{k} (L_i r + M_i)\Omega_i + |a| + \left| \frac{\Lambda_2}{\Lambda_1} \right| \\
=\ & \left(L\Omega + \sum_{i=1}^{k} L_i\Omega_i \right) r + M\Omega + \sum_{i=1}^{k} M_i\Omega_i + |a| + \left| \frac{\Lambda_2}{\Lambda_1} \right| \le r,
\end{aligned}$$

which shows that $\mathcal{F}B_r \subset B_r$.

Step 2. We show that $\mathcal{F}$ is a contraction. For $x, y \in C$ and for each $t \in [0,1]$, we obtain

$$\begin{aligned}
&\|\mathcal{F}x - \mathcal{F}y\| \\
\leq\ & \sup_{t\in[0,1]} \Bigg\{ \int_0^t \frac{(t-s)^{q-1}}{\Gamma(q)} |f(s,x(s)) - f(s,y(s))| ds \\
&+ \sum_{i=1}^{k} \int_0^t \frac{(t-s)^{q+p_i-1}}{\Gamma(q+p_i)} |g_i(s,x(s)) - g_i(s,y(s))| ds \\
&+ \frac{t^{m-1}}{|\Lambda_1|} \Bigg[|\alpha| \int_0^1 \Bigg(\frac{(1-s)^{q-1}}{\Gamma(q)} |f(s,x(s)) - f(s,y(s))| \\
&+ \sum_{i=1}^{k} \frac{(1-s)^{q+p_i-1}}{\Gamma(q+p_i)} |g_i(s,x(s)) - g_i(s,y(s))| \Bigg) ds \\
&+ |\beta| \int_0^1 \Bigg(\frac{(1-s)^{q-2}}{\Gamma(q-1)} |f(s,x(s)) - f(s,y(s))| \\
&+ \sum_{i=1}^{k} \frac{(1-s)^{q+p_i-2}}{\Gamma(q+p_i-1)} |g_i(s,x(s)) - g_i(s,y(s))| \Bigg) ds \\
&+ |\gamma_1| \int_0^{\zeta} \int_0^s \frac{(s-u)^{q-1}}{\Gamma(q)} |f(u,x(u)) - f(u,y(u))|\, du\, ds \\
&+ |\gamma_1| \sum_{i=1}^{k} \int_0^{\zeta} \int_0^s \frac{(s-u)^{q-1}}{\Gamma(q)} \int_0^u \frac{(u-w)^{p_i-1}}{\Gamma(p_i)} |g_i(w,x(w)) - g_i(w,y(w))| dw du ds \\
&+ \sum_{j=1}^{p} |\alpha_j| \Bigg(\int_0^{\eta_j} \frac{(\eta_j-s)^{q-1}}{\Gamma(q)} |f(s,x(s)) - f(s,y(s))| ds \\
&+ \sum_{i=1}^{k} \int_0^{\eta_i} \frac{(\eta_i-s)^{q+p_i-1}}{\Gamma(q+p_i)} |g_i(s,x(s)) - g_i(x,y(s))| ds \Bigg) \\
&+ |\gamma_2| \int_{\xi}^{1} \int_0^s \frac{(s-u)^{q-1}}{\Gamma(q)} |f(u,x(u)) - f(u,y(u))|\, du\, ds \\
&+ |\gamma_2| \sum_{i=1}^{k} \int_{\xi}^{1} \int_0^s \frac{(s-u)^{q-1}}{\Gamma(q)} \int_0^u \frac{(u-w)^{p_i-1}}{\Gamma(p_i)} |g_i(w,x(w)) - g_i(w,y(w))|\, dw\, du\, ds \Bigg] \Bigg\} \\
\leq\ & L \left[\sup_{t\in[0,1]} \left\{ \frac{t^q}{\Gamma(q+1)} + \frac{t^{m-1}}{|\Lambda_1|} \left(\frac{|\alpha|}{\Gamma(q+1)} + \frac{|\beta|}{\Gamma(q)} + \frac{|\gamma_1|\zeta^{q+1}}{\Gamma(q+2)} + \frac{\sum_{j=1}^{p} |\alpha_j| \eta_j^q}{\Gamma(q+1)} \right.\right.\right. \\
&\left.\left.\left. + \frac{|\gamma_2|(1-\xi^{q+1})}{\Gamma(q+2)} \right) \right\} \right] \|x-y\| + \sum_{i=1}^{k} L_i \left[\frac{t^{q+p_i}}{\Gamma(q+p_i+1)} + \frac{t^{m-1}}{|\Lambda_1|} \left(\frac{|\alpha|}{\Gamma(q+p_i+1)} \right.\right. \\
&\left.\left. + \frac{|\beta|}{\Gamma(q+p_i)} + \frac{|\gamma_1|\zeta^{q+p_i+1}}{\Gamma(q+p_i+2)} + \frac{\sum_{j=1}^{p} |\alpha_j| \eta_j^{q+p_i}}{\Gamma(q+p_i+1)} + \frac{|\gamma_2|(1-\xi^{q+p_i+1})}{\Gamma(q+p_i+2)} \right) \right] \|x-y\| \\
\leq\ & \left(L\Omega + \sum_{i=1}^{k} L_i \Omega_i \right) \|x-y\|,
\end{aligned}$$

which, by the condition (15), implies that $\mathcal{F}$ is a contraction. Thus the conclusion of the Banach contraction mapping principle applies and hence the operator $\mathcal{F}$ has a unique fixed point. Therefore, there exists a unique solution for the boundary value problem (1) on $[0,1]$. □

Next, we prove an existence result for the boundary value problem (1), which relies on Krasnosel'skiĭ's fixed point theorem [46].

Theorem 2. *Let $f, g_i : [0,1] \times \mathbb{R} \to \mathbb{R}$, $(i = 1, \ldots, k)$ be continuous functions satisfying the condition (A_1). In addition, we assume that:*

(A_2) $|f(t,x)| \leq \mu(t)$, $|g_i(t,x)| \leq \mu_i(t)$, *for all* $(t,x) \in [0,1] \times \mathbb{R}$, $\mu, \mu_i \in C([0,1], \mathbb{R}^+)$.

Then, the boundary value problem (1) has at least one solution on $[0,1]$, *provided that*

$$L\left[\Omega - \left(\frac{1}{\Gamma(q+1)}\right)\right] + \sum_{i=1}^{k} L_i \left[\Omega_i - \left(\frac{1}{\Gamma(q+p_i+1)}\right)\right] < 1, \tag{17}$$

where Ω, $\Omega_i, i = 1, 2, \ldots, k$ *are given by (13) and (14), respectively.*

Proof. Consider $B_\rho = \{x \in C : \|x\| \leq \rho\}$, $\|\mu\| = \sup_{t\in[0,1]} |\mu(t)|$, $\|\mu_i\| = \sup_{t\in[0,1]} |\mu_i(t)|$, $i = 1, 2, \ldots, k$ with $\rho \geq \|\mu\|\Omega + \sum_{i=1}^{k} \|\mu_i\|\Omega_i + |a| + |\Lambda_2|/|\Lambda_1|$. Then, we define the operators Φ and Ψ on B_ρ as

$$\begin{aligned}
\Phi x(t) &= \int_0^t \frac{(t-s)^{q-1}}{\Gamma(q)} f(s,x(s))ds + \sum_{i=1}^{k} \int_0^t \frac{(t-s)^{q+p_i-1}}{\Gamma(q+p_i)} g_i(s,x(s))ds, \ t \in [0,1], \\
\Psi x(t) &= a - \frac{t^{m-1}}{\Lambda_1} \left[\alpha \int_0^1 \left(\frac{(1-s)^{q-1}}{\Gamma(q)} f(s,x(s)) - \sum_{i=1}^{k} \frac{(1-s)^{q+p_i-1}}{\Gamma(q+p_i)} g_i(s,x(s)) \right) ds \right. \\
&\quad + \beta \int_0^1 \left(\frac{(1-s)^{q-2}}{\Gamma(q-1)} f(s,x(s)) - \sum_{i=1}^{k} \frac{(1-s)^{q+p_i-2}}{\Gamma(q+p_i-1)} g_i(s,x(s)) \right) ds \\
&\quad - \gamma_1 \int_0^\zeta \int_0^s \frac{(s-u)^{q-1}}{\Gamma(q)} f(u,x(u))\,du\,ds \\
&\quad + \gamma_1 \sum_{i=1}^{k} \int_0^\zeta \int_0^s \frac{(s-u)^{q-1}}{\Gamma(q)} \int_0^u \frac{(u-w)^{p_i-1}}{\Gamma(p_i)} g_i(w,x(w))dw\,du\,ds \\
&\quad - \sum_{j=1}^{p} \alpha_j \left(\int_0^{\eta_j} \frac{(\eta_j - s)}{\Gamma(q)} f(s,x(s))ds - \sum_{i=1}^{k} \int_0^{\eta_j} \frac{(\eta_j-s)^{q+p_i-1}}{\Gamma(q+p_i)} g_i(s,x(s))ds \right) \\
&\quad - \gamma_2 \int_\zeta^1 \int_0^s \frac{(s-u)^{q-1}}{\Gamma(q)} f(u,x(u))\,du\,ds \\
&\quad \left. + \gamma_2 \sum_{i=1}^{k} \int_\zeta^1 \int_0^s \frac{(s-u)^{q-1}}{\Gamma(q)} \int_0^u \frac{(u-w)^{p_i-1}}{\Gamma(p_i)} g_i(w,x(w))dwduds + \Lambda_2 \right], t \in [0,1].
\end{aligned}$$

We complete the proof in three steps.

Step 1. We show that $\Phi x + \Psi y \in B_\rho$. For $x, y \in B_\rho$, we find that

$$\begin{aligned}
&\|\Phi x + \Psi y\| \\
\leq\ & \sup_{t\in[0,1]} \left\{ \int_0^t \frac{(t-s)^{q-1}}{\Gamma(q)} |f(s,x(s))|ds + \sum_{i=1}^{k} \int_0^t \frac{(t-s)^{q+p_i-1}}{\Gamma(q+p_i)} |g_i(s,x(s))|ds + |a| \right. \\
&+ \frac{t^{m-1}}{|\Lambda_1|} \left[|\alpha| \int_0^1 \left(\frac{(1-s)^{q-1}}{\Gamma(q)} |f(s,y(s))| + \sum_{i=1}^{k} \frac{(1-s)^{q+p_i-1}}{\Gamma(q+p_i)} |g_i(s,y(s))| \right) ds \right. \\
&+ |\beta| \int_0^1 \left(\frac{(1-s)^{q-2}}{\Gamma(q-1)} |f(s,y(s))| + \sum_{i=1}^{k} \frac{(1-s)^{q+p_i-2}}{\Gamma(q+p_i-1)} |g_i(s,y(s))| \right) ds \\
&+ |\gamma_1| \int_0^\zeta \int_0^s \frac{(s-u)^{q-1}}{\Gamma(q)} |f(u,y(u))|\,du\,ds
\end{aligned}$$

$$
\begin{aligned}
& +|\gamma_1|\sum_{i=1}^{k}\int_0^{\zeta}\int_0^{s}\frac{(s-u)^{q-1}}{\Gamma(q)}\int_0^{u}\frac{(u-w)^{p_i-1}}{\Gamma(p_i)}|g_i(w,y(w))|\,dw\,du\,ds \\
& +\sum_{j=1}^{p}|\alpha_j|\left(\int_0^{\eta_j}\frac{(\eta_j-s)^{q-1}}{\Gamma(q)}|f(s,y(s))|ds+\sum_{i=1}^{k}\int_0^{\eta_j}\frac{(\eta_j-s)^{q+p_i-1}}{\Gamma(q+p_i)}|g_i(s,y(s))|ds\right) \\
& +|\gamma_2|\int_{\zeta}^{1}\int_0^{s}\frac{(s-u)^{q-1}}{\Gamma(q)}|f(u,y(u))|\,du\,ds \\
& +|\gamma_2|\sum_{i=1}^{k}\int_{\zeta}^{1}\int_0^{s}\frac{(s-u)^{q-1}}{\Gamma(q)}\int_0^{u}\frac{(u-w)^{p_i-1}}{\Gamma(p_i)}|g_i(w,y(w))|\,dw\,du\,ds+|\Lambda_2|\Bigg]\Bigg\} \\
\leq\ & \|\mu\|\left[\sup_{t\in[0,1]}\left\{\frac{t^q}{\Gamma(q+1)}+\frac{t^{m-1}}{|\Lambda_1|}\left(\frac{|\alpha|}{\Gamma(q+1)}+\frac{|\beta|}{\Gamma(q)}+\frac{|\gamma_1|\zeta^{q+1}}{\Gamma(q+2)}+\frac{\sum_{j=1}^{p}|\alpha_j|\eta_j^q}{\Gamma(q+1)}\right.\right.\right. \\
& \left.\left.\left.+\frac{|\gamma_2|(1-\zeta^{q+1})}{\Gamma(q+2)}\right)\right\}\right]+\sum_{i=1}^{k}\|\mu_i\|\left[\frac{t^{q+p_i}}{\Gamma(q+p_i+1)}+\frac{t^{m-1}}{|\Lambda_1|}\left(\frac{|\alpha|}{\Gamma(q+p_i+1)}+\frac{|\beta|}{\Gamma(q+p_i)}\right.\right. \\
& \left.\left.+\frac{|\gamma_1|\zeta^{q+p_i+1}}{\Gamma(q+p_i+2)}+\frac{\sum_{j=1}^{p}|\alpha_j|\eta_j^{q+p_i}}{\Gamma(q+p_i+1)}+\frac{|\gamma_2|(1-\zeta^{q+p_i+1})}{\Gamma(q+p_i+2)}\right)\right]+\left(|a|+\left|\frac{\Lambda_2}{\Lambda_1}\right|\right) \\
=\ & \|\mu\|\Omega+\sum_{i=1}^{k}\|\mu_i\|\Omega_i+|a|+\left|\frac{\Lambda_2}{\Lambda_1}\right|\leq\rho.
\end{aligned}
$$

Thus, $\Phi x+\Psi y\in B_\rho$.

Step 2. We show that Ψ is a contraction mapping. For that, let $x,y\in\mathcal{C}$. Then, for each $t\in[0,1]$, we have

$$
\begin{aligned}
& \|\Psi x-\Psi y\| \\
\leq\ & \sup_{t\in[0,1]}\left\{\frac{t^{m-1}}{|\Lambda_1|}\left[|\alpha|\int_0^1\left(\frac{(1-s)^{q-1}}{\Gamma(q)}|f(s,x(s))-f(s,y(s))|\right.\right.\right. \\
& \left.+\sum_{i=1}^{k}\frac{(1-s)^{q+p_i-1}}{\Gamma(q+p_i)}|g_i(s,x(s))-g_i(s,y(s))|\right)ds \\
& +|\beta|\int_0^1\left(\frac{(1-s)^{q-2}}{\Gamma(q-1)}|f(s,x(s))-f(s,y(s))|\right. \\
& \left.+\sum_{i=1}^{k}\frac{(1-s)^{q+p_i-2}}{\Gamma(q+p_i-1)}|g_i(s,x(s))-g_i(s,y(s))|\right)ds \\
& +|\gamma_1|\int_0^{\zeta}\int_0^{s}\frac{(s-u)^{q-1}}{\Gamma(q)}|f(u,x(u))-f(u,y(u))|\,du\,ds \\
& +|\gamma_1|\sum_{i=1}^{k}\int_0^{\zeta}\int_0^{s}\frac{(s-u)^{q-1}}{\Gamma(q)}\int_0^{u}\frac{(u-w)^{p_i-1}}{\Gamma(p_i)}|g_i(w,x(w))-g_i(w,y(w))|\,dw\,du\,ds \\
& +\sum_{j=1}^{p}|\alpha_j|\left(\int_0^{\eta_j}\frac{(\eta_j-s)^{q-1}}{\Gamma(q)}|f(s,x(s))-f(s,y(s))|ds\right. \\
& \left.+\sum_{i=1}^{k}\int_0^{\eta_j}\frac{(\eta_j-s)^{q+p_i-1}}{\Gamma(q+p_i)}|g_i(s,x(s))-g_i(x,y(s))|ds\right) \\
& +|\gamma_2|\int_{\zeta}^{1}\int_0^{s}\frac{(s-u)^{q-1}}{\Gamma(q)}|f(u,x(u))-f(u,y(u))|\,du\,ds
\end{aligned}
$$

$$+|\gamma_2|\sum_{i=1}^{k}\int_{\xi}^{1}\int_{0}^{s}\frac{(s-u)^{q-1}}{\Gamma(q)}\int_{0}^{u}\frac{(u-w)^{p_i-1}}{\Gamma(p_i)}|g_i(w,x(w))-g_i(w,y(w))|\,dw\,du\,ds\Bigg]\Bigg\}$$

$$\leq L\left[\sup_{t\in[0,1]}\left\{\frac{t^{m-1}}{|\Lambda_1|}\left(\frac{|\alpha|}{\Gamma(q+1)}+\frac{|\beta|}{\Gamma(q)}+\frac{|\gamma_1|\zeta^{q+1}}{\Gamma(q+2)}+\frac{\sum_{j=1}^{p}|\alpha_j|\eta_j^q}{\Gamma(q+1)}+\frac{|\gamma_2|(1-\xi^{q+1})}{\Gamma(q+2)}\right)\right\}\right]$$

$$\times\|x-y\|+\sum_{i=1}^{k}L_i\left[\frac{t^{m-1}}{|\Lambda_1|}\left(\frac{|\alpha|}{\Gamma(q+p_i+1)}+\frac{|\beta|}{\Gamma(q+p_i)}+\frac{|\gamma_1|\zeta^{q+p_i+1}}{\Gamma(q+p_i+2)}\right.\right.$$

$$\left.\left.+\frac{\sum_{j=1}^{p}|\alpha_j|\eta_j^{q+p_i}}{\Gamma(q+p_i+1)}+\frac{|\gamma_2|(1-\xi^{q+p_i+1})}{\Gamma(q+p_i+2)}\right)\right]\|x-y\|$$

$$\leq \left(L\left[\Omega-\left(\frac{1}{\Gamma(q+1)}\right)\right]+\sum_{i=1}^{k}L_i\left[\Omega_i-\left(\frac{1}{\Gamma(q+p_i+1)}\right)\right]\right)\|x-y\|,$$

which is a contraction by the condition (17).

Step 3. We show that Φ is compact and continuous.

(i) Observe that the continuity of the operator Φ follows from that of f and $g_i, i=1,\ldots,k$.
(ii) Φ is uniformly bounded on B_ρ as:

$$\begin{aligned}\|\Phi x\| &\leq \sup_{t\in[0,1]}\left\{\int_0^t\frac{(t-s)^{q-1}}{\Gamma(q)}|f(s,x(s))|ds+\sum_{i=1}^{k}\int_0^t\frac{(t-s)^{q+p_i-1}}{\Gamma(q+p_i)}|g_i(s,x(s))|ds\right\}\\ &\leq \|\mu\|\sup_{t\in[0,1]}\left\{\int_0^t\frac{(t-s)^{q-1}}{\Gamma(q)}ds\right\}+\|\mu_i\|\sup_{t\in[0,1]}\left\{\sum_{i=1}^{k}\int_0^t\frac{(t-s)^{q+p_i-1}}{\Gamma(q+p_i)}ds\right\}\\ &\leq \frac{\|\mu\|}{\Gamma(q+1)}+\sum_{i=1}^{k}\frac{\|\mu_i\|}{\Gamma(q+p_i+1)}.\end{aligned}$$

(iii) Φ is equicontinuous.

Let us set $\max_{(t,x)\in[0,1]\times B_\rho}|f(t,x)|=\widehat{f}$ and $\max_{(t,x)\in[0,1]\times B_\rho}|g_i(t,x)|=\widehat{g}_i$, $i=1,2,\ldots,m$. Then, for $t_1,t_2\in[0,1], t_1>t_2$, we have

$$\begin{aligned}&|\Phi x(t_1)-\Phi x(t_2)|\\ &= \left|\int_0^{t_1}\frac{(t_1-s)^{q-1}}{\Gamma(q)}f(s,x(s))ds-\sum_{i=1}^{k}\int_0^{t_1}\frac{(t_1-s)^{q+p_i-1}}{\Gamma(q+p_i)}g_i(s,x(s))ds\right.\\ &\quad\left.-\int_0^{t_2}\frac{(t_2-s)^{q-1}}{\Gamma(q)}f(s,x(s))ds+\sum_{i=1}^{k}\int_0^{t_2}\frac{(t_2-s)^{q+p_i-1}}{\Gamma(q+p_i)}g_i(s,x(s))ds\right|\\ &\leq \left|\int_0^{t_2}\frac{(t_1-s)^{q-1}-(t_2-s)^{q-1}}{\Gamma(q)}f(s,x(s))ds\right|+\left|\int_{t_2}^{t_1}\frac{(t_1-s)^{q-1}}{\Gamma(q)}f(s,x(s))ds\right|\\ &\quad+\left|\sum_{i=1}^{k}\int_0^{t_2}\frac{(t_2-s)^{q+p_i-1}-(t_1-s)^{q+p_i-1}}{\Gamma(q+p_i)}g_i(s,x(s))ds\right|\\ &\quad+\left|\int_{t_2}^{t_1}\frac{(t_1-s)^{q+p_i-1}}{\Gamma(q+p_i)}g_i(s,x(s))ds\right|\\ &\leq \frac{\widehat{f}}{\Gamma(q+1)}\left\{|(t_1-t_2)^q|+|t_1^q-t_2^q|+|(t_1-t_2)^q|\right\}\end{aligned}$$

$$+\frac{\widehat{g}_i}{\Gamma(q+p_i+1)}\left\{|t_2^{q+p_i}-t_1^{q+p_i}|+|(t_1-t_2)^{q+p_i}|+|(t_1-t_2)^{q+p_i}|\right\}$$
$$\leq \frac{\widehat{f}}{\Gamma(q+1)}\left|2\,(t_1-t_2)^q+t_1^q-t_2^q\right|+\frac{\widehat{g}_i}{\Gamma(q+p_i+1)}|2\,(t_1-t_2)^{q+p_i}-t_1^{q+p_i}+t_2^{q+p_i}|,$$

which tends to zero, independent of x, as $t_1-t_2\to 0$. So, Φ is equicontinuous. Hence, we deduce by the Arzelá–Ascoli Theorem that Φ is compact on B_r. So, the hypothesis (Steps 1–3) of Krasnosel'skiĭ's fixed point theorem [46] holds true. Consequently, there exists at least one solution for the boundary value problem (1) on $[0,1]$. The proof is completed. □

4. Examples

Here, we illustrate the applicability of our results by constructing numerical examples.

Example 1. *Consider the following boundary value problem:*

$$\begin{cases} {}^cD^{13/4}x(t)+\sum\limits_{i=1}^{2} I^{p_i}g_i(t,x(t))=f(t,x(t)),\ t\in[0,1],\\ x(0)=0,\ x'(0)=0,\ x''(0)=0,\\ \alpha x(1)+\beta x'(1)=\gamma_1\displaystyle\int_0^{1/4}x(s)ds+\sum_{j=1}^{3}\alpha_j\,x(\eta_j)+\gamma_2\int_{1/2}^{1}x(s)ds. \end{cases} \tag{18}$$

Here, $m=4,\ q=13/14,\ p_1=10/14,\ p_2=11/14,\ p_3=12/14,\ \alpha=\beta=\gamma_1=\gamma_2=1,\ \zeta=1/4,$ $\alpha_1=1/2,\ \alpha_2=3/4,\ \alpha_3=1,\ \eta_1=1/7,\ \eta_2=2/7,\ \eta_3=3/7,\ \xi=1/2.$

(i) Let $f(t,x)=\dfrac{e^{2t}}{70}(\arctan x+\sin 5t),\ g_1(t,x)=\dfrac{1}{4}\left(\dfrac{e^{-t}\cos x+t^2+1}{\sqrt{t^2+49}}\right)$, and $g_2(t,x)=\dfrac{1}{34}(\sin x+e^{-t}\sqrt{57})$. It is easy to see that (A_1) is satisfied with $L=e^2/70,\ L_1=1/28$, and $L_2=1/34$.

Using the given data, we have $\Omega\approx 1.932128,\ \Omega_1\approx 0.677039,\ \Omega_2\approx 1.237301$, and

$$\Lambda_1=\left(\alpha+\beta(m-1)-\gamma_1\left(\frac{\zeta^m}{m}\right)-\sum_{j=1}^{3}\alpha_j\eta_j^{m-1}-\gamma_2\left(\frac{1-\xi^m}{m}\right)\right)\approx 3.666981.$$

Then, $L\Omega+\sum\limits_{i=1}^{2}L_i\Omega_i\approx 0.203951+0.060571<1$. Thus, by Theorem 1, the boundary value problem (18) has a unique solution on $[0,1]$.

(ii) We choose the following functions in problem (18) for illustrating Theorem 2:

$$f(t,x)=\frac{2}{17}(\sin x+e^{-t}\cos 7t),\ g_1(t,x)=\frac{3}{32}\left(\frac{|x|}{1+|x|}\right)+2t,\ g_2(t,x)=\frac{1}{34}(\sin x+e^{-t}\sqrt{32}). \tag{19}$$

Here $L=2/17, L_1=3/32$ and $L_2=1/34$ as $|f(t,x)-f(t,y)|\leq\dfrac{2}{17}|x-y|$, $|g_1(t,x)-g_1(t,y)|\leq\dfrac{3}{32}|x-y|$ and $|g_2(t,x)-g_2(t,y)|\leq\dfrac{1}{34}|x-y|$.

Further,

$$\begin{aligned}\|f(t,x)\| &\leq \frac{2}{17}|\sin x|+e^{-t}|\cos 7t|\leq\frac{2}{17}+e^{-t}\cos 7t=\mu(t),\\ \|g_1(t,x)\| &\leq \frac{3}{32}+2t=\mu_1(t),\end{aligned}$$

and

$$\|g_2(t,x)\|\leq\frac{1}{34}|\sin x|+e^{-t}\sqrt{32}\leq\frac{1}{34}+e^{-t}\sqrt{32}=\mu_2(t).$$

Obviously, $\|\mu\| = 19/17$, $\|\mu_1\| = 67/32$ and $\|\mu_2\| = 5.686266$. Moreover, we have

$$\left(L \left[\Omega - \left(\frac{1}{\Gamma(q+1)} \right) \right] + \sum_{i=1}^{2} L_i \left[\Omega_i - \left(\frac{1}{\Gamma(q+p_i+1)} \right) \right] \right) \approx 0.1824334 < 1.$$

As the hypothesis of Theorem 2 holds true, so there exists least one solution for problem (18) with the functions given by (19).

5. Conclusions

We have studied a nonlinear fractional integro-differential equation involving many finitely Riemann–Liouville fractional integral type nonlinearities together with a non-integral nonlinearity complemented by multi-point sub-strip boundary conditions. In fact, we considered a more general situation by considering the fractional order nonlinear integral terms in the integro-differential equation at hand, which reduce to the usual nonlinear integral terms for $p_i = 1, \forall i = 1, \ldots, k$. Under appropriate assumptions, the existence and uniqueness results for the given problem are proved by applying the standard tools of the fixed point theory. The results obtained in this paper are not only new, but they also lead to some new results associated with the particular choices of the parameters involved in the problem. For example, our results correspond to the two-strip aperture (ζ, ξ) boundary value problem when $\alpha_j = 0, \forall j = 1, \ldots, p$. On the other hand, by letting $\gamma_1 = 0 = \gamma_2$ in the the results of this paper, we obtain the ones for a nonlinear Caputo–Riemann–Liouville type fractional integro-differential equation with multi-point boundary conditions. Thus, the work presented in this paper significantly contributes to the existing literature on the topic.

Author Contributions: Conceptualization, B.A.; formal analysis, A.A., A.F.A., S.K.N. and B.A.; funding acquisition, A.A.; methodology, A.A., A.F.A., S.K.N. and B.A. All authors have read and agreed to the published version of the manuscript.

Funding: The Deanship of Scientific Research (DSR) at King Abdulaziz University, Jeddah, Saudi Arabia, funded this project, under grant no. (FP-20-42).

Acknowledgments: The Deanship of Scientific Research (DSR) at King Abdulaziz University, Jeddah, Saudi Arabia, funded this project, under grant no. (FP-20-42). The authors, therefore, acknowledge the DSR, with thanks for the technical and financial support provided. The authors thank the reviewers for their constructive remarks on our work.

Conflicts of Interest: The authors declare no conflicts of interest.

References

1. Magin, R.L. *Fractional Calculus in Bioengineering*; Begell House Publishers: Redding, CT, USA, 2006.
2. Javidi, M.; Ahmad, B. Dynamic analysis of time fractional order phytoplankton-toxic phytoplankton-zooplankton system. *Ecol. Model.* **2015**, *318*, 8–18. [CrossRef]
3. Fallahgoul, H.A.; Focardi, S.M.; Fabozzi, F.J. *Fractional Calculus and Fractional Processes with Applications to Financial Economics. Theory and Application*; Elsevier/Academic Press: London, UK, 2017.
4. Zaslavsky, G.M. *Hamiltonian Chaos and Fractional Dynamics*; Oxford University Press: Oxford, UK, 2005.
5. Kilbas, A.A.; Srivastava, H.M.; Trujillo, J.J. *Theory and Applications of Fractional Differential Equations*; North-Holland Mathematics Studies, 204; Elsevier Science B.V.: Amsterdam, The Netherlands, 2006.
6. Sabatier, J.; Agrawal, O.P.; Machado, J.A.T. (Eds.) *Advances in Fractional Calculus: Theoretical Developments and Applications in Physics and Engineering*; Springer: Dordrecht, The Netherlands, 2007.
7. Ahmad, B.; Alsaedi, A.; Ntouyas, S.K.; Tariboon, J. *Hadamard-Type Fractional Differential Equations, Inclusions and Inequalities*; Springer: Cham, Switzerland, 2017.
8. Henderson, J.; Kosmatov, N. Eigenvalue comparison for fractional boundary value problems with the Caputo derivative. *Fract. Calc. Appl. Anal.* **2014**, *17*, 872–880. [CrossRef]
9. Peng, L.; Zhou, Y. Bifurcation from interval and positive solutions of the three-point boundary value problem for fractional differential equations. *Appl. Math. Comput.* **2015**, *257*, 458–466. [CrossRef]

10. Agarwal, R.P.; Ahmad, B.; Alsaedi, A. Fractional-order differential equations with anti-periodic boundary conditions: A survey. *Bound. Value Probl.* **2017**, *173*, 27. [CrossRef]
11. Cui, Y.; Ma, W.; Sun, Q.; Su, X. New uniqueness results for boundary value problem of fractional differential equation. *Nonlinear Anal. Model. Control* **2018**, *23*, 31–39. [CrossRef]
12. Baghani, H.; Nieto, J.J. On fractional Langevin equation involving two fractional orders in different intervals. *Nonlinear Anal. Model. Control* **2019**, *24*, 884–897. [CrossRef]
13. Alsaedi, A.; Ahmad, B.; Alghanmi, M. Extremal solutions for generalized Caputo fractional differential equations with Steiltjes-type fractional integro-initial conditions. *Appl. Math. Lett.* **2019**, *91*, 113–120. [CrossRef]
14. Ahmad, B.; Alsaedi, A.; Alruwaily, Y.; Ntouyas, S.K. Nonlinear multi-term fractional differential equations with Riemann-Stieltjes integro-multipoint boundary conditions. *AIMS Math.* **2020**, *5*, 1446–1461. [CrossRef]
15. Liang, S.; Wang, L.; Yin, G. Fractional differential equation approach for convex optimization with convergence rate analysis. *Optim. Lett.* **2020**, *14*, 145–155. [CrossRef]
16. Iskenderoglu, G.; Kaya, D. Symmetry analysis of initial and boundary value problems for fractional differential equations in Caputo sense. *Chaos Solitons Fractals* **2020**, *134*, 109684. [CrossRef]
17. Cen, Z.; Liu, L.-B.; Huang, J. A posteriori error estimation in maximum norm for a two-point boundary value problem with a Riemann-Liouville fractional derivative. *Appl. Math. Lett.* **2020**, *102*, 106086. [CrossRef]
18. Laitinen, M.; Tiihonen, T. Heat transfer in conducting and radiating bodies. *Appl. Math. Lett.* **1997**, *10*, 5–8. [CrossRef]
19. Laitinen, M.; Tiihonen, T. Integro-differential equation modelling heat transfer in conducting, radiating and semitransparent materials. *Math. Methods Appl. Sci.* **1998**, *21*, 375–392. [CrossRef]
20. Tarasov, V.E. Fractional integro-differential equations for electromagnetic waves in dielectric media. *Theor. Math. Phys.* **2009**, *158*, 355–359. [CrossRef]
21. Ahmad, B.; Nieto, J.J. Boundary value problems for a class of sequential integrodifferential equations of fractional order. *J. Funct. Spaces Appl.* **2013**, 149659. [CrossRef]
22. Debbouche, A.; Nieto, J.J. Relaxation in controlled systems described by fractional integro-differential equations with nonlocal control conditions. *Electron. J. Differ. Equ.* **2015**, *89*, 18.
23. Ahmad, B.; Luca, R. Existence of solutions for sequential fractional integro-differential equations and inclusions with nonlocal boundary conditions. *Appl. Math. Comput.* **2018**, *339*, 516–534. [CrossRef]
24. Alsaedi, A.; Ahmad, B.; Aljoudi, S.; Ntouyas, S.K. A study of a fully coupled two-parameter system of sequential fractional integro-differential equations with nonlocal integro-multipoint boundary conditions. *Acta Math. Sci. Ser. B* **2019**, *39*, 927–944. [CrossRef]
25. Zhou, Y.; Suganya, S.; Arjunan, M.M.; Ahmad, B. Approximate controllability of impulsive fractional integro-differential equation with state-dependent delay in Hilbert spaces. *IMA J. Math. Control Inform.* **2019**, *36*, 603–622. [CrossRef]
26. Ahmad, B.; Alsaedi, A.; Aljoudi, S.; Ntouyas, S.K. A six-point nonlocal boundary value problem of nonlinear coupled sequential fractional integro-differential equations and coupled integral boundary conditions. *J. Appl. Math. Comput.* **2018**, *56*, 367–389. [CrossRef]
27. Liu, J.; Zhao, K. Existence of mild solution for a class of coupled systems of neutral fractional integro-differential equations with infinite delay in Banach space. *Adv. Differ. Equ.* **2019**, *2019*, 284. [CrossRef]
28. Ravichandran, C.; Valliammal, N.; Nieto, J.J. New results on exact controllability of a class of fractional neutral integro-differential systems with state-dependent delay in Banach spaces. *J. Franklin Inst.* **2019**, *356*, 1535–1565. [CrossRef]
29. Ahmad, B.; Broom, A.; Alsaedi, A.; Ntouyas, S.K. Nonlinear integro-differential equations involving mixed right and left fractional derivatives and integrals with nonlocal boundary data. *Mathematics* **2020**, *8*, 336. [CrossRef]
30. Huang, L.; Li, X.-F.; Zhao, Y.; Duan, X.-Y. Approximate solution of fractional integro-differential equations by Taylor expansion method. *Comput. Math. Appl.* **2011**, *62*, 1127–1134. [CrossRef]
31. Zhang, L.; Ahmad, B.; Wang, G.; Agarwal, R.P. Nonlinear fractional integro-differential equations on unbounded domains in a Banach space. *J. Comput. Appl. Math.* **2013**, *249*, 51–56. [CrossRef]
32. Tate, S.; Kharat, V.V.; Dinde, H.T. On nonlinear mixed fractional integro-differential equations with positive constant coefficient. *Filomat* **2019**, *33*, 5623–5638. [CrossRef]
33. Jalilian, Y.; Ghasemi, M. On the solutions of a nonlinear fractional integro-differential equation of pantograph type. *Mediterr. J. Math.* **2017**, *14*, 194. [CrossRef]

34. Ahmad, B. A quasilinearization method for a class of integro-differential equations with mixed nonlinearities. *Nonlinear Anal. Real World Appl.* **2006**, *7*, 997–1004. [CrossRef]
35. Hussain, K.H.; Hamoud, A.A.; Mohammed, N.M. Some new uniqueness results for fractional integro-differential equations. *Nonlinear Funct. Anal. Appl.* **2019**, *24*, 827–836.
36. Hamoud, A.A.; Ghadle, K.; Atshan, S. The approximate solutions of fractional integro-differential equations by using modified Adomian decomposition method. *Khayyam J. Math.* **2019**, *5*, 21–39.
37. Ahmad, B.; Alruwaily, Y.; Alsaedi, A.; Nieto, J.J. Fractional integro-differential equations with dual anti-periodic boundary conditions. *Differ. Integral Equ.* **2020**, *33*, 181–206.
38. Ahmad, B.; Alsaedi, A.; Alghamdi, B.S. Analytic approximation of solutions of the forced Duffing equation with integral boundary conditions. *Nonlinear Anal. Real World Appl.* **2008**, *9*, 1727–1740. [CrossRef]
39. Ciegis, R.; Bugajev, A. Numerical approximation of one model of the bacterial self-organization. *Nonlinear Anal. Model. Control* **2012**, *17*, 253–270. [CrossRef]
40. Ahmad, B.; Asghar, S.; Hayat, T. Diffraction of a plane wave by an elastic knife-edge adjacent to a rigid strip. *Canad. Appl. Math. Quart.* **2001**, *9*, 303–316.
41. Yusufoglu, E.; Turhan, I. A mixed boundary value problem in orthotropic strip containing a crack. *J. Franklin Inst.* **2012**, *349*, 2750–2769. [CrossRef]
42. Bitsadze, A.; Samarskii, A. On some simple generalizations of linear elliptic boundary problems. *Russ. Acad. Sci. Dokl. Math.* **1969**, *10*, 398–400.
43. Byszewski, L.; Lakshmikantham, V. Theorem about the existence and uniqueness of a solution of a nonlocal abstract Cauchy problem in a Banach space. *Appl. Anal.* **1991**, *40*, 11–19. [CrossRef]
44. Byszewski, L. Theorems about the existence and uniqueness of solutions of a semilinear evolution nonlocal Cauchy problem. *J. Math. Anal. Appl.* **1991**, *162*, 494–505. [CrossRef]
45. Groza, G.; Pop, N. Approximate solution of multipoint boundary value problems for linear differential equations by polynomial functions. *J. Differ. Equ. Appl.* **2008**, *14*, 1289–1309. [CrossRef]
46. Smart, D.R. *Fixed Point Theorems*; Cambridge University Press: Cambridge, UK, 1974.

mathematics

MDPI

Article

Dissipativity of Fractional Navier–Stokes Equations with Variable Delay

Lin F. Liu [1] **and Juan J. Nieto** [2,*]

[1] School of Mathematics, Northwest University, Xi'an 710075, China; liulinfang2020@nwu.edu.cn
[2] Instituto de Matemáticas, Universidade de Santiago de Compostela, 15782 Santiago de Compostela, Spain
* Correspondence: juanjose.nieto.roig@usc.es

Received: 14 October 2020; Accepted: 9 November 2020; Published: 16 November 2020

Abstract: We use classical Galerkin approximations, the generalized Aubin–Lions Lemma as well as the Bellman–Gronwall Lemma to study the asymptotical behavior of a two-dimensional fractional Navier–Stokes equation with variable delay. By modifying the fractional Halanay inequality and the comparison principle, we investigate the dissipativity of the corresponding system, namely, we obtain the existence of global absorbing set. Besides, some available results are improved in this work. The existence of a global attracting set is still an open problem.

Keywords: fractional Navier–Stokes equations; variable delay; modified fractional Halanay inequality; generalized comparison principle; dissipativity

1. Introduction

We study the longtime behavior of the following two-dimensional Navier–Stokes equation of fractional order with variable delay on a bounded domain $\Omega \subset \mathbb{R}^2$,

$$D_t^\alpha u - \nu\Delta u + (u \cdot \nabla)u + \nabla p = f(t) + g(t, u_t), \text{ in } (0,T) \times \Omega, \tag{1}$$

$$div\ u \equiv 0, \text{ in } (0,T) \times \Omega, \tag{2}$$

$$u = 0, \text{ on } (0,T) \times \partial\Omega, \tag{3}$$

$$u(t,x) = \phi(t,x),\ t \in [-h, 0],\ x \in \Omega, \tag{4}$$

where D_t^α is a fractional derivative of order $\alpha \in (0,1)$, $T > 0$, $\Omega \subset \mathbb{R}^2$ is a bounded open set with regular boundary $\partial\Omega$, $\nu > 0$ is the kinematic viscosity, u is the velocity field of the fluid, p is the pressure, ϕ is the initial datum, $h > 0$ is a constant, f is an external force field without delay, and g is the external force containing some functional delay. We will refer to (1)–(4) as problem (P).

In fact, hereditary characteristics are ubiquitous in engineering, biology and physics. For example, feedback control problem, immune systems, soft matter with viscoelasticity [1] could all have hereditary properties (including memory, variable delay or distributed delay, constant delay, etc). The delay term is very often denoted by a function $u_t(\cdot)$ defined on some interval $[-h, 0]$ (here h could be $-\infty$). The memory effect is modeled by using fractional calculus, which actually has been widely applied in many sciences [2–5]. We would like to mention that the concept of fractional calculus was raised by L'Hospital, who wrote to Leibniz in the year 1695, seeking the meaning of $\frac{d^n y}{dx^n}$ when $n = \frac{1}{2}$. However, it only became popular in practical applications in the past few decades. Several kinds of definitions of fractional derivatives have been introduced [2], but maybe the most commonly used nowadays are the so-called Riemann–Liouville derivative and Caputo derivative. More definitions for Riemann–Liouville and Caputo derivative can be found in [3,6,7].

It is worth pointing out that using a convolution group, Li and Liu [8] introduced a generalized definition of Caputo derivative of order $\alpha \in (0,1)$, and built a convenient framework for studying initial value problems of time fractional differential equations. Compared with Riemann–Liouville derivative, the Caputo derivative defined in [8] removes the singularity at $t = 0$ and characterizes memory from $t = 0_+$. It is probably this character that makes the Caputo derivative share many similarities with the corresponding ordinary derivative and then more manageable for Cauchy problems. In this work, we use the Caputo derivative introduced in [8] to investigate the fractional dynamic system (1).

On the other hand, there are many results about time-fractional Navier–Stokes equations, which can be used to simulate anomalous diffusion in fractal media. For instance, applying Laplace and finite Hankel transforms, Chaurasia and Kumar [9] obtained the solution of a time-fractional Navier–Stokes equation. In [10], Zhou and Peng studied the mild solutions of Navier–Stokes equations with a time-fractional derivative, meanwhile Nieto and Planas [11] investigated the existence and uniqueness of mild solutions to the Navier–Stokes equations with time fractional differential operators, and obtained several interesting properties about the solution, such as regularity and decay rate in Lebesgue spaces. Nevertheless, most of the available works including the mentioned ones did not take into account the delay in the external forcing term, and are concerned mainly with the existence of solution/mild solution or the regularity. There is no result on the limit behavior of solutions, even less work about fractional Navier–Stokes equations with delay, such as the existence of weak solution and asymptotical behavior of solutions. Actually, for general fractional PDEs, this discussion is limited due to the lack of tools although some special cases have been studied [12–14].

The traditional method used to study solutions of classic nonlinear PDEs is to find some "a priori" estimates of approximate solutions, then to apply some compactness criteria—i.e., the Arzelà-Ascoli theorem, etc. However, this method seems not to work for fractional PDEs with variable delay. Because of the appearance of variable delay term, the generalized fractional Gronwall inequality [15] (Theorem 1) is not enough to find some "a priori" estimates of Lyapunov functions. Even though Ye and Gao [16] obtained the Henry-Gronwall type retarded integral inequalities, this only works for fractional differential equations with constant delay but not for variable delay. Fortunately, Li and Liu [17] (Theorem 4.1–4.2), generalized the classic Aubin–Lions lemma and some convergence theorem to the fractional case, respectively. To our purpose, we first improve [17] (Proposition 3.5) and [8] (Theorem 4.10). Then, under the condition that $\alpha \in (\frac{1}{2}, 1)$, we investigate the solutions of our system by combing the Galerkin approximation and the generalized Aubin–Lions lemma as well as the Bellman–Gronwall Lemma.

We would like to mention that Wen, Yu and Wang [18] analyzed the dissipativity of Volterra functional differential equations by using the generalized Halanay inequalities, while Wang and Zou [19] studied the dissipativity and contractivity analysis for fractional functional differential equations and their numerical approximations via a fractional Halanay inequality. However, to analyze the dissipativity of fractional PDEs with variable delay, the fractional Halanay inequality [19] alone is not enough any more, in fact, it cannot be applied directly for our case, either. We modify the fractional Halanay inequality [19] (Lemma 4) to a more general case, and then improve the comparison principle [20] (Lemma 3.4) and combine the fractional Halanay inequality to overcome this difficulty.

Motivated by [19], we study the long time behavior of fractional Navier–Stokes equations with variable delay. More precisely, we first prove the existence and uniqueness of weak solutions by Galerkin approximation, and then analyze the dissipativity of system (P), namely, we obtain the existence of an absorbing set by fractional Halanay inequalities and generalized comparison principle. We would like to mention that similar results about the classic model of problem (P) can be found in [21].

The organization of this work is as follows. In the next Section, we recall some basic concepts about fractional calculus, and present some auxiliary lemmas which will be useful in later study. In Section 3, we focus on the existence and uniqueness of weak solutions, and the dissipativity of the fractional dynamic system (P) is shown in Section 4. Throughout the work, C, c are positive constants, which can be different from line to line, even in the same line.

2. Preliminaries

In this Section, we first recollect the generalize definitions of fractional calculus to functions valued in general Banach spaces as studied in [8,17]. Then we prefer to recall some notations and abstract spaces for the sake of completeness and to make the reading of the paper easier, although the notations and results included in this section may seem somehow repetitive, since they can be found in several already published monographs or articles [22–24]. Besides, two examples of delay are presented and some lemmas, propositions that will be used in our later discussion are stated.

Now, we start with the definition of fractional integral, readers are referred to [2,3,8] for more details.

Definition 1. *([3,17]) The fractional Riemann–Liouville integral of order $\alpha \in (0,1)$ for a function $u : \mathbb{R}^+ \to \mathbb{R}$ locally integrable is defined by*

$$[I_\alpha u](t) = \frac{1}{\Gamma(\alpha)} \int_0^t (t-s)^{\alpha-1} u(s) ds, \ \ t > 0,$$

where $\Gamma(\alpha) = \int_0^\infty x^{\alpha-1} e^{-x} dx$ is the classical Gamma function.

Definition 2. *([8]) Let X be a Banach space. For a locally integrable function $u \in L^1_{loc}((0,T);X)$, if there exists $u_0 \in X$ such that*

$$\lim_{t \to 0_+} \frac{1}{t} \int_0^t \|u(s) - u_0\|_X ds = 0,$$

then u_0 is called the right limit of u at $t = 0$, denote as $u(0_+) = u_0$. Similarly, we define $u(T^-) = u_T$ to be the left limit of u at $t = T$—i.e., $u_T \in X$ such that

$$\lim_{t \to T^-} \frac{1}{T-t} \int_t^T \|u(s) - u_T\|_X ds = 0.$$

As pointed out in [8], this fractional integral can be expressed as the convolution between the kernel $g_\alpha(t) = \frac{H(t)t^{\alpha-1}}{\Gamma(\alpha)}$ and $H(t)u(t)$ on $\mathbb{R}$, where

$$H(t) = \begin{cases} 1, & t \geq 0, \\ 0, & t < 0. \end{cases}$$

is the standard Heaviside step function. By this fact, it is not difficult to verify that the integral operators I_α form a semigroup, and I_α is a bounded linear operator from $L^1(0,T)$ to $L^1(0,T)$. Inspired by [25] (Section 5, Chapter 1), Li and Liu [8] proposed a generalized definition of Caputo derivative. The new definition is consistent with various definitions in the literature while revealing the underlying group structure. The underlying group property makes many properties of the Caputo derivative natural.

Before introducing this generalized Caputo derivative, we need to use the distributions $\{g_\alpha\}$ as the convolution kernels for $\alpha \in (-1,1)$:

$$g_\alpha(t) := \begin{cases} \frac{H(t)t^{\alpha-1}}{\Gamma(\alpha)}, & \alpha \in (0,1), \\ \delta(t), & \alpha = 0, \\ \frac{D(H(t)t^{\alpha})}{\Gamma(1+\alpha)}, & \alpha \in (-1,0), \end{cases}$$

where δ is the usual Dirac distribution, and D means the distributional derivative. As in [8], the fractional integral operator I_α can be expressed as

$$[I_\alpha u](t) := g_\alpha * [H(t)u(t)].$$

Given $f, g \in L^1_{loc}(0,T)$, we define the convolution between f and g as

$$f(t) * g(t) = \int_0^t f(s)g(t-s)ds.$$

Now, we introduce the generalized Caputo derivative as

Definition 3. *([8]) Let $\alpha \in (0,1)$. Suppose that $u \in L^1_{loc}(0,T)$ has a right limit $u(0_+) = u_0$ at $t = 0$ in the sense of Definition 2. The Caputo derivative of fractional order α of u is a distribution in $\mathcal{D}'(-\infty, T)$ with support in $[0,T)$, given by*

$$D_t^\alpha u := I_{-\alpha}u - u_0 g_{1-\alpha} = g_{-\alpha} * [(u-u_0)H(t)].$$

The right fractional Caputo derivative is defined as

Definition 4. *([17]) Let $\alpha \in (0,1)$. Consider that $u \in L^1_{loc}(-\infty, T)$ has a left limit u_T at $t = T$ in the sense of Definition 2. The right Caputo derivative fractional order α of u is a distribution in $\mathcal{D}'(\mathbb{R})$ with support in $(-\infty, T]$, given by*

$$\tilde{D}_{c;T}^\alpha u := \tilde{g}_{-\alpha} * [H(T-t)(u-u_T)].$$

To introduce the Caputo derivatives for functions valued in general Banach spaces, for fix $T > 0$, we present the following sets:

$$\mathcal{D}' := \{v | v : C_c^\infty((-\infty, T); \mathbb{R}) \to X \text{ is a bounded linear operator}\},$$

which is analogous of the distribution $\mathcal{D}'$ used in [17]. We would like to point out that $\mathcal{D}'$ can be understood as the generalization of distribution. In fact, if $X = \mathbb{R}$, then it is reduced to the usual distribution as in [17].

The weak fractional Caputo derivative of the functions valued in Banach spaces is given by

Definition 5. *([17]) Let X be a Banach space and $u \in L^1_{loc}([0,T); X)$. Let $u_0 \in X$. We define the weak Caputo derivative of fractional order α of u associated with initial value u_0 to be $D_t^\alpha u \in \mathcal{D}'$ such that for any test function $v \in C_c^\infty((-\infty, T); \mathbb{R})$,*

$$\langle v, D_t^\alpha u \rangle := \int_{-\infty}^T (\tilde{D}_{c;T}^\alpha v)(u-u_0)H(t)dt = \int_0^T (\tilde{D}_{c;T}^\alpha v)(u-u_0)dt.$$

Next, let us consider the following usual abstract spaces:

$$\mathcal{V} = \left\{ u \in (C_0^\infty(\Omega))^2 : div\, u = 0 \right\}.$$

$H =$ the closure of $\mathcal{V}$ in $(L^2(\Omega))^2$ with norm $|\cdot|$, and inner product $(\cdot,\cdot)$, where for $u, v \in (L^2(\Omega))^2$,

$$(u,v) = \sum_{j=1}^{2} \int_\Omega u_j(x) v_j(x) dx.$$

$V =$ the closure of $\mathcal{V}$ in $(H_0^1(\Omega))^2$ with norm $\|\cdot\|$, and inner product $((\cdot,\cdot))$, where for $u, v \in (H_0^1(\Omega))^2$,

$$((u,v)) = \sum_{i,j=1}^{2} \int_\Omega \frac{\partial u_j}{\partial x_i} \frac{\partial v_j}{\partial x_i} dx.$$

It follows that $V \subset H \equiv H' \subset V'$, where the injections are dense and compact. We will use $\|\cdot\|_*$ for the norm in V', and $\langle \cdot,\cdot \rangle$ for the duality pairing between V and V'. Now we define $A : V \to V'$ by $\langle Au, v\rangle = ((u,v))$, and the trilinear form B on $V \times V \times V$ by

$$B(u,v,w) = \sum_{i,j=1}^{2} \int_\Omega u_i \frac{\partial v_j}{\partial x_i} w_j dx, \ \forall\, u,v,w \in V.$$

Note that the trilinear form B satisfies the following inequalities which will be used later in proofs (see [23] (p. 2015)).

$$|B(u,v,u)| \leq \|u\|^2_{(L^4(\Omega))^2} \|v\| \leq 2^{-1/2} |u| \|u\| \|v\|, \ \forall\, u,v \in V. \tag{5}$$

The phase space used in this paper is defined as $C_H = C([-h,0];H)$ with the norm

$$\|u_t\|_{C_H} = \sup_{-h \leq \theta \leq 0} |u(t+\theta)|, \text{ for } u_t \in C_H \text{ and } t \geq 0,$$

where u_t is a function defined on $[-h,0]$—i.e., $u_t := u_t(\theta) = u(t+\theta), \theta \in [-h,0]$.

We now enumerate the assumptions on the delay term g. For $g : [0,T] \times C_H \to (L^2(\Omega))^2$, we assume:

(g1) For any $\xi \in C_H$, the mapping $[0,T] \ni t \mapsto g(t,\xi) \in (L^2(\Omega))^2$ is measurable.

(g2) $g(\cdot,0) = 0$.

(g3) There exists $L_g > 0$ such that, for any $t \in [0,T]$ and all $\xi, \eta \in C_H$,

$$|g(t,\xi) - g(t,\eta)| \leq L_g \|\xi - \eta\|_{C_H}.$$

Remark 1. *(i) As pointed out in [23], condition $(g2)$ is not a restriction. Indeed, if $|g(\cdot,0)| \in L^2(0,T)$, we could redefine $\hat{l}(t) = l(t) + g(t,0)$ and $\hat{g}(t,\cdot) = g(t,\cdot) - g(t,0)$. In this way the problem is exactly the same, $\hat{l}$ and $\hat{g}$ satisfy the required assumptions.*
(ii) Conditions $(g2)$ and $(g3)$ imply that

$$|g(t,\xi)| \leq L_g \|\xi\|_{C_H},$$

whence $|g(t,\xi)| \in L^\infty(0,T)$.

Example 1. ***A forcing term with bounded variable delay.*** *Let $G : [0,T] \times \mathbb{R}^2 \to \mathbb{R}^2$ be a measurable function satisfying $G(t,0) = 0$ for all $t \in [0,T]$, and assume that there exists $L_G > 0$ such that*

$$|G(t,u) - G(t,v)|_{\mathbb{R}^2} \leq L_G |u - v|_{\mathbb{R}^2}, \ \ \forall u, v \in \mathbb{R}^2.$$

Consider a function $\rho(\cdot) : [0,+\infty) \to [0,h]$, which plays the role of the variable delay. Assume that $\rho(\cdot)$ is measurable and define $g(t,\xi)(x) = G(t,\xi(-\rho(t))(x))$ for each $\xi \in C_H$, $x \in \Omega$ and $t \in [0,T]$. Notice that, in this case, the delayed term g in our problem becomes

$$g(t,\xi) = G(t,\xi(-\rho(t))).$$

Example 2. ***A forcing term with finite distributed delay.*** *Let $G : [0,T] \times [-h,0] \times \mathbb{R}^2 \to \mathbb{R}^2$ be a measurable function satisfying $G(t,s,0) = 0$ for all $(t,s) \in [0,T] \times [-h,0]$, and there exists a function $\beta(s) \in L^1(-h,0)$ such that*

$$|G(t,s,u) - G(t,s,v)|_{\mathbb{R}^2} \leq \beta(s)|u - v|_{\mathbb{R}^2}, \ \ \forall u, v \in \mathbb{R}^2, \ \forall (t,s) \in [0,T] \times [-h,0].$$

Define $g(t,\xi)(x) = \int_{-h}^{0} G(t,s,\xi(s)(x))ds$ for each $\xi \in C_H$, $t \in [0,T]$, and $x \in \Omega$. Then, the delayed term g in our problem becomes

$$g(t,\xi) = \int_{-h}^{0} G(t,s,\xi(s))ds.$$

After introducing the above operators, an equivalent abstract formulation to problem (P) is

$$D_t^\alpha u + \nu A u + B(u) = f(t) + g(t,u_t), \ \ \forall t > 0, \tag{6}$$

$$u(t) = \phi(t), \ t \in [-h,0]. \tag{7}$$

The definition of weak solution to problem (6) and (7) is defined as

Definition 6. *([17]) Given an initial datum $\phi \in C_H$, a weak solution u to (6) and (7) in the interval $[-h,T]$ is a function $u \in C([-h,T];H) \cap L^2(0,T;V)$ with $u_0 = \phi(0)$ such that, for all $v \in V$,*

$$(D_t^\alpha u(t), v) + \nu((u(t),v)) + B(u(t),u(t),v) = \langle f(t), v\rangle + (g(t,u_t),v),$$

where the equation must be understood in the sense of distribution.

The following auxiliary Lemmas will be needed in this work.

Lemma 1. *(See [17,26]) For any function $u(t)$ absolutely continuous on $[0,T]$, one has the inequality*

$$u(t)D_t^\alpha u(t) \geq \frac{1}{2} D_t^\alpha u^2(t), \ \alpha \in (0,1).$$

The following result is a generalization of the Aubin–Lions Lemma [27].

Lemma 2. *([17] (Theorem 4.2)) Let $T > 0$, $\alpha \in (0,1)$ and $p \in [1,\infty)$. Let M, X, Y be Banach spaces. The inclusion $M \hookrightarrow X$ compact and the inclusion $X \hookrightarrow Y$ continuous. Suppose $W \subset L^1_{loc}((0,T);M)$ satisfies:*

(i) *There exists* $r_1 \in [1,\infty)$ *and* $C > 0$ *such that* $\forall u \in W$,

$$\sup_{t\in(0,T)} I^{\alpha}(\|u\|_M^{r_1}) = \sup_{t\in(0,T)} \frac{1}{\Gamma(\alpha)}\int_0^t (t-s)^{\alpha-1}\|u\|_M^{r_1}(s)ds \le C.$$

(ii) *There exists* $p_1 \in (p,\infty]$, *such that,* W *is bounded in* $L^{p_1}((0,T);X)$.

(iii) *There exists* $r_2 \in [1,\infty)$, $C > 0$ *such that for any* $u \in W$ *with right limit* u_0 *at* $t = 0$, *it holds that*

$$\|D_t^{\alpha}u\|_{L^{r_2}((0,T);Y)} \le C.$$

Then, W *is relatively compact in* $L^p((0,T);X)$.

Proposition 1. *(An improvement in [17] (Proposition 3.5)) Suppose* Y *is a reflexive Banach space,* $\alpha \in (0,1)$ *and* $T > 0$. *Assume the sequence* $\{u^n\}$ *converges to* u *in* $L^p((0,T);Y)$, $p \ge 1$. *If there is an assignment of initial values* $u_{0,n}$ *for* u^n *such that the weak Caputo derivatives* $D_t^{\alpha}u^n$ *are bounded in* $L^r((0,T);Y)$ $(r \in [1,\infty))$, *then*

(i) *There is a subsequence such that* $u_{0,n}$ *converges weakly to some value* $u_0 \in Y$.

(ii) *If* $r > 1$, *there exists a subsequence such that* $D_t^{\alpha}u^{nk}$ *converges weakly to* v *and* u_{0,n_k} *converges weakly to* u_0. *Moreover,* v *is the Caputo derivative of* u *with initial value* u_0 *so that*

$$u(t) = u_0 + \frac{1}{\Gamma(\alpha)}\int_0^t (t-s)^{\alpha-1}v(s)ds.$$

Further, if $r \ge 1$, *then,* $u(0_+) = u_0$ *in* Y *is the sense of Definition 2.*

Proof. We would like to mention that this Proposition is just a slightly improvement of [17] (Proposition 3.5), in which, the final conclusion—i.e., $u(0_+) = u_0$ in Y—holds true for $r \ge \frac{1}{\alpha}$. However, this conclusion holds for $r \ge 1$.

So, we just need to prove that for $r \ge 1$, if $D_t^{\alpha}u \in L^1_{loc}([0,T),Y)$, then $u(0_+) = u_0$ in Y under the sense of Definition 2. By a similar argument in [17] (Corollary 2.16) and Young's inequality with the conjugate index $p = \infty$, $q = 1$, $\frac{1}{p} + \frac{1}{q} = 1$, we find

$$\begin{aligned}\frac{1}{t}\int_0^t \|u-u_0\|_Y dt &\le \frac{1}{t\Gamma(\alpha)}\int_0^t\int_0^{\tau}(\tau-s)^{\alpha-1}\|D_t^{\alpha}u\|_Y ds d\tau\\ &\le \frac{1}{t\Gamma(\alpha+1)}\int_0^t (t-s)^{\alpha}\|D_t^{\alpha}u\|_Y ds\\ &\le \frac{1}{\Gamma(\alpha+2)}t^{\alpha}\|D_t^{\alpha}u\|_{L^1((0,t),Y)} \to 0 \text{ as } t \to 0_+,\end{aligned}$$

Since $\|D_t^{\alpha}u\|_Y$ is integrable on $[0, T-\delta]$ for some $\delta > 0$. The proof is finished immediately. □

Remark 2. *Li and Liu in [17] (Theorem 5.2) proved the existence of weak solution for a time fractional incompresible Navier–Stokes equation for* $\alpha \in [\frac{1}{2},1)$, *because* $u(0_+) = u_0$ *is obtained under this condition. However, by using this Proposition 1, we also can prove* $u(0_+) = u_0$ *for* $\alpha \in (0,1)$. *Therefore, the existence result of [17] (Theorem 5.2) still holds for* $\alpha \in (0,1)$. *In this extent, we say that Proposition 1 improves [17] (Proposition 3.5).*

Proposition 2. *(Modified Fractional Halanay Inequality) Assume that the non-negative continuous function v satisfies*

$$D_t^\alpha v(t) \leq \gamma + av(t) + b \sup_{t-\tau(t)\leq s\leq t} v(s), \; 0 < t \leq T,$$
$$v(t) = |\varphi(t)|, \; -\sigma \leq t \leq 0, \tag{8}$$

where γ is a positive constant and $a + b \neq 0$, $\sigma = -\inf_{t\geq 0}(t - \tau(t)) > 0$, and the delay function $\tau(t) \geq 0$. If $a + b < 0$, then the following estimates holds

$$v(t) \leq -\frac{\gamma}{a+b} + M' E_\alpha(\lambda^* t^\alpha), \; \textit{for all } t \geq \tau(t), \tag{9}$$

where $M' = \sup_{-h\leq t\leq 0} |\varphi(t)|$, and the parameter λ^* is defined by

$$\lambda^* = \sup_{t-\tau(t)\geq 0} \{\lambda : \lambda - a - b\frac{E_\alpha(\lambda(t-\tau(t))^\alpha)}{E_\alpha(\lambda t^\alpha)} = 0\},$$

and it holds that $\lambda^* \in [a + b, 0]$.

Further, if the delay is bounded—i.e., $\tau(t) \leq \tau_0$ for some constant τ_0—then the parameter λ^* defined by

$$\lambda^* = \sup_{t-\tau(t)\geq 1} \{\lambda : \lambda - a - b\frac{E_\alpha(\lambda(t-\tau(t))^\alpha)}{E_\alpha(\lambda t^\alpha)} = 0\},$$

is strictly negative, namely, there exists some positive constants ϵ_0 satisfying $a + b < -\epsilon_0$ such that $\lambda^* \in [a + b, -\epsilon_0]$, and the estimate in (9) holds for all t such that $t \geq \tau(t) + 1$.

Proof. Actually, Proposition 2 is a slightly modification of [19] (Lemma 4), in which $\tau(t) > 0$ strictly for the first conclusion (9). However, in our case, (9) holds true for $\tau(t) \geq 0$. So, we only need to prove (9) is true when $\tau(t) = 0$. We prove this by comparison principle.

If $\tau(t) = 0$, then the original system (8) becomes.

$$D_t^\alpha v(t) \leq \gamma + (a+b)v(t), \; 0 < t \leq T,$$
$$v(0) = |\varphi(0)|, \; t = 0,$$

where γ is a positive constant and $a + b < 0$.

From system (8), there exists a nonnegative function $m(t)$ satisfying

$$D_t^\alpha v(t) = \gamma + (a+b)v(t) + m(t), \; 0 < t \leq T,$$
$$v(0) = |\varphi(0)|, \; t = 0.$$

According to [2] (Theorem 4.3), the initial value problem (8) has a unique solution that can be represented by

$$\begin{aligned} v(t) &= |\phi(0)|E_\alpha((a+b)t^\alpha) + \int_0^t (t-s)^{\alpha-1}E_{\alpha,\alpha}((a+b)(t-s)^\alpha)(\gamma+m(s))ds \\ &\le |\phi(0)|E_\alpha((a+b)t^\alpha) + \gamma\int_0^t (t-s)^{\alpha-1}E_{\alpha,\alpha}((a+b)(t-s)^\alpha)ds \\ &\le |\phi(0)|E_\alpha((a+b)t^\alpha) - \frac{\gamma}{a+b} \\ &\le M'E_\alpha(\lambda^* t^\alpha) - \frac{\gamma}{a+b}, \end{aligned}$$

where we used that $t^{\alpha-1}$ and $E_{\alpha,\alpha}((a+b)t^\alpha)$ are nonnegative and $\lambda^* \in [a+b, 0]$, as well as the fact that $E_\alpha(\lambda t^\alpha)$ is non-decreasing respect to λ. The proof is complete. □

Remark 3. *It turns out that the modified fractional Halanay inequality holds true not only for delay fractional dynamical system but also for the nondelay case, which means that it could be applied to more fractional differential equations. In this sense, we say it improves [19] (Lemma 4).*

Proposition 3. *(The generalized comparison principle.) Assume that for any function u and w are absolutely continuous on $[0,T]$, one has the inequality*

$$\begin{aligned} &D_t^\alpha u(t) \le -au(t) + bu(t-\tau(t)) + c, \;\; 0 < t < T, \\ &u(t) = \varphi(t), \;\; -h \le t \le 0, \end{aligned} \tag{10}$$

and the following fractional differential equation

$$\begin{aligned} &D_t^\alpha w(t) = -aw(t) + bw(t-\tau(t)) + c, \;\; 0 < t < T, \\ &w(t) = \varphi(t), \;\; -h \le t \le 0, \end{aligned} \tag{11}$$

where a, b, c are positive constants. Then it holds that

$$u(t) \le w(t), \text{ for all } t \ge -h. \tag{12}$$

Proof. Obviously, (12) holds true for any $t \in [-h, 0]$. Hence, we only need to verify that (12) is correct for $t \in [0, T]$. We will prove this through two steps.

Step 1. We first prove that (12) holds for $t > \tau(t)$. By contradiction, if it is not true, then there exists some $t > \tau(t)$ such that $u(t) \ge w(t)$. Denote t_* by

$$t_* = \inf\{t > \tau(t) : u(t) \ge w(t)\}.$$

Now, set $z(t) = u(t) - w(t)$. Then we know from the definition that $z(t_*) = 0$, and $z(t) < 0$ for $0 < t_* - \tau(t_*) \le t < t_*$. Then by the fractional comparison principle in [19] (Lemma 3), we have that

$$D_t^\alpha z(t_*) \ge 0. \tag{13}$$

However,

$$\begin{aligned} D_t^\alpha z(t_*) &= D_t^\alpha u(t_*) - D_t^\alpha w(t_*) \\ &\le -a(u(t_*) - w(t_*)) + b(u(t_* - \tau(t_*)) - w(t_* - \tau(t_*))) \\ &= -az(t_*) + bz(t_* - \tau(t_*)) \le bz(t_* - \tau(t_*)) < 0, \end{aligned}$$

which contradicts (13); therefore, $u(t) \le w(t)$ for $t > \tau(t)$.

Step 2. On the other hand, when $0 < t \le \tau(t)$, then $-h \le t - \tau(t) \le 0$, since $\tau(t) \in [0, h]$. So (14) and (15) can be rewritten as, respectively,

$$\begin{aligned} &D_t^\alpha u(t) \le -au(t) + b\varphi(t - \tau(t)) + c, \ 0 < t \le \tau(t), \\ &u(t) = \varphi(t), \ -h \le t \le 0, \end{aligned} \tag{14}$$

and the following fractional differential equation

$$\begin{aligned} &D_t^\alpha w(t) = -aw(t) + b\varphi(t - \tau(t)) + c, \ 0 < t \le \tau(t), \\ &w(t) = \varphi(t), \ -h \le t \le 0, \end{aligned} \tag{15}$$

Then there is a nonnegative function $m(t)$, such that

$$\begin{aligned} &D_t^\alpha u(t) = -au(t) + b\varphi(t - \tau(t)) - m(t) + c, \ 0 < t \le \tau(t), \\ &u(t) = \varphi(t), \ -h \le t \le 0, \end{aligned} \tag{16}$$

Then, by [28] (Theorem 1), system (16) has a unique solution on $[0, h]$ that can be represented as

$$u(t) = \int_0^t (t-s)^{\alpha-1} E_{\alpha,\alpha}(-a(t-s)^\alpha)(b\varphi(s - \tau(s)) - m(s) + c)ds + c' E_\alpha(-at^\alpha), \ 0 < t \le \tau(t).$$

Similarly, the solution of system (15) can be written as

$$w(t) = \int_0^t (t-s)^{\alpha-1} E_{\alpha,\alpha}(-a(t-s)^\alpha)(b\varphi(s - \tau(s)) + c)ds + c' E_\alpha(-at^\alpha), \ 0 < t \le \tau(t).$$

Notice that t^α and $E_{\alpha,\alpha}(-at^\alpha)$ are non-negative for $a > 0$, then we have $u(t) \le w(t)$ for all $0 < t \le \tau(t)$. In summary, $u(t) \le w(t)$ for all $t \ge -h$.

Therefore the proof is complete. □

Remark 4. *We would like to point out that Proposition 3 generalizes the conclusion in [8] (Theorem 4.10) to some extent. Proposition 3 also improves the comparison principle in [20] (Lemma 3.4), which is proven only for constant delay—i.e., $\tau(t) = \tau$. However, in our case, the delay term $\tau(t)$ is a function taking values in $[0, h]$. In this way, we could say that Lemma 3.4 of [20] is a special case of Proposition 3.*

Lemma 3. *(Bellman–Gronwall Lemma [29] (p. 252)) Let $T > 0$, $g \in L^1(0, T)$ and $g \ge 0$ a.e., C_1, C_2 be positive constants. If $\varphi \in L^1(0, T)$, $\varphi \ge 0$ a.e., satisfying $g\varphi \in L^1(0, T)$ and*

$$\varphi(t) \le C_1 + C_2 \int_0^t g(s)\varphi(s)ds, \ \ a.e. \ t \in (0, T),$$

then

$$\varphi(t) \le C_1 \exp\{C_2 \int_0^t g(s)ds\}, \ \ a.e.\ t \in (0,T).$$

Remark 5. *Actually, the positive constants C_1, C_2 can be replaced by functions $C_1(t)$ or $C_2(t)$, but a similar result can be obtained—readers are referred to [29] (p. 252) for more information.*

3. Existence and Uniqueness of Weak Solutions

In this section, we prove the existence and uniqueness of weak solutions to problem (6) and (7) by Galerkin approximations. Denote

$$\lambda_1 = \inf_{v \in V \backslash \{0\}} \frac{\|v\|^2}{|v|^2} > 0.$$

Moreover,

$$\int_0^t \|f(s)\|_*^{2(1+\frac{1}{\alpha})} ds < \infty, \ \ \alpha \in (0,1), \ \ \text{for any } t \ge 0. \tag{17}$$

We have the following result:

Theorem 1. *Suppose that $(g1)-(g3)$ and (17) hold true, then for any $\phi \in C_H$ and $\alpha \in (\frac{1}{2},1)$ system, (6) and (7) has a unique weak solution.*

Proof. We split it into several steps.

Step 1. The Galerkin approximation. By the definition of $A = -\Delta$ and the classical spectral theory of elliptic operators, it follows that A possesses a sequence of eigenvalues $\{\lambda_j\}_{j\ge1}$ and a corresponding family of eigenfunctions $\{w_j\}_{j\ge1} \subset V$, which form a Hilbert basis of H, dense on V. We consider the subspace $V_m = \text{span}\{w_1, w_2, \cdots, w_m\}$, and the projector $P_m : H \to V_m$ given by $P_m u = \sum_{j=1}^m (u, w_j) w_j$, and define $u^{(m)}(t) = \sum_{j=1}^m \gamma_{m,j}(t) w_j$, where the superscript m will be used instead of (m), for short, since no confusion is possible with powers of u, and where the coefficients $\gamma_{m,j}(t)$ are required to satisfy the Cauchy problem

$$\begin{aligned} &(D_t^\alpha u^m(t), w_j) + \nu((u^m(t), w_j)) + B(u^m(t), u^m(t), w_j) = \langle f(t), w_j\rangle + (g(t, u_t^m), w_j), \ \ 1 \le j \le m, \\ &u^m(t) = P_m \phi(t), \ t \in [-h, 0]. \end{aligned} \tag{18}$$

The above system of fractional order functional differential equations with finite delay fulfills the conditions for the existence and uniqueness of a local solution (e.g., cf. [30] (Theorem 3.1)). Hence, we conclude that (18) has a unique local solution defined in $[0, t_m)$ with $0 \le t_m \le T$. Next, we will obtain a priori estimates and ensure that the solutions u^m do exist in the whole interval $[0, \infty)$. Assume that $M = \|\phi\|_{C_H}^2 = \sup_{-h \le t \le 0} |\phi(t)|^2$.

Step 2. A priori estimates. Multiplying (18) by $\gamma_{m,j}(t)$, $j = 1, \ldots, m$, summing up, and using Lemma 1, Cauchy–Schwartz and Young's inequalities, we obtain

$$\begin{aligned} \frac{1}{2} D_t^\alpha |u^m(t)|^2 + \nu \|u^m(t)\|^2 &\le \|f(t)\|_* \|u^m(t)\| + |g(u_t^m)||u^m(t)| \\ &\le \frac{\nu}{2}\|u^m(t)\|^2 + \frac{\|f(t)\|_*^2}{2\nu} + L_g \|u_t^m\|_{C_H}^2. \end{aligned}$$

Hence,

$$D_t^{\alpha}|u^m(t)|^2 + \nu\|\nabla u^m(t)\|^2 \le \frac{\|f(t)\|_*^2}{\nu} + 2L_g\|u_t^m\|_{C_H}^2. \tag{19}$$

Multiplying (19) by I_α, and let $p = 1+\alpha$, $q = 1+\frac{1}{\alpha}$, we find

$$\begin{aligned}
&|u^m(t)|^2 + \frac{\nu}{\Gamma(\alpha)}\int_0^t (t-s)^{\alpha-1}\|\nabla u^m(s)\|^2 ds \\
&\le |u^m(0)|^2 + \frac{1}{\nu\Gamma(\alpha)}\int_0^t (t-s)^{\alpha-1}\|f(s)\|_*^2 ds + \frac{2L_g}{\Gamma(\alpha)}\int_0^t (t-s)^{\alpha-1}\|u_s^m\|_{C_H}^2 ds \\
&\le |u^m(0)|^2 + \frac{1}{\nu\Gamma(\alpha)}\Big(\int_0^t (t-s)^{p(\alpha-1)}ds\Big)^{\frac{1}{p}}\Big(\int_0^t \|f(s)\|_*^{2q}ds\Big)^{1/q} + \frac{2L_g}{\Gamma(\alpha)}\int_0^t (t-s)^{\alpha-1}\|u_s^m\|_{C_H}^2 ds \\
&\le |u^m(0)|^2 + \frac{1}{\nu\alpha^2\Gamma(\alpha)}t^{\alpha^2}F(t) + \frac{2L_g}{\Gamma(\alpha)}\Big(\int_0^t (t-s)^{p(\alpha-1)}e^{ps}ds\Big)^{\frac{1}{p}}\Big(\int_0^t e^{-qs}\|u_s^m\|_{C_H}^{2q}ds\Big)^{1/q} \\
&\le |u^m(0)|^2 + \frac{1}{\nu\alpha^2\Gamma(\alpha)}t^{\alpha^2}F(t) + \frac{2L_g\Gamma(\alpha^2)}{\Gamma(\alpha)}(1+\alpha)^{-\frac{\alpha^2}{1+\alpha}}e^t\Big(\int_0^t e^{-qs}\|u_s^m\|_{C_H}^{2q}ds\Big)^{1/q}.
\end{aligned} \tag{20}$$

Denote by $A(t) = |u^m(0)|^2 + \frac{1}{\nu\alpha^2\Gamma(\alpha)}t^{\alpha^2}F(t)$, $B(t) = \frac{2L_g\Gamma(\alpha^2)}{\Gamma(\alpha)}(1+\alpha)^{-\frac{\alpha^2}{1+\alpha}}e^t$. Then, we have

$$\|u_t^m\|_{C_H}^2 + \nu\int_0^{t+\theta}(t+\theta-s)^{\alpha-1}\|\nabla u^m(s)\|^2 ds \le A(t) + B(t)\Big(\int_0^t \|u_s^m\|^{2q}ds\Big)^{1/q}.$$

Therefore,

$$\|u_t^m\|_{C_H}^{2q} \le 2^q A^q(t) + 2^q B^q(t)\Big(\int_0^t \|u_s^m\|_{C_H}^{2q}ds\Big).$$

Using the Gronwall Lemma, we obtain that

$$\|u_t^m\|_{C_H}^2 \le c\Big(A(t) + B(t)\int_0^t A(s)e^{c\int_s^t B(r)dr}ds\Big), \text{ for all } t\in[0,T] \text{ and } \theta\in[-h,0].$$

Hence, we conclude that for any $T>0$, $\|u_t^m\|_{C_H}$ is finite, which means the local solution $u^m(t;\phi)$ is actually a global one. We also can have that there exists a constant $C>0$, depending on some constants of the problem (namely, ν, L_g and f), and on T and $M>0$, such that

$$\|u_t^m\|_{C_H}^2 \le C(T,M)\ \ \forall t\in[0,T],\ \ \|\phi\|_{C_H}\le M,\ \ \forall m\ge 1,$$

which also implies that $\{u^m\}$ is bounded in $L^\infty(-h,T;H)$.

Now it follows from (20) and the above uniform estimates that

$$\begin{aligned}
\nu\|u^m\|_{L^2(0,T;V)}^2 &\le \nu t^{1-\alpha}\int_0^t (t-s)^{\alpha-1}\|u^m(s)\|^2 ds \\
&\le t^{1-\alpha}\left(\Gamma(\alpha)|u^m(0)|^2 + \int_0^t (t-s)^{\alpha-1}\left(\frac{1}{\nu}\|f(s)\|_*^2 + 2L_gC(T,M)\right)ds\right) \\
&\le C(T,M),\ \ \forall m\ge 1.
\end{aligned}$$

Therefore, we conclude that

$$\{u^m\} \text{ is bounded in } L^2((0,T);V) \cap L^\infty((-h,T);H). \tag{21}$$

From (5) and (18), it holds $\|D_t^\alpha u^m\|_* \le \nu\|u^m\| + 2^{1/2}|u^m| \cdot \|u^m\| + \|f\|_* + \lambda_1^{-1/2}|g(t,u_t^m)|$, which implies that

$$\{D_t^\alpha u^m\} \text{ is bounded in } L^2(0,T;V'). \tag{22}$$

Step 3. Approximation of initial datum in C_H. Let us check

$$P_m\phi \to \phi \ \text{ in } C_H. \tag{23}$$

Assume that $\theta_m \to \theta \in [-h,0]$, then $P_m\phi(\theta_m) \to \phi(\theta)$, since $\|P_m\phi(\theta_m) - \phi(\theta)\| \le \|P_m\phi(\theta_m) - P_m\phi(\theta)\| + \|P_m\phi(\theta) - \phi(\theta)\| \to 0$ as $m \to \infty$. So (23) holds true.

Step 4. Compactness results. By (21) and (22), the compact imbedding $V \hookrightarrow H$, and the generalized Aubin–Lions Lemma 2 as well as Proposition 1, for any $\alpha \in (0,1)$, we obtain there exist a subsequence still relabeled as $\{u^m\}$ and a function $u \in C([-h,T);H) \cap L^2((0,T);V)$ for all $T > 0$, with $u(t) = \phi(t)$ in $[-h,0]$, $u(0_+) = u_0$, and $D_t^\alpha u \in L^2((0,T);V')$ for all $T > 0$, and an element $\chi \in L^\infty((0,T);H)$ such that

$$\begin{aligned}
&u^m \overset{*}{\rightharpoonup} u \ \text{ weakly-star in } L^\infty((0,T);H),\\
&u^m \rightharpoonup u \ \text{ weakly in } L^2((0,T);V),\\
&D_t^\alpha u^m \rightharpoonup D_t^\alpha u \ \text{ weakly in } L^2((0,T);V'),\\
&u^m \to u \ \text{ strongly in } L^2((0,T);H),\\
&g(\cdot,u_t^m) \overset{*}{\rightharpoonup} \chi \ \text{ weakly-star in } L^\infty((0,T);H).
\end{aligned} \tag{24}$$

Observe that if $\alpha \in (\frac{1}{2},1)$, for all $s,t \in [0,T]$, by

$$\begin{aligned}
u^m(t) - u^m(s) &= \frac{1}{\Gamma(\alpha)} \int_s^t (t-r)^{\alpha-1} (D_r^\alpha u^m)(r)dr\\
&\le \frac{1}{\Gamma(\alpha)(2\alpha-1)} (t-s)^{2\alpha-1} \|D_t^\alpha u^m\|_{L^2((s,t),V')}, \text{ in } V',
\end{aligned}$$

and combing (22) we find that u^m is equi-continuous on $[0,T]$ with values in V'. Notice that the inclusion $H \hookrightarrow V'$ compact, so using Ascoli-Arzelà and (24), we conclude that

$$u^m \to u \ \text{ in } C([0,T];V'), \ \forall\, T > 0. \tag{25}$$

Combining (21) and (25), then for any $\{t_m\} \subset [0,T)$ with $t_m \to t$, one obtains

$$u^m(t_m) \rightharpoonup u(t) \ \text{ weakly in } H. \tag{26}$$

Now we prove that

$$u^m \to u \ \text{ in } C([0,T];H), \ \forall\, T > 0. \tag{27}$$

By contradiction, if (27) is not true, then there would exists a $\epsilon_1 > 0$, $\{t_m\}$ and t_* with $t_m \to t_*$ such that

$$|u^m(t_m) - u(t_*)| \ge \epsilon_1. \tag{28}$$

On one hand, by (26), we have $|u(t_*)| \leq \lim_{n\to\infty} \inf |u^m(t_m)|$. Therefore, if we could prove $|u^m(t_m)| \to |u(t_*)|$, then (28) is contradictory, in other words, (27) is obtained immediately. To this end, it is enough to show that

$$\lim_{m\to\infty} \sup |u^m(t_m)| \leq |u(t_*)|. \tag{29}$$

On the other hand, for system (18), we have the following energy inequality,

$$\begin{aligned} &|u^m(t)|^2 + \frac{\nu}{\Gamma(\alpha)} \int_0^t (t-s)^{\alpha-1} \|u^m(s)\|^2 ds \\ &\leq |u^m(0)|^2 + \frac{1}{\Gamma(\alpha)} \int_0^t (t-s)^{\alpha-1} \|f(s)\|_*^2 ds + \frac{2L_g C(T,M)}{\Gamma(\alpha+1)} (t-s)^\alpha. \end{aligned} \tag{30}$$

Besides, by (24), passing to the limit in (18), we have that $u \in C([0,T];H)$ is a solution of a similar problem to (6)—i.e.,

$$(D_t^\alpha u, v) + \nu(Au, v) + (B(u), v) = \langle f(t), v\rangle + (\chi, v), \quad \forall v \in V',$$

which also has the energy inequality,

$$\begin{aligned} &|u(t)|^2 + \frac{\nu}{\Gamma(\alpha)} \int_0^t (t-s)^{\alpha-1} \|u(s)\|^2 ds \\ &\leq |u(0)|^2 + \frac{1}{\Gamma(\alpha)} \int_0^t (t-s)^{\alpha-1} \|f(s)\|_*^2 ds + \frac{1}{\Gamma(\alpha)} \int_0^t (t-s)^{\alpha-1} (\chi(s), u(s)) ds. \end{aligned}$$

Combing the last convergence in (24) and the dominate convergence theorem, we find

$$\int_0^t (t-s)^{\alpha-1} |\chi(s)|^2 ds \leq \lim_{m\to\infty} \inf \int_0^t (t-s)^{\alpha-1} |g(u_s^m)|^2 ds \leq \frac{2L_g}{\alpha} C(T,M)(t-s)^\alpha.$$

Therefore, u also satisfies inequality (30) with the same last term on the right-hand side.
Consider now two continuous functions defined as

$$\begin{aligned} J(t) &= |u(t)|^2 - \frac{1}{\Gamma(\alpha)} \int_0^t (t-s)^{\alpha-1} \|f(s)\|_*^2 ds - \frac{2L_g C(T,M)}{\Gamma(\alpha+1)} (t-s)^\alpha, \\ J_m(t) &= |u^m(t)|^2 - \frac{1}{\Gamma(\alpha)} \int_0^t (t-s)^{\alpha-1} \|f(s)\|_*^2 ds - \frac{2L_g C(T,M)}{\Gamma(\alpha+1)} (t-s)^\alpha. \end{aligned}$$

J and J_m are non-increasing in t. Moreover, again from (24), we have

$$J_m(t) \to J(t), \quad a.e.\ t \in (0,T). \tag{31}$$

Assume that $t_* > 0$, consider $\{t_k\} \subset (0, t_*)$ with $t_k \to t_*$, by the continuity of J,

$$\exists k_\epsilon : \ |J(t_k) - J(t_*)| < \frac{\epsilon}{2}, \ \forall k \geq k_\epsilon.$$

Take now $m(k_\epsilon)$ such that

$$t_m \geq t_{k_\epsilon} : \ |J_m(t_{k_\epsilon}) - J(t_{k_\epsilon})| < \frac{\epsilon}{2}, \ \forall n \geq n(k_\epsilon).$$

Then, we conclude that for all $m \geq n(k_\epsilon)$

$$J_m(t_m) - J(t_*) \leq |J_m(t_{k_\epsilon}) - J(t_{k_\epsilon})| + |J(t_{k_\epsilon}) - J(t_*)| < \epsilon,$$

which gives (29).

Therefore, we find that

$$u^m \to u \ \text{ in } C([0,T];H).$$

Then, steps 3 and 4 imply that

$$u_t^m \to u_t \ \text{ in } C_H, \ \forall\, 0 \leq t \leq T.$$

Therefore, combining $(g3)$, we can prove that

$$g(\cdot, u_\cdot^m) \to g(\cdot, u_\cdot) \ \text{ in } L^2(0,T;H).$$

Thus, we can finally pass to the limit in (18), concluding that u solves (P).

Step 5. Uniqueness of solution. Let $u(t;\phi), v(t;\phi)$ be two solutions of (P) with the same initial values—i.e., $u(t) = v(t) = \phi(t), t \in [-h,0]$. Set $w(t) = u(t) - v(t), t \geq 0$, then $w(t) = 0$, for all $t \in [-h,0]$. For $w(t)$, we have

$$D_t^\alpha w - \nu \Delta w + B(u) - B(v) = g(t,u_t) - g(t,v_t).$$

Multiplying above equation by $w(t)$, and integral over Ω, we obtain

$$\begin{aligned} D_t^\alpha |w|^2 + \nu \|w\|^2 &= -(B(u) - B(v), w) + (g(t,u_t) - g(t,v_t), w) \\ &\leq c|w|^2\|v\|^2 + 2L_g \sup_{0\leq s\leq t} |w(s)|^2 \\ &\leq c(\|v\|^2 + 1) \sup_{0\leq s\leq t} |w(s)|^2, \quad \text{for all } t \in [0,T]. \end{aligned}$$

The above inequality holds true for any $t \in [0,T]$, then we have

$$\sup_{0\leq s\leq t} |w(s)|^2 \leq |w(0)|^2 + c\int_0^t (\|v\|^2 + 1)(t-s)^{\alpha-1} \sup_{0\leq r\leq s} |w(r)|^2 ds.$$

Using the Bellman–Gronwall Lemma 3 and (21), we have

$$\sup_{0\leq s\leq t} |w(s)|^2 \leq |w(0)|^2 \exp\left\{ c\int_0^t (\|v\|^2 + 1)(t-s)^{\alpha-1} ds \right\} = 0, \quad \text{for all } t \in [0,T].$$

Therefore, $|w(t)| = 0$ on $[-h,T]$. The proof is finished. □

Remark 6. *We prove the existence of solution for a general delay case, namely, $g(t,u_t)$ could be variable delay or distributed delay. In Section 4, we take $g(t,u_t) = g(u(t-\tau(t)))$—i.e., the delay function $\tau(t) \in C(\mathbb{R}_+;[0,h])$, to study the dissipativity.*

Remark 7. *It is worth mentioning that only the existence result is proved under the condition that $\alpha \in (\frac{1}{2}, 1)$, which is due to the phase space $C([-h,0];H)$. If $C([-h,0];H)$ is replaced by some Sobolev space, such as $L^2((-h,0);H)$. Then the existence of solution can be established for any $\alpha \in (0,1)$ and without additional conditions.*

Theorem 2. *Suppose that* $(g1)-(g3)$ *hold true, then the solutions of system* (P) *are continuous with respect of initial values—i.e.,*

$$\|u_t - v_t\|_{C_H}^2 \leq c\|\phi - \varphi\|_{C_H}^2 \exp\left\{c\int_0^t (\|v\|^2+1)(t-s)^{\alpha-1}ds\right\},\ t\in[0,T].$$

Proof. Let $u(t;\phi)$, $v(t;\varphi)$ be the solutions of (1)–(4) with initial values, ϕ and φ, respectively. Set $w(t) = u(t)-v(t)$ for $t>0$, and $w(t)=\phi(t)-\varphi(t)$ for $t\in[-h,0]$. Then we have

$$D_t^\alpha w - \nu\Delta w + B(u) - B(v) = g(t,u_t) - g(t,v_t),\ \ t>0.$$

Multiplying above equation by $w(t)$, and integral over Ω, we obtain

$$\begin{aligned}D_t^\alpha |w|^2 + \nu\|w\|^2 &= -(B(u)-B(v),w) + (g(t,u_t)-g(t,v_t),w)\\ &\leq c|w|^2\|v\|^2 + 2L_g\|w_t\|_{C_H}^2 \leq c|w|^2\|v\|^2 + 2L_g(\|\phi-\varphi\|_{C_H}^2 + \sup_{0\leq t\leq T}|w|^2)\\ &\leq (c\|v\|^2+2L_g)\sup_{0\leq t\leq T}|w(t)|^2 + 2L_g\|\phi-\varphi\|_{C_H}^2.\end{aligned}$$

Hence, we have

$$\sup_{0\leq t\leq T}|w(t)|^2 \leq |w(0)|^2 + \frac{2L_gT^\alpha}{\Gamma(\alpha+1)}\|\phi-\varphi\|_{C_H}^2 + \frac{1}{\Gamma(\alpha)}\int_0^t (c\|v\|^2+2L_g)(t-s)^{\alpha-1}\sup_{0\leq s\leq t}|w(s)|^2ds$$

Again using the Bellman–Gronwall Lemma 3 and (21), we find that

$$\|w_t\|_{C_H}^2 \leq c\|\phi-\varphi\|_{C_H}^2 \exp\left\{c\int_0^t (\|v\|^2+1)(t-s)^{\alpha-1}ds\right\},\ \ t\in[0,T].$$

The proof is complete. □

4. Dissipativity

In this section, we derive some uniform estimates of solutions to problem (P) by using Proposition 2. Besides, in this section, we assume that $g(t,u_t) = g(u(t-\tau(t)))$.

Definition 7. *The system* (P) *is said to be dissipative in* C_H *if there exists a bounded set* $B\subset C_H$*, such that for any given bounded set* $A\subset C_H$*, there is a time* $t^* = t^*(A)$*, such that for any given initial function* $\phi\in A$*, for all* $t\in[-h,0]$*, the values of the corresponding solution* $u(t)$ *of the problem* (P) *are contained in* B *for all* $t\geq t^*$*. The set* B *is called an absorbing set of the system* (P).

We assume that

$$\lambda_1\nu > \sqrt{2}L_g. \tag{32}$$

Theorem 3. *(Existence of absorbing sets in* C_H*) Assume that* $(g1)-(g3)$*, (17) and (32) hold. Then there exists* $T>0$*, such that for all* $t\geq T$*, the solution of problem* (P) *satisfies*

$$\|u_t\|_{C_H}^2 \leq \frac{\lambda_1\nu f_0}{(\lambda_1\nu)^2 - 2L_g^2} + 1,\ \forall\, t\geq T,$$

where $f_0 = \nu^{-1}\sup_{t\geq 0}\|f(t)\|_*^2$.

Proof. Multiplying (1) by u, integrating over Ω, we have

$$\begin{aligned} D_t^\alpha |u(t)|^2 + \nu\|u(t)\|^2 &\le \|f(t)\|_*\|u(t)\| + |g(u(t-\tau(t)))||u(t)| \\ &\le \frac{\|f(t)\|_*^2}{\nu} + \frac{2L_g^2}{\lambda_1\nu} \sup_{t-\tau(t)\le s\le t} |u(s)|^2. \end{aligned} \tag{33}$$

Then we obtain

$$\begin{aligned} D_t^\alpha |u(t)|^2 &\le f_0 - \lambda_1\nu|u(t)|^2 + \frac{2L_g^2}{\lambda_1\nu} \sup_{t-\tau(t)\le s\le t} |u(s)|^2, \;\; t\in(0,T], \\ |u(t)|^2 &= |\phi(t)|^2, \;\; t\in[-h,0], \end{aligned}$$

where $f_0 = \nu^{-1}\sup_{t\ge 0}\|f(t)\|_*^2$. Using Proposition 2, we find that

$$|u(t)|^2 \le \frac{\lambda_1\nu f_0}{(\lambda_1\nu)^2 - 2L_g^2} + ME_\alpha(\lambda^* t^\alpha)$$

for all $t \ge \tau(t)$, where $M = \|\phi\|_{C_H}^2 = \sup_{-h\le t\le 0} |\phi(t)|^2$, and the parameter λ^* is defined by

$$\lambda^* = \sup_{t-\tau(t)\ge 1} \{\lambda : \lambda - (-\lambda_1\nu) - \frac{2L_g^2}{\lambda_1\nu}\frac{E_\alpha(\lambda(t-\tau(t))^\alpha)}{E_\alpha(\lambda t^\alpha)} = 0\},$$

is strictly negative, namely, there exists some positive constants ϵ_0 satisfying $-\lambda_1\nu + \frac{2L_g^2}{\lambda_1\nu} < -\epsilon_0$ such that $\lambda^* \in [-\lambda_1\nu + \frac{2L_g^2}{\lambda_1\nu}, -\epsilon_0]$, and the estimate in (9) holds for all t such that $t \ge \tau(t)+1$. In other words, for $\lambda^* \in [-\lambda_1\nu + \frac{2L_g^2}{\lambda_1\nu}, -\epsilon_0]$, we have

$$|u(t)|^2 \le \frac{\lambda_1\nu f_0}{(\lambda_1\nu)^2 - 2L_g^2} + ME_\alpha(\lambda^* t^\alpha), \;\; \forall\, t \ge \tau(t)+1.$$

For the case of $t < \tau(t)+1$, in order to analyze the dissipativity of problem (P) in phase space C_H by Proposition 3, we first need to consider the following fractional differential equation,

$$\begin{aligned} D_t^\alpha w(t) + \lambda_1\nu w(t) &= f_0 + \frac{2L_g^2}{\lambda_1\nu} w(t-\tau(t)), \;\; 0 < t \le h+1, \\ w(t) &= |\phi(t)|^2, \;\; t\in[-h,0], \end{aligned} \tag{34}$$

Then, by using the method of steps [28] (Theorem 1), we have that the initial value problem (34) has, on the interval $[0,kh]$, a unique solution that can be represented by $w(t) = w_{ih}(t)$, if $(i-1)h \le t \le ih$,

$$w_{ih}(t) = \int_0^t E_{\alpha,\alpha}(-\lambda_1\nu(t-s)^\alpha) f_{ih}(s)ds + c_{ih}E_\alpha(-\lambda_1\nu t^\alpha), \;\; t\in[(i-1)h, ih],$$

where c_{ih} is a constant, $i = 1, 2, \cdots, k$.

$$f_{kh}(t) := \begin{cases} \frac{2L_g^2}{\lambda_1 \nu} w_{0h}(t-h) + f_0, & 0 < t \leq h, \\ \frac{2L_g^2}{\lambda_1 \nu} w_{1h}(t-h) + f_0, & h < t \leq 2h, \\ \cdots \\ \frac{2L_g^2}{\lambda_1 \nu} w_{(k-1)h}(t-h) + f_0, & (k-1)h < t \leq kh, \end{cases}$$

is continuous and $w_{0h}(t) = |\phi(t)|^2$. k is a smallest integer such that $kh \geq h + 1$. Therefore, we obtain that

$$|w(t)| \leq \sum_{0 \leq i \leq k} |w_{ih}(t)| \leq C E_\alpha(-\lambda_1 \nu t^\alpha), \ \forall\, 0 \leq t \leq \tau(t) + 1. \tag{35}$$

Now, we estimate the solution of (1), for $t < \tau(t) + 1$. By (33), we have

$$\begin{aligned} D_t^\alpha |u(t)|^2 + \lambda_1 \nu |u(t)|^2 &\leq \frac{\|f(t)\|_*^2}{\nu} + \frac{2L_g^2}{\lambda_1 \nu} |u(t - \tau(t))|, \ 0 \leq t < \tau(t) + 1, \\ |u(t)|^2 &= |\phi(t)|^2, \ \ t \in [-h, 0]. \end{aligned} \tag{36}$$

Then, by Proposition 3 and (34)–(36), we have

$$|u(t)|^2 \leq C E_\alpha(-\lambda_1 \nu t^\alpha), \ \ 0 \leq t < \tau(t) + 1.$$

So, we find that

$$|u(t)|^2 \leq \frac{\lambda_1 \nu f_0}{(\lambda_1 \nu)^2 - 2L_g^2} + M E_\alpha(\lambda^* t^\alpha) + C E_\alpha(-\lambda_1 \nu t^\alpha), \text{ for all } t \geq 0.$$

By the norm of C_H, we conclude that

$$\|u_t\|_{C_H}^2 \leq \frac{\lambda_1 \nu f_0}{(\lambda_1 \nu)^2 - 2L_g^2} + M E_\alpha(\lambda^* t^\alpha) + C E_\alpha(-\lambda_1 \nu t^\alpha), \text{ for all } t \geq 0, \theta \in [-h, 0].$$

Since λ^* and $-\lambda_1 \nu$ are strictly negative, by the property of Mittag–Leffler function [2], we obtain

$$\|u_t\|_{C_H}^2 \leq \frac{\lambda_1 \nu f_0}{(\lambda_1 \nu)^2 - 2L_g^2} + C \frac{C_\alpha}{t^\alpha}, \text{ as } t \to +\infty,$$

where $C_\alpha > 0$ is a constant independent of t. Therefore, there exists $T > 0$ large enough, such that for all $t \geq T$, the solution of problem (P) satisfies

$$\|u_t\|_{C_H}^2 \leq \frac{\lambda_1 \nu f_0}{(\lambda_1 \nu)^2 - 2L_g^2} + 1, \ t \geq T.$$

Denote by $B_{C_H} = B\left(0, \sqrt{\frac{\lambda_1 \nu f_0}{(\lambda_1 \nu)^2 - 2L_g^2} + 1}\right)$ the absorbing set in phase space C_H, which implies that system (P) is dissipative. The proof is complete. □

5. Discussion

In this work, we prove the existence and uniqueness of solution for fractional Navier–Stokes equations with variable delays for $\alpha \in (\frac{1}{2}, 1)$, and we show that this system is dissipative in the phase space C_H, namely, there exists a global absorbing set in C_H. Different from the classic Navier–Stokes equations with variable delays [22–24], in which the existence of pullback absorbing set and pullback attractors were established. Here, we obtained the forward absorbing set, which is more meaningful from the view of applications. Besides, the existence of global attracting set as well as the existence of solution for $\alpha \in (0, 1)$ in phase space C_H are still open problems. These will be considered in the future.

Author Contributions: Conceptualization, L.F.L. and J.J.N.; methodology, L.F.L. and J.J.N.; writing—original draft preparation, L.F.L.; writing—review and editing, J.J.N. All authors have read and agreed to the published version of the manuscript.

Funding: The work of Lin F. Liu has been partially supported by NSF of China (Nos. 11901448, 11871022 and 11671142) as well as by China Postdoctoral Science Foundation Grant (Nos. 2018M643610). The work of Juan J. Nieto has been partially supported by the Agencia Estatal de Investigación (AEI) of Spain, co-financed by the European Fund for Regional Development (FEDER) corresponding to the 2014-2020 multiyear financial framework, project MTM2016-75140-P, Xunta de Galicia under grant ED431C 2019/02; by Instituto de Salud Carlos III (Spain), grant COV20/00617.

Conflicts of Interest: The authors declare no conflict of interest.

References

1. Piero, G.; Deseri, L. On the concepts of state and free energy in linear viscoelasticity. *Arch. Rat. Mech. Anal.* **1997**, *138*, 1–35. [CrossRef]
2. Kilbas, A.A.; Srivastava, H.M.; Trujillo, J.J. *Theory and Applications of Fractional Differential Equations*; Elsevier Science B.V.: Amsterdam, The Netherlands, 2006.
3. Podlubny, I. Fractional Differential Equations. In *An Introduction to Fractional Derivatives, Fractional Differential Equations, To Methods of Their Solution and Some of Their Applications*; Academic Press Inc.: San Diego, CA, USA, 1999.
4. Mainardi, F.; Paradisi, P.; Gorenflo, R. Probability distributions generated by fractional diffusion equations. *arXiv* **2007**, arXiv:0704.0320.
5. Diethelm, K. *The Analysis of Fractional Differential Equations: An Application-Oriented Exposition Using Differential Operators of Caputo Type*; Springer: Berlin, Germany, 2010.
6. Caputo, M. Linear models of dissipation whose Q is almost frequency independent-II. *Geophys. J. Intern.* **1967**, *13*, 529–539. [CrossRef]
7. Allen, M.; Caffarelli, L.; Vasseur, A. A parabolic problem with a fractional-time derivative. *Arch. Ration. Mech. Anal.* **2016**, *221*, 603–630. [CrossRef]
8. Li, L.; Liu, J. A Generalized Definition of Caputo Derivatives and Its Application to Fractional ODEs. *SIAM J. Math. Anal.* **2018**, *50*, 2867–2900. [CrossRef]
9. Chaurasia, V.; Kumar, D. Solution of the Time-Fractional Navier-Stokes Equation. *Gen. Math. Notes* **2011**, *4*, 49–59.
10. Zhou, Y.; Peng, L. On the time-fractional Navier-Stokes equations. *Comp. Math. Appl.* **2017**, *73*, 874–891. [CrossRef]
11. de Carvalho-Neto, P.M.; Planas, G. Mild solutions to the time fractional Navier-Stokes equations in R^n. *J. Differ. Equ.* **2015**, *259*, 2948–2980. [CrossRef]
12. Gorenflo, R.; Luchko, Y.; Yamamoto, M. Operator theoretic approach to the caputo derivative and the fractional diffusion equations. *arXiv* **2014**, arXiv:1411.7289.
13. Chen, F.; Nieto, J.J.; Zhou, Y. Global attractivity for nonlinear fractional differential equations. *Nonlinear Anal. RWA* **2012**, *13*, 287–298. [CrossRef]
14. Harikrishnan, S.; Prakash, P.; Nieto, J.J. Forced oscillation of solutions of a nonlinear fractional partial differential equation. *Appl. Math. Comput.* **2015**, *254*, 14–19. [CrossRef]

15. Ye, H.; Gao, J.; Ding, Y. A generalized Gronwall inequality and its application to a fractional differential equation. *J. Math. Anal. Appl.* **2007**, *328*, 1075–1081. [CrossRef]
16. Ye, H.; Gao, J. Henry-Gronwall type retarded integral inequalities and their applications to fractional differential equations with delay. *Appl. Math. Comp.* **2011**, *218*, 4152–4160. [CrossRef]
17. Li, L.; Liu, J. Some compactness criteria for weak solutions of time fractional PDEs. *SIAM J. Math. Anal.* **2018**, *50*, 3963–3995. [CrossRef]
18. Wen, L.; Yu, W.; Wang, W. Generalized Halanay inequalities for dissipativity of Volterra functional differential equations. *J. Math. Anal. Appl.* **2008**, *347*, 169–178. [CrossRef]
19. Wang, D.; Zou, J. Dissipativity and contractivity analysis for fractional functional differential equations and their numerical approximations. *SIAM J. Numer. Anal.* **2019**, *57*, 1445–1470. [CrossRef]
20. Wang, H.; Yu, Y.; Wen, G.; Zhang, S.; Yu, J. Global stability analysis of fractional-order Hopfield neural networks with time delay. *Neurocomputing* **2015**, *154*, 15–23. [CrossRef]
21. Liu, L.F.; Caraballo, T.; Marín-Rubio, P. Stability results for 2D Navier–Stokes equations with unbounded delay. *J. Differ. Equ.* **2018**, *265*, 5685–5708. [CrossRef]
22. Marín-Rubio, P.; Real, J. Pullback attractors for 2D-Navier-Stokes equations with delays in continuous and sub-linear operators. *Discrete Cont. Dyn. Syst. A* **2012**, *26*, 989–1006. [CrossRef]
23. Marín-Rubio, P.; Real, J.; Valero, J. Pullback attractors for a two-dimensional Navier-Stokes model in an infinite delay case. *Nonlinear Anal.* **2011**, *74*, 2012–2030. [CrossRef]
24. Caraballo, T.; Real, J. Attractors for 2D-Navier-Stokes models with delays. *J. Differ. Equ.* **2004**, *205*, 271–297. [CrossRef]
25. Gel'fand, I.M.; Shilov, G.E. *Generalized Functions*; Academic Press: New York, NY, USA, 1964; Volume 1.
26. Alikhanov, A. A priori estimates for solutions of boundary vale problems for fractional-order equations. *Differ. Equ.* **2010**, *46*, 660–666. [CrossRef]
27. Simon, J. Compact sets in the space $L^p(0,T;B)$. *Ann. Mat. Pure Appl.* **1987**, *146*, 65–96. [CrossRef]
28. Morgadoa, M.L.; Ford, N.J.; Lima, P.M. Analysis and numerical methods for fractional differential equations with delay. *J. Comput. Appl. Math.* **2013**, *252*, 159–168. [CrossRef]
29. Desoer, C.A.; Vidyasagar, M. *Feedback Systems: Input-Output Properties*; Academic Press: New York, NY, USA, 1975.
30. Lakshmikantham, V. Theory of fractional functional differential equations. *Nonlinear Anal.* **2008**, *69*, 3337–3343. [CrossRef]

MDPI
St. Alban-Anlage 66
4052 Basel
Switzerland
Tel. +41 61 683 77 34
Fax +41 61 302 89 18
www.mdpi.com

Mathematics Editorial Office
E-mail: mathematics@mdpi.com
www.mdpi.com/journal/mathematics

www.ingramcontent.com/pod-product-compliance
Lightning Source LLC
LaVergne TN
LVHW070709100526
838202LV00013B/1055
* 9 7 8 3 0 3 6 5 0 0 7 4 4 *